Yamaha RD350 YPVS Twins Owners Workshop Manual

by Pete Shoemark

Models covered

RD350 LC II. 347cc. April 1983 to June 1985
RD350 N. 347cc. February 1985 to October 1986
RD350 N II. 347cc. March 1986 to July 1987
RD350 F. 347cc. February 1985 to October 1986
RD350 F II. 347cc. March 1986 to December 1991
RD350 R. 347cc. March 1992 on

(1158-4Z9)

Haynes Group Limited
Haynes North America, Inc

www.haynes.com

Acknowledgements

Our thanks are due to APS Motorcycles of Wells, Somerset, who supplied the motorcycles featured in this manual, and Mitsui Machinery Sales (UK) Ltd who supplied the necessary service information and gave permission to reproduce many of the line drawings used.

The Avon Rubber Company supplied information on tyre care and fitting, and NGK Spark Plugs (UK) Ltd provided information on plug maintenance and electrode conditions. Renold Ltd advised on chain maintenance and renewal.

A book in the **Haynes Owners Workshop Manual Series**

ISBN 978 1 85010 879 5

British Library Cataloguing in Publication Data
A catalogue record for this book is available from the British Library

Disclaimer

There are risks associated with automotive repairs. The ability to make repairs depends on the individual's skill, experience and proper tools. Individuals should act with due care and acknowledge and assume the risk of performing automotive repairs.

The purpose of this manual is to provide comprehensive, useful and accessible automotive repair information, to help you get the best value from your vehicle. However, this manual is not a substitute for a professional certified technician or mechanic.

This repair manual is produced by a third party and is not associated with an individual vehicle manufacturer. If there is any doubt or discrepancy between this manual and the owner's manual or the factory service manual, please refer to the factory service manual or seek assistance from a professional certified technician or mechanic.

Even though we have prepared this manual with extreme care and every attempt is made to ensure that the information in this manual is correct, neither the publisher nor the author can accept responsibility for loss, damage or injury caused by any errors in, or omissions from, the information given.

Contents

The RD350 F model

Engine/gearbox unit

About this manual

The purpose of this manual is to present the owner with a concise and graphic guide which will enable him to tackle any operation from basic routine maintenance to a major overhaul. It has been assumed that any work would be undertaken without the luxury of a well-equipped workshop and a range of manufacturer's service tools.

To this end, the machine featured in the manual was stripped and rebuilt in our own workshop, by a team comprising a mechanic, a photographer and the author. The resulting photographic sequence depicts events as they took place, the hands shown being those of the author and the mechanic.

The use of specialised, and expensive, service tools was avoided unless their use was considered to be essential due to risk of breakage or injury. There is usually some way of improvising a method of removing a stubborn component, providing that a suitable degree of care is exercised.

The author learnt his motorcycle mechanics over a number of years, faced with the same difficulties and using similar facilities to those encountered by most owners. It is hoped that this practical experience can be passed on through the pages of this manual.

Where possible, a well-used example of the machine is chosen for the workshop project, as this highlights any areas which might be particularly prone to giving rise to problems. In this way, any such difficulties are encountered and resolved before the text is written, and the techniques used to deal with them can be incorporated in the relevant section. Armed with a working knowledge of the machine, the author undertakes a considerable amount of research in order that the maximum amount of data can be included in the manual.

A comprehensive section, preceding the main part of the manual, describes procedures for carrying out the routine maintenance of the machine at intervals of time and mileage. This section is included particularly for those owners who wish to ensure the efficient day-to-day running of their motorcycle, but who choose not to undertake overhaul or renovation work.

Each Chapter is divided into numbered sections. Within these sections are numbered paragraphs. Cross reference throughout the manual is quite straightforward and logical. When reference is made 'See Section 6.10' it means Section 6, paragraph 10 in the same Chapter. If another Chapter were intended, the reference would read, for example, 'See Chapter 2, Section 6.10'. All the photographs are captioned with a section/paragraph number to which they refer and are relevant to the Chapter text adjacent.

Figures (usually line illustrations) appear in a logical but numerical order, within a given Chapter. Fig. 1.1 therefore refers to the first figure in Chapter 1.

Left-hand and right-hand descriptions of the machines and their components refer to the left and right of a given machine when the rider is seated normally.

Motorcycle manufacturers continually make changes to specifications and recommendations, and these, when notified, are incorporated into our manuals at the earliest opportunity.

Introduction to the RD350 YPVS models

The machines covered in this manual represent the further development of Yamaha's earlier LC models. The ancestry of the current RD350 range can be traced back to the YDS2 produced in the late 1960s. From this model evolved both the TZ racer and the road-going (and air-cooled) RD series.

By the later 1970s the appeal of a simple air-cooled two-stroke was beginning to wane somewhat, and the motorcycling public began to look for a similar level of technology to that being displayed in the larger four-strokes. At the time, the increasingly swingeing emission laws in most countries seemed to augur badly for the two-stroke, and many felt that its time was nearly up. This impression was backed up by Yamaha themselves; for many years at the forefront of two-stroke production, even they seemed to be turning their attention to the more civilised four-stroke.

In 1980 Yamaha caused an appreciable stir by introducing the new RDs (sold as RZs in some countries). The new models were based on the proven air-cooled RD machines, but also bore a close resemblance to the highly-respected TZ racing models. The water-cooled LC models soon became popular with road riders and production racers alike.

The RD350 LC II was first introduced in 1983, and though broadly similar to its predecessors, it exibited more than cosmetic changes. The most significant of these is Yamaha's YPVS power valve system. This arrangement, consisting of a microprocessor-controlled servomotor controlling a spool-shaped exhaust port valve permits continually variable port timing. This means that more useful power can be extracted at peak engine speeds without the usual detrimental effects on low speed running normally associated with tuned two-strokes.

As well as the fitting of the power valve, attention was turned to the chassis. New front forks were fitted, these employing air assistance and variable damping. The combination provided progressive fork action, the fork moving compliantly in response to minor bumps, yet still able to cope with larger obstacles without bottoming.

At the rear of the machine, Yamaha's ageing cantilever suspension was replaced with a more sophisticated rising rate monoshock design. Confusingly, the new system is still called 'Monocross', even though it has nothing in common with the earlier system.

During 1985 the RD350 F and RD350 N models (the LC suffix having by now been dropped) were introduced. They were basically similar to the RD350 LC II, with additional detail and cosmetic alterations. The F model was equipped with a full-length fairing in place of the previous model's half fairing and belly pan. The RD350 N retains the general layout of the II model, but is unfaired.

Both the RD350 F and N models were superseded by the F II and N II models in April 1986, followed by the RD350R in March 1992. These later models are covered in Chapter 8.

Model dimensions and weights

Overall length	2095 mm (82.5 in)
Overall width	
RD350 LC II	710 mm (28.0 in)
RD350 N	690 mm (27.2 in)
RD350 F	670 mm (26.4 in)
Overall height	
RD350 LC II	1175 mm (46.3 in)
RD350 F, N	1190 mm (46.9 in)
Seat height	800 mm (31.5 in)
Wheelbase	1385 mm (54.5 in)
Ground clearance	
RD350 LC II	175 mm (6.89 in)
RD350 F, N	165 mm (6.50 in)
Weight (with oil and full fuel tank)	
RD350 LC II	164 kg (361.6 lb)
RD350 N	161 kg (355.0 lb)
RD350 F	165 kg (364.0 lb)

Ordering spare parts

When ordering spare parts for any Yamaha model it is advisable to deal direct wth an official Yamaha dealer who will be able to supply most items ex-stock. Where parts have to be ordered, an authorised dealer will be able to obtain them as quickly as possible. The engine and frame numbers must always be quoted in full. This avoids the risk of incorrect parts being supplied and is particularly important where detail modifications have been made in the middle of production runs. In some instances it will be necessary for the dealer to check compatability of later parts designs with earlier models. The frame number is stamped into the right-hand side of the steering head boss, and the engine number on a raised boss on the crankcase below the left-hand carburettor.

It is recommended that genuine Yamaha parts are used. Although pattern parts are often cheaper, remember that there is no guarantee that they are of the same specification as the original, and in some instances may be positively dangerous. Note also that the use of non-standard parts may invalidate the warranty in the event of a subsequent failure.

Some of the more expendable parts such as oils, greases, spark plugs, tyres and bulbs, can safely be obtained from auto accessory shops. These are often more conveniently positioned and may open during weekends. It is also possible to obtain parts on a mail order basis from specialists who advertise in the motorcycle magazines.

Each model has a model code which relates to a range of engine and frame numbers, and these are summarised below:

Model name	RD350 LC II	RD350 F	RD350 N
Model code	31K	57V	1JF
Initial engine No	31K-000101	31K-053101	31K-077101
Initial frame No	31K-000101	31K-053101	31K-077101

Engine number location

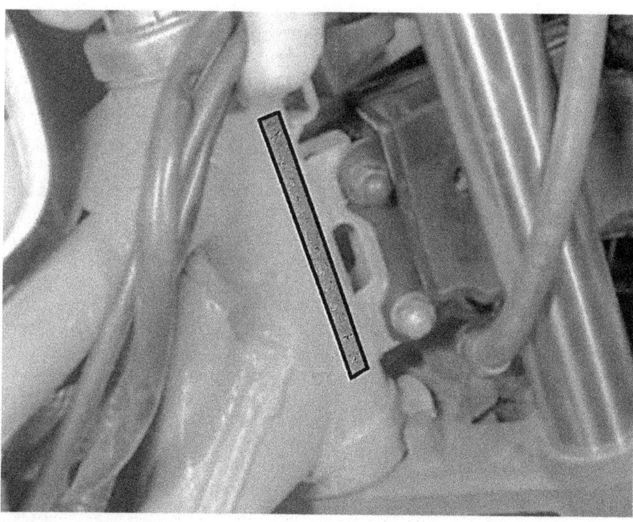

Frame number location

Safety first!

Professional motor mechanics are trained in safe working procedures. However enthusiastic you may be about getting on with the job in hand, do take the time to ensure that your safety is not put at risk. A moment's lack of attention can result in an accident, as can failure to observe certain elementary precautions.

There will always be new ways of having accidents, and the following points do not pretend to be a comprehensive list of all dangers; they are intended rather to make you aware of the risks and to encourage a safety-conscious approach to all work you carry out on your vehicle.

Essential DOs and DON'Ts

DON'T start the engine without first ascertaining that the transmission is in neutral.

DON'T suddenly remove the filler cap from a hot cooling system – cover it with a cloth and release the pressure gradually first, or you may get scalded by escaping coolant.

DON'T attempt to drain oil until you are sure it has cooled sufficiently to avoid scalding you.

DON'T grasp any part of the engine, exhaust or silencer without first ascertaining that it is sufficiently cool to avoid burning you.

DON'T allow brake fluid or antifreeze to contact the machine's paintwork or plastic components.

DON'T syphon toxic liquids such as fuel, brake fluid or antifreeze by mouth, or allow them to remain on your skin.

DON'T inhale dust – it may be injurious to health (see *Asbestos* heading).

DON'T allow any spilt oil or grease to remain on the floor – wipe it up straight away, before someone slips on it.

DON'T use ill-fitting spanners or other tools which may slip and cause injury.

DON'T attempt to lift a heavy component which may be beyond your capability – get assistance.

DON'T rush to finish a job, or take unverified short cuts.

DON'T allow children or animals in or around an unattended vehicle.

DON'T inflate a tyre to a pressure above the recommended maximum. Apart from overstressing the carcase and wheel rim, in extreme cases the tyre may blow off forcibly.

DO ensure that the machine is supported securely at all times. This is especially important when the machine is blocked up to aid wheel or fork removal.

DO take care when attempting to slacken a stubborn nut or bolt. It is generally better to pull on a spanner, rather than push, so that if slippage occurs you fall away from the machine rather than on to it.

DO wear eye protection when using power tools such as drill, sander, bench grinder etc.

DO use a barrier cream on your hands prior to undertaking dirty jobs – it will protect your skin from infection as well as making the dirt easier to remove afterwards; but make sure your hands aren't left slippery. Note that long-term contact with used engine oil can be a health hazard.

DO keep loose clothing (cuffs, tie etc) and long hair well out of the way of moving mechanical parts.

DO remove rings, wristwatch etc, before working on the vehicle – especially the electrical system.

DO keep your work area tidy – it is only too easy to fall over articles left lying around.

DO exercise caution when compressing springs for removal or installation. Ensure that the tension is applied and released in a controlled manner, using suitable tools which preclude the possibility of the spring escaping violently.

DO ensure that any lifting tackle used has a safe working load rating adequate for the job.

DO get someone to check periodically that all is well, when working alone on the vehicle.

DO carry out work in a logical sequence and check that everything is correctly assembled and tightened afterwards.

DO remember that your vehicle's safety affects that of yourself and others. If in doubt on any point, get specialist advice.

IF, in spite of following these precautions, you are unfortunate enough to injure yourself, seek medical attention as soon as possible.

Asbestos

Certain friction, insulating, sealing, and other products – such as brake linings, clutch linings, gaskets, etc – contain asbestos. *Extreme care must be taken to avoid inhalation of dust from such products since it is hazardous to health.* If in doubt, assume that they *do* contain asbestos.

Fire

Remember at all times that petrol (gasoline) is highly flammable. Never smoke, or have any kind of naked flame around, when working on the vehicle. But the risk does not end there – a spark caused by an electrical short-circuit, by two metal surfaces contacting each other, by careless use of tools, or even by static electricity built up in your body under certain conditions, can ignite petrol vapour, which in a confined space is highly explosive.

Always disconnect the battery earth (ground) terminal before working on any part of the fuel or electrical system, and never risk spilling fuel on to a hot engine or exhaust.

It is recommended that a fire extinguisher of a type suitable for fuel and electrical fires is kept handy in the garage or workplace at all times. Never try to extinguish a fuel or electrical fire with water.

Note: *Any reference to a 'torch' appearing in this manual should always be taken to mean a hand-held battery-operated electric lamp or flashlight. It does* **not** *mean a welding/gas torch or blowlamp.*

Fumes

Certain fumes are highly toxic and can quickly cause unconsciousness and even death if inhaled to any extent. Petrol (gasoline) vapour comes into this category, as do the vapours from certain solvents such as trichloroethylene. Any draining or pouring of such volatile fluids should be done in a well ventilated area.

When using cleaning fluids and solvents, read the instructions carefully. Never use materials from unmarked containers – they may give off poisonous vapours.

Never run the engine of a motor vehicle in an enclosed space such as a garage. Exhaust fumes contain carbon monoxide which is extremely poisonous; if you need to run the engine, always do so in the open air or at least have the rear of the vehicle outside the workplace.

The battery

Never cause a spark, or allow a naked light, near the vehicle's battery. It will normally be giving off a certain amount of hydrogen gas, which is highly explosive.

Always disconnect the battery earth (ground) terminal before working on the fuel or electrical systems.

If possible, loosen the filler plugs or cover when charging the battery from an external source. Do not charge at an excessive rate or the battery may burst.

Take care when topping up and when carrying the battery. The acid electrolyte, even when diluted, is very corrosive and should not be allowed to contact the eyes or skin.

If you ever need to prepare electrolyte yourself, always add the acid slowly to the water, and never the other way round. Protect against splashes by wearing rubber gloves and goggles.

Mains electricity and electrical equipment

When using an electric power tool, inspection light etc, always ensure that the appliance is correctly connected to its plug and that, where necessary, it is properly earthed (grounded). Do not use such appliances in damp conditions and, again, beware of creating a spark or applying excessive heat in the vicinity of fuel or fuel vapour. Also ensure that the appliances meet the relevant national safety standards.

Ignition HT voltage

A severe electric shock can result from touching certain parts of the ignition system, such as the HT leads, when the engine is running or being cranked, particularly if components are damp or the insulation is defective. Where an electronic ignition system is fitted, the HT voltage is much higher and could prove fatal.

Tools and working facilities

The first priority when undertaking maintenance or repair work of any sort on a motorcycle is to have a clean, dry, well-lit working area. Work carried out in peace and quiet in the well-ordered atmosphere of a good workshop will give more satisfaction and much better results than can usually be achieved in poor working conditions. A good workshop must have a clean flat workbench or a solidly constructed table of convenient working height. The workbench or table should be equipped with a vice which has a jaw opening of at least 4 in (100 mm). A set of jaw covers should be made from soft metal such as aluminium alloy or copper, or from wood. These covers will minimise the marking or damaging of soft or delicate components which may be clamped in the vice. Some clean, dry, storage space will be required for tools, lubricants and dismantled components. It will be necessary during a major overhaul to lay out engine/gearbox components for examination and to keep them where they will remain undisturbed for as long as is necessary. To this end it is recommended that a supply of metal or plastic containers of suitable size is collected. A supply of clean, lint-free, rags for cleaning purposes and some newspapers,. other rags, or paper towels for mopping up spillages should also be kept. If working on a hard concrete floor note that both the floor and one's knees can be protected from oil spillages and wear by cutting open a large cardboard box and spreading it flat on the floor under the machine or workbench. This also helps to provide some warmth in winter and to prevent the loss of nuts, washers, and other tiny components which have a tendency to disappear when dropped on anything other than a perfectly clean, flat, surface.

Unfortunately, such working conditions are not always available to the home mechanic. When working in poor conditions it is essential to take extra time and care to ensure that the components being worked on are kept scrupulously clean and to ensure that no components or tools are lost or damaged.

A selection of good tools is a fundamental requirement for anyone contemplating the maintenance and repair of a motor vehicle. For the owner who does not possess any, their purchase will prove a considerable expense, offsetting some of the savings made by doing-it-yourself. However, provided that the tools purchased meet the relevant national safety standards and are of good quality, they will last for many years and prove an extremely worthwhile investment.

To help the average owner to decide which tools are needed to carry out the various tasks detailed in this manual, we have compiled three lists of tools under the following headings: *Maintenance and minor repair*, *Repair and overhaul*, and *Specialized*. The newcomer to practical mechanics should start off with the simpler jobs around the vehicle. Then, as his confidence and experience grow, he can undertake more difficult tasks, buying extra tools as and when they are needed. In this way, a *Maintenance and minor repair* tool kit can be built-up into a *Repair and overhaul* tool kit over a considerable period of time without any major cash outlays. The experienced home mechanic will have a tool kit good enough for most repair and overhaul procedures and will add tools from the specialized category when he feels the expense is justified by the amount of use these tools will be put to.

It is obviously not possible to cover the subject of tools fully here. For those who wish to learn more about tools and their use there is a book entitled *Motorcycle Workshop Practice Manual* (Book no 1454) available from the publishers of this manual

As a general rule, it is better to buy the more expensive, good quality tools. Given reasonable use, such tools will last for a very long time, whereas the cheaper, poor quality, item will wear out faster and need to be renewed more often, thus nullifying the original saving. There is also the risk of a poor quality tool breaking while in use, causing personal injury or expensive damage to the component being worked on.

For practically all tools, a tool factor is the best source since he will have a very comprehensive range compared with the average garage or accessory shop. Having said that, accessory shops often offer excellent quality tools at discount prices, so it pays to shop around. There are plenty of tools around at reasonable prices, but always aim to purchase items which meet the relevant national safety standards. If in doubt, seek the advice of the shop proprietor or manager before making a purchase.

The basis of any toolkit is a set of spanners. While open-ended spanners with their slim jaws, are useful for working on awkwardly-positioned nuts, ring spanners have advantages in that they grip the nut far more positively. There is less risk of the spanner slipping off the nut and damaging it, for this reason alone ring spanners are to be preferred. Ideally, the home mechanic should acquire a set of each, but if expense rules this out a set of combination spanners (open-ended at one end and with a ring of the same size at the other) will provide a good compromise. Another item which is so useful it should be considered an essential requirement for any home mechanic is a set of socket spanners. These are available in a variety of drive sizes. It is recommended that the ½-inch drive type is purchased to begin with as although bulkier and more expensive than the ⅜-inch type, the larger size is far more common and will accept a greater variety of torque wrenches, extension pieces and socket sizes. The socket set should comprise sockets of sizes between 8 and 24 mm, a reversible ratchet drive, an extension bar of about 10 inches in length, a spark plug socket with a rubber insert, and a universal joint. Other attachments can be added to the set at a later date.

Maintenance and minor repair tool kit

Set of spanners 8 – 24 mm
Set of sockets and attachments
Spark plug spanner with rubber insert – 10, 12, or 14 mm as appropriate
Adjustable spanner
C-spanner/pin spanner
Torque wrench (same size drive as sockets)
Set of screwdrivers (flat blade)
Set of screwdrivers (cross-head)
Set of Allen keys 4 – 10 mm
Impact screwdriver and bits
Ball pein hammer – 2 lb
Hacksaw (junior)
Self-locking pliers – Mole grips or vice grips
Pliers – combination
Pliers – needle nose
Wire brush (small)
Soft-bristled brush
Tyre pump
Tyre pressure gauge
Tyre tread depth gauge
Oil can
Fine emery cloth
Funnel (medium size)
Drip tray
Grease gun
Set of feeler gauges
Brake bleeding kit
Strobe timing light
Continuity tester (dry battery and bulb)
Soldering iron and solder
Wire stripper or craft knife
PVC insulating tape
Assortment of split pins, nuts, bolts, and washers

Repair and overhaul toolkit

The tools in this list are virtually essential for anyone undertaking major repairs to a motorcycle and are additional to the tools listed above. Concerning Torx driver bits, Torx screws are encountered on some of the more modern machines where their use is restricted to fastening certain components inside the engine/gearbox unit. It is therefore recommended that if Torx bits cannot be borrowed from a local dealer, they are purchased individually as the need arises. They are not in regular use in the motor trade and will therefore only be available in specialist tool shops.

Plastic or rubber soft-faced mallet
Torx driver bits
Pliers – electrician's side cutters
Circlip pliers – internal (straight or right-angled tips are available)
Circlip pliers – external
Cold chisel
Centre punch
Pin punch
Scriber
Scraper (made from soft metal such as aluminium or copper)
Soft metal drift
Steel rule/straight edge
Assortment of files
Electric drill and bits
Wire brush (large)
Soft wire brush (similar to those used for cleaning suede shoes)
Sheet of plate glass
Hacksaw (large)
Stud extractor set (E-Z out)

Specialized tools

This is not a list of the tools made by the machine's manufacturer to carry out a specific task on a limited range of models. Occasional references are made to such tools in the text of this manual and, in general, an alternative method of carrying out the task without the manufacturer's tool is given where possible. The tools mentioned in this list are those which are not used regularly and are expensive to buy in view of their infrequent use. Where this is the case it may be possible to hire or borrow the tools against a deposit from a local dealer or tool hire shop. An alternative is for a group of friends or a motorcycle club to join in the purchase.

Piston ring compressor
Universal bearing puller
Cylinder bore honing attachment (for electric drill)
Micrometer set
Vernier calipers
Dial gauge set
Cylinder compression gauge
Vacuum gauge set
Multimeter
Dwell meter/tachometer

Care and maintenance of tools

Whatever the quality of the tools purchased, they will last much longer if cared for. This means in practice ensuring that a tool is used for its intended purpose; for example screwdrivers should not be used as a substitute for a centre punch, or as chisels. Always remove dirt or grease and any metal particles but remember that a light film of oil will prevent rusting if the tools are infrequently used. The common tools can be kept together in a large box or tray but the more delicate, and more expensive, items should be stored separately where they cannot be damaged. When a tool is damaged or worn out, be sure to renew it immediately. It is false economy to continue to use a worn spanner or screwdriver which may slip and cause expensive damage to the component being worked on.

Fastening systems

Fasteners, basically, are nuts, bolts and screws used to hold two or more parts together. There are a few things to keep in mind when working with fasteners. Almost all of them use a locking device of some type; either a lock washer, lock nut, locking tab or thread adhesive. All threaded fasteners should be clean, straight, have undamaged threads and undamaged corners on the hexagon head where the spanner fits. Develop the habit of replacing all damaged nuts and bolts with new ones.

Rusted nuts and bolts should be treated with a rust penetrating fluid to ease removal and prevent breakage. After applying the rust penetrant, let it 'work' for a few minutes before trying to loosen the nut or bolt. Badly rusted fasteners may have to be chiseled off or removed with a special nut breaker, available at tool shops.

Flat washers and lock washers, when removed from an assembly should always be replaced exactly as removed. Replace any damaged washers with new ones. Always use a flat washer between a lock washer and any soft metal surface (such as aluminium), thin sheet metal or plastic. Special lock nuts can only be used once or twice before they lose their locking ability and must be renewed.

If a bolt or stud breaks off in an assembly, it can be drilled out and removed with a special tool called an E-Z out. Most dealer service departments and motorcycle repair shops can perform this task, as well as others (such as the repair of threaded holes that have been stripped out).

Spanner size comparison

Jaw gap (in)	Spanner size
0.250	$\frac{1}{4}$ in AF
0.276	7 mm
0.313	$\frac{5}{16}$ in AF
0.315	8 mm
0.344	$\frac{11}{32}$ in AF; $\frac{1}{8}$ in Whitworth
0.354	9 mm
0.375	$\frac{3}{8}$ in AF
0.394	10 mm
0.433	11 mm
0.438	$\frac{7}{16}$ in AF
0.445	$\frac{3}{16}$ in Whitworth; $\frac{1}{4}$ in BSF
0.472	12 mm
0.500	$\frac{1}{2}$ in AF
0.512	13 mm
0.525	$\frac{1}{4}$ in Whitworth; $\frac{5}{16}$ in BSF
0.551	14 mm
0.563	$\frac{9}{16}$ in AF
0.591	15 mm
0.600	$\frac{5}{16}$ in Whitworth; $\frac{3}{8}$ in BSF
0.625	$\frac{5}{8}$ in AF
0.630	16 mm
0.669	17 mm
0.686	$\frac{11}{16}$ in AF
0.709	18 mm
0.710	$\frac{3}{8}$ in Whitworth; $\frac{7}{16}$ in BSF
0.748	19 mm
0.750	$\frac{3}{4}$ in AF
0.813	$\frac{13}{16}$ in AF
0.820	$\frac{7}{16}$ in Whitworth; $\frac{1}{2}$ in BSF
0.866	22 mm
0.875	$\frac{7}{8}$ in AF
0.920	$\frac{1}{2}$ in Whitworth; $\frac{9}{16}$ in BSF
0.938	$\frac{15}{16}$ in AF

Jaw gap (in)	Spanner size
0.945	24 mm
1.000	1 in AF
1.010	$\frac{9}{16}$ in Whitworth; $\frac{5}{8}$ in BSF
1.024	26 mm
1.063	$1\frac{1}{16}$ in AF; 27 mm
1.100	$\frac{5}{8}$ in Whitworth; $\frac{11}{16}$ in BSF
1.125	$1\frac{1}{8}$ in AF
1.181	30 mm
1.200	$\frac{11}{16}$ in Whitworth; $\frac{3}{4}$ in BSF
1.250	$1\frac{1}{4}$ in AF
1.260	32 mm
1.300	$\frac{3}{4}$ in Whitworth; $\frac{7}{8}$ in BSF
1.313	$1\frac{5}{16}$ in AF
1.390	$\frac{13}{16}$ in Whitworth; $\frac{15}{16}$ in BSF
1.417	36 mm
1.438	$1\frac{7}{16}$ in AF
1.480	$\frac{7}{8}$ in Whitworth; 1 in BSF
1.500	$1\frac{1}{2}$ in AF
1.575	40 mm; $\frac{15}{16}$ in Whitworth
1.614	41 mm
1.625	$1\frac{5}{8}$ in AF
1.670	1 in Whitworth; $1\frac{1}{8}$ in BSF
1.688	$1\frac{11}{16}$ in AF
1.811	46 mm
1.813	$1\frac{13}{16}$ in AF
1.860	$1\frac{1}{8}$ in Whitworth; $1\frac{1}{4}$ in BSF
1.875	$1\frac{7}{8}$ in AF
1.969	50 mm
2.000	2 in AF
2.050	$1\frac{1}{4}$ in Whitworth; $1\frac{3}{8}$ in BSF
2.165	55 mm
2.362	60 mm

Standard torque settings

Specific torque settings will be found at the end of the specifications section of each Chapter. Where no figure is given, bolts and nuts should be tightened according to the table shown below. Note that the size of nuts given is that measured across its flats, whilst bolt sizes refer to the diameter of the bolt:

Nut size	Bolt size	kgf m	lbf ft
10 mm	6 mm	0.6	4.3
12 mm	8 mm	1.5	11.0
14 mm	10 mm	3.0	22.0
17 mm	12 mm	5.5	40.0
19 mm	14 mm	8.5	61.0
22 mm	16 mm	13.0	94.0

Choosing and fitting accessories

The range of accessories available to the modern motorcyclist is almost as varied and bewildering as the range of motorcycles. This Section is intended to help the owner in choosing the correct equipment for his needs and to avoid some of the mistakes made by many riders when adding accessories to their machines. It will be evident that the Section can only cover the subject in the most general terms and so it is recommended that the owner, having decided that he wants to fit, for example, a luggage rack or carrier, seeks the advice of several local dealers and the owners of similar machines. This will give a good idea of what makes of carrier are easily available, and at what price. Talking to other owners will give some insight into the drawbacks or good points of any one make. A walk round the motorcycles in car parks or outside a dealer will often reveal the same sort of information.

The first priority when choosing accessories is to assess exactly what one needs. It is, for example, pointless to buy a large heavy-duty carrier which is designed to take the weight of fully laden panniers and topbox when all you need is a place to strap on a set of waterproofs and a lunchbox when going to work. Many accessory manufacturers have ranges of equipment to cater for the individual needs of different riders and this point should be borne in mind when looking through a dealer's catalogues. Having decided exactly what is required and the use to which the accessories are going to be put, the owner will need a few hints on what to look for when making the final choice. To this end the Section is now sub-divided to cover the more popular accessories fitted. Note that it is in no way a customizing guide, but merely seeks to outline the practical considerations to be taken into account when adding aftermarket equipment to a motorcycle.

Fairings and windscreens

A fairing is possibly the single, most expensive, aftermarket item to be fitted to any motorcycle and, therefore, requires the most thought before purchase. Fairings can be divided into two main groups: front fork mounted handlebar fairings and windscreens, and frame mounted fairings.

The first group, the front fork mounted fairings, are becoming far more popular than was once the case, as they offer several advantages over the second group. Front fork mounted fairings generally are much easier and quicker to fit, involve less modification to the motorcycle, do not as a rule restrict the steering lock, permit a wider selection of handlebar styles to be used, and offer adequate protection for much less money than the frame mounted type. They are also lighter, can be swapped easily between different motorcycles, and are available in a much greater variety of styles. Their main disadvantages are that they do not offer as much weather protection as the frame mounted types, rarely offer any storage space, and, if poorly fitted or naturally incompatible, can have an adverse effect on the stability of the motorcycle.

The second group, the frame mounted fairings, are secured so rigidly to the main frame of the motorcycle that they can offer a substantial amount of protection to motorcycle and rider in the event of a crash. They offer almost complete protection from the weather and, if double-skinned in construction, can provide a great deal of useful storage space. The feeling of peace, quiet and complete relaxation encountered when riding behind a good full fairing has to be experienced to be believed. For this reason full fairings are considered essential by most touring motorcyclists and by many people who ride all year round. The main disadvantages of this type are that fitting can take a long time, often involving removal or modification of standard motorcycle components, they restrict the steering lock and they can add up to about 40 lb to the weight of the machine. They do not usually affect the stability of the machine to any great extent once the front tyre pressure and suspension have been adjusted to compensate for the extra weight, but can be affected by sidewinds.

The first thing to look for when purchasing a fairing is the quality of the fittings. A good fairing will have strong, substantial brackets constructed from heavy-gauge tubing; the brackets must be shaped to fit the frame or forks evenly so that the minimum of stress is imposed on the assembly when it is bolted down. The brackets should be properly painted or finished – a nylon coating being the favourite of the better manufacturers – the nuts and bolts provided should be of the same thread and size standard as is used on the motorcycle and be properly plated. Look also for shakeproof locking nuts or locking washers to ensure that everything remains securely tightened down. The fairing shell is generally made from one of two materials: fibreglass or ABS plastic. Both have their advantages and disadvantages, but the main consideration for the owner is that fibreglass is much easier to repair in the event of damage occurring to the fairing. Whichever material is used, check that it is properly finished inside as well as out, that the edges are protected by beading and that the fairing shell is insulated from vibration by the use of rubber grommets at all mounting points. Also be careful to check that the windscreen is retained by plastic bolts which will snap on impact so that the windscreen will break away and not cause personal injury in the event of an accident.

Having purchased your fairing or windscreen, read the manufacturer's fitting instructions very carefully and check that you have all the necessary brackets and fittings. Ensure that the mounting brackets are located correctly and bolted down securely. Note that some manufacturers use hose clamps to retain the mounting brackets; these should be discarded as they are convenient to use but not strong enough for the task. Stronger clamps should be substituted; car exhaust pipe clamps of suitable size would be a good alternative. Ensure that the front forks can turn through the full steering lock available without fouling the fairing. With many types of frame-mounted fairing the handlebars will have to be altered or a different type fitted and the steering lock will be restricted by stops provided with the fittings. Also check that the fairing does not foul the front wheel or mudguard, in any steering position, under full fork compression. Re-route any cables, brake pipes or electrical wiring which may snag on the fairing and take great care to protect all electrical connections, using insulating tape. If the manufacturer's instructions are followed carefully at every stage no serious problems should be encountered. Remember that hydraulic pipes that have been disconnected must be carefully re-tightened and the hydraulic system purged of air bubbles by bleeding.

Two things will become immediately apparent when taking a motorcycle on the road for the first time with a fairing – the first is the tendency to underestimate the road speed because of the lack of wind pressure on the body. This must be very carefully watched until one has grown accustomed to riding behind the fairing. The second thing is the alarming increase in engine noise which is an unfortunate but inevitable by-product of fitting any type of fairing or windscreen, and is caused by normal engine noise being reflected, and in some cases amplified, by the flat surface of the fairing.

Luggage racks or carriers

Carriers are possibly the commonest item to be fitted to modern motorcycles. They vary enormously in size, carrying capacity, and durability. When selecting a carrier, always look for one which is made specifically for your machine and which is bolted on with as few separate brackets as possible. The universal-type carrier, with its mass of brackets and adaptor pieces, will generally prove too weak to be of any real use. A good carrier should bolt to the main frame, generally using the two suspension unit top mountings and a mudguard mounting bolt as attachment points, and have its luggage platform as low and as far forward as possible to minimise the effect of any load on the machine's stability. Look for good quality, heavy gauge tubing, good welding and good finish. Also ensure that the carrier does not prevent opening of the seat, sidepanels or tail compartment, as appropriate. When using a carrier, be very careful not to overload it. Excessive weight placed so high and so far to the rear of any motorcycle will have an adverse effect on the machine's steering and stability.

Luggage

Motorcycle luggage can be grouped under two headings: soft and hard. Both types are available in many sizes and styles and have advantages and disadvantages in use.

Soft luggage is now becoming very popular because of its lower cost and its versatility. Whether in the form of tankbags, panniers, or strap-on bags, soft luggage requires in general no brackets and no modification to the motorcycle. Equipment can be swapped easily from one motorcycle to another and can be fitted and removed in seconds. Awkwardly shaped loads can easily be carried. The disadvantages of soft luggage are that the contents cannot be secure against the casual thief, very little protection is afforded in the event of a crash, and waterproofing is generally poor. Also, in the case of panniers, carrying capacity is restricted to approximately 10 lb, although this amount will vary considerably depending on the manufacturer's recommendation. When purchasing soft luggage, look for good quality material, generally vinyl or nylon, with strong, well-stitched attachment points. It is always useful to have separate pockets, especially on tank bags, for items which will be needed on the journey. When purchasing a tank bag, look for one which has a separate, well-padded, base. This will protect the tank's paintwork and permit easy access to the filler cap at petrol stations.

Hard luggage is confined to two types: panniers, and top boxes or tail trunks. Most hard luggage manufacturers produce matching sets of these items, the basis of which is generally that manufacturer's own heavy-duty luggage rack. Variations on this theme occur in the form of separate frames for the better quality panniers, fixed or quickly-detachable luggage, and in size and carrying capacity. Hard luggage offers a reasonable degree of security against theft and good protection against weather and accident damage. Carrying capacity is greater than that of soft luggage, around 15 – 20 lb in the case of panniers, although top boxes should never be loaded as much as their apparent capacity might imply. A top box should only be used for lightweight items, because one that is heavily laden can have a serious effect on the stability of the machine. When purchasing hard luggage look for the same good points as mentioned under fairings and windscreens, ie good quality mounting brackets and fittings, and well-finished fibreglass or ABS plastic cases. Again as with fairings, always purchase luggage made specifically for your motorcycle, using as few separate brackets as possible, to ensure that everything remains securely bolted in place. When fitting hard luggage, be careful to check that the rear suspension and brake operation will not be impaired in any way and remember that many pannier kits require re-siting of the indicators. Remember also that a non-standard exhaust system may make fitting extremely difficult.

Handlebars

The occupation of fitting alternative types of handlebar is extremely popular with modern motorcyclists, whose motives may vary from the purely practical, wishing to improve the comfort of their machines, to the purely aesthetic, where form is more important than function. Whatever the reason, there are several considerations to be borne in mind when changing the handlebars of your machine. If fitting lower bars, check carefully that the switches and cables do not foul the petrol tank on full lock and that the surplus length of cable, brake pipe, and electrical wiring are smoothly and tidily disposed of. Avoid tight kinks in cable or brake pipes which will produce stiff controls or the premature and disastrous failure of an overstressed component. If necessary, remove the petrol tank and re-route the cable from the engine/gearbox unit upwards, ensuring smooth gentle curves are produced. In extreme cases, it will be necessary to purchase a shorter brake pipe to overcome this problem. In the case of higher handlebars than standard it will almost certainly be necessary to purchase extended cables and brake pipes. Fortunately, many standard motorcycles have a custom version which will be equipped with higher handlebars and, therefore, factory-built extended components will be available from your local dealer. It is not usually necessary to extend electrical wiring, as switch clusters may be used on several different motorcycles, some being custom versions. This point should be borne in mind however when fitting extremely high or wide handlebars.

When fitting different types of handlebar, ensure that the mounting clamps are correctly tightened to the manufacturer's specifications and that cables and wiring, as previously mentioned, have smooth easy runs and do not snag on any part of the motorcycle throughout the full steering lock. Ensure that the fluid level in the front brake master cylinder remains level to avoid any chance of air entering the hydraulic system. Also check that the cables are adjusted correctly and that all handlebar controls operate correctly and can be easily reached when riding.

Crashbars

Crashbars, also known as engine protector bars, engine guards, or case savers, are extremely useful items of equipment which can contribute protection to the machine's structure if a crash occurs. They do not, as has been inferred in the US, prevent the rider from crashing, or necessarily prevent rider injury should a crash occur.

It is recommended that only the smaller, neater, engine protector type of crashbar is considered. This type will offer protection while restricting, as little as is possible, access to the engine and the machine's ground clearance. The crashbars should be designed for use specifically on your machine, and should be constructed of heavy-gauge tubing with strong, integral mounting brackets. Where possible, they should bolt to a strong lug on the frame, usually at the engine mounting bolts.

The alternative type of crashbar is the larger cage type. This type is not recommended in spite of their appearance which promises some protection to the rider as well as to the machine. The larger amount of leverage imposed by the size of this type of crashbar increases the risk of severe frame damage in the event of an accident. This type also decreases the machine's ground clearance and restricts access to the engine. The amount of protection afforded the rider is open to some doubt as the design is based on the premise that the rider will stay in the normally seated position during an accident, and the crash bar structure will not itself fail. Neither result can in any way be guaranteed.

As a general rule, always purchase the best, ie usually the most expensive, set of crashbars you an afford. The investment will be repaid by minimising the amount of damage incurred, should the machine be involved in an accident. Finally, avoid the universal type of crashbar. This should be regarded only as a last resort to be used if no alternative exists. With its usual multitude of separate brackets and spacers, the universal crashbar is far too weak in design and construction to be of any practical value.

Exhaust systems

The fitting of aftermarket exhaust systems is another extremely popular pastime amongst motorcyclists. The usual motive is to gain more performance from the engine but other considerations are to gain more ground clearance, to lose weight from the motorcycle, to obtain a more distinctive exhaust note or to find a cheaper alternative to the manufacturer's original equipment exhaust system. Original equipment exhaust systems often cost more and may well have a relatively short life. It should be noted that it is rare for an aftermarket exhaust system alone to give a noticeable increase in the engine's power output. Modern motorcycles are designed to give the highest power output possible allowing for factors such as quietness, fuel economy, spread of power, and long-term reliability. If there were a magic formula which allowed the exhaust system to produce more power without affecting these other considerations you can be sure that the manufacturers, with their large research and development facilities, would have found it and made use of it. Performance increases of a worthwhile and noticeable nature only come from well-tried and properly matched modifications to the entire engine, from the air filter, through the carburettors, port timing or camshaft and valve design, combustion chamber shape, compression ratio, and the exhaust system. Such modifications are well outside the scope of this manual but interested owners might refer to specialist books produced by the publisher of this manual which go into the whole subject in great detail.

Whatever your motive for wishing to fit an alternative exhaust system, be sure to seek expert advice before doing so. Changes to the carburettor jetting will almost certainly be required for which you must consult the exhaust system manufacturer. If he cannot supply adequately specific information it is reasonable to assume that insufficient development work has been carried out, and that particular make should be avoided. Other factors to be borne in mind are whether the exhaust system allows the use of both centre and side stands, whether it allows sufficient access to permit oil and filter changing and whether modifications are necessary to the standard exhaust system. Many two-stroke expansion chamber systems require the use of the standard exhaust pipe; this is all very well if the standard exhaust pipe and silencer are separate units but can cause problems if the two, as with so many modern two-strokes, are a one-piece unit. While the exhaust pipe can be removed easily by means of a hacksaw it is not so easy to refit the original silencer should you at any time wish to return the machine to standard trim. The same applies to several four-stroke systems.

On the subject of the finish of aftermarket exhausts, avoid black-painted systems unless you enjoy painting. As any trail-bike owner will tell you, rust has a great affinity for black exhausts and re-painting or rust removal becomes a task which must be carried out with monotonous regularity. A bright chrome finish is, as a general rule, a far better proposition as it is much easier to keep clean and to prevent rusting. Although the general finish of aftermarket exhaust systems is not always up to the standard of the original equipment the lower cost of such systems does at least reflect this fact.

When fitting an alternative system always purchase a full set of new exhaust gaskets, to prevent leaks. Fit the exhaust first to the cylinder head or barrel, as appropriate, tightening the retaining nuts or bolts by hand only and then line up the exhaust rear mountings. If the new system is a one-piece unit and the rear mountings do not line up exactly, spacers must be fabricated to take up the difference. Do not force the system into place as the stress thus imposed will rapidly cause cracks and splits to appear. Once all the mountings are loosely fixed, tighten the retaining nuts or bolts securely, being careful not to overtighten them. Where the motorcycle manufacturer's torque settings are available, these should be used. Do not forget to carry out any carburation changes recommended by the exhaust system's manufacturer.

Electrical equipment

The vast range of electrical equipment available to motorcyclists is so large and so diverse that only the most general outline can be given here. Electrical accessories vary from electronic ignition kits fitted to replace contact breaker points, to additional lighting at the front and rear, more powerful horns, various instruments and gauges, clocks, anti-theft systems, heated clothing, CB radios, radio-cassette players, and intercom systems, to name but a few of the more popular items of equipment.

As will be evident, it would require a separate manual to cover this subject alone and this section is therefore restricted to outlining a few basic rules which must be borne in mind when fitting electrical equipment. The first consideration is whether your machine's electrical system has enough reserve capacity to cope with the added demand of the accessories you wish to fit. The motorcycle's manufacturer or importer should be able to furnish this sort of information and may also be able to offer advice on uprating the electrical system. Failing this, a good dealer or the accessory manufacturer may be able to help. In some cases, more powerful generator components may be available, perhaps from another motorcycle in the manufacturer's range. The second consideration is the legal requirements in force in your area. The local police may be prepared to help with this point. In the UK for example, there are strict regulations governing the position and use of auxiliary riding lamps and fog lamps.

When fitting electrical equipment always disconnect the battery first to prevent the risk of a short-circuit, and be careful to ensure that all connections are properly made and that they are waterproof. Remember that many electrical accessories are designed primarily for use in cars and that they cannot easily withstand the exposure to vibration and to the weather. Delicate components must be rubber-mounted to insulate them from vibration, and sealed carefully to prevent the entry of rainwater and dirt. Be careful to follow exactly the accessory manufacturer's instructions in conjunction with the wiring diagram at the back of this manual.

Accessories – general

Accessories fitted to your motorcycle will rapidly deteriorate if not cared for. Regular washing and polishing will maintain the finish and will provide an opportunity to check that all mounting bolts and nuts are securely fastened. Any signs of chafing or wear should be watched for, and the cause cured as soon as possible before serious damage occurs.

As a general rule, do not expect the re-sale value of your motorcycle to increase by an amount proportional to the amount of money and effort put into fitting accessories. It is usually the case that an absolutely standard motorcycle will sell more easily at a better price than one that has been modified. If you are in the habit of exchanging your machine for another at frequent intervals, this factor should be borne in mind to avoid loss of money.

Fault diagnosis

Contents

1 Introduction

This Section provides an easy reference-guide to the more common ailments that are likely to afflict your machine. Obviously, the opportunities are almost limitless for faults to occur as a result of obscure failures, and to try and cover all eventualities would require a book. Indeed, a number have been written on the subject.

Successful fault diagnosis is not a mysterious 'black art' but the application of a bit of knowledge combined with a systematic and logical approach to the problem. Approach any fault diagnosis by first accurately identifying the symptom and then checking through the list of possible causes, starting with the simplest or most obvious and progressing in stages to the most complex. Take nothing for granted, but above all apply liberal quantities of common sense.

The main symptom of a fault is given in the text as a major heading below which are listed, as Section headings, the various systems or areas which may contain the fault. Details of each possible cause for a fault and the remedial action to be taken are given, in brief, in the paragraphs below each Section heading. Further information should be sought in the relevant Chapter.

Engine does not start when turned over

2 No fuel flow to carburettors

- Float bowls have run dry. Prime float bowls by turning fuel tap to the "PRI" position. If this fails to cure the problem, pull off the fuel pipe and check that fuel flows from the tap outlet when set to "PRI".
- Tank filler cap air vent obstructed. This can prevent fuel from flowing into the carburettor float bowl because air cannot enter the fuel tank to replace it. The problem is more likely to appear when the machine is being ridden. Check by listening close to the filler cap and releasing it. A hissing noise indicates that a blockage is present. Remove the cap and clear the vent hole with wire or by using an air line from the inside of the cap.
- Fuel tap or filter blocked. Blockage may be due to accumulation of rust or paint flakes from the tank's inner surface or of foreign matter from contaminated fuel. Remove the tap and clean it and the filter. Look also for water droplets in the fuel.
- Fuel line blocked. Blockage of the fuel line is more likely to result from a kink in the line rather than the accumulation of debris.

3 Fuel not reaching cylinder

- Float chambers not filling. Caused by float needles or floats sticking in up position. This may occur after the machine has been left standing for an extended length of time allowing the fuel to evaporate. When this occurs a gummy residue is often left which hardens to a varnish-like substance. This condition may be worsened by corrosion and crystaline deposits produced prior to the total evaporation of contaminated fuel. Sticking of the float needles may also be caused by wear. In any case removal of the float chambers will be necessary for inspection and cleaning.
- Blockage in starting circuit, slow running circuit or jets. Blockage of these items may be attributable to debris from the fuel tank by-passing the filter system or to gumming up as described in paragraph 1. Water droplets in the fuel will also block jets and passages. The carburettor should be dismantled for cleaning.
- Fuel level too low. The fuel level in the float chambers is controlled by float height. The fuel level may increase with wear or damage but will never reduce, thus a low fuel level is an inherent rather than developing condition. Check the float height, renewing the float or needle if required.

4 Engine flooding

- Float valve needle worn or stuck open. A piece of rust or other debris can prevent correct seating of the needle against the valve seat thereby permitting an uncontrolled flow of fuel. Similarly, a worn needle or needle seat will prevent valve closure. Dismantle the carburettor float bowl for cleaning and, if necessary, renewal of the worn components.
- Fuel level too high. The fuel level is controlled by the float height which may increase due to wear of the float needle, pivot pin or operating tang. Check the float height, and make any necessary adjustments. A leaking float will cause an increase in fuel level, and thus should be renewed.
- Cold starting mechanism. Check the choke (starter mechanism) for correct operation. If the mechanism jams in the 'On' position subsequent starting of a hot engine will be difficult.
- Blocked air filter. A badly restricted air filter will cause flooding. Check the filter and clean or renew as required. A collapsed inlet hose will have a similar effect. Check that the air filter inlet has not become blocked by a rag or similar item.

5 No spark at plugs

- Ignition switch not on.
- Engine stop switch off.
- Fuse blown. Check fuse for ignition circuit. See wiring diagram.
- Spark plugs dirty, oiled or 'whiskered'. Because the induction mixture of a two-stroke engine is inclined to be of a rather oily nature it is comparatively easy to foul the plug electrodes, especially where there have been repeated attempts to start the engine. A machine used for short journeys will be more prone to fouling because the engine may never reach full operating temperature, and the deposits will not burn off. On rare occasions a change of plug grade may be required but the advice of a dealer should be sought before making such a change. 'Whiskering' is a comparatively rare occurrence on modern machines but may be encountered where pre-mixed petrol and oil (petroil) lubrication is employed. An electrode deposit in the form of a barely visible filament across the plug electrodes can short circuit the plug and prevent its sparking. On all two-stroke machines it is a sound precaution to carry a pair of new spare spark plugs for substitution in the event of fouling problems.
- Spark plug failure. Clean the spark plugs thoroughly and reset their electrode gaps. Refer to the spark plug section If the spark plugs short internally or have sustained visible damage to the electrodes, core or ceramic insulator they should be renewed. On rare occasions a plug that appears to spark vigorously will fail to do so when refitted to the engine and subjected to the compression pressure in the cylinder.
- Spark plug caps or high tension (HT) leads faulty. Check condition and security. Replace if deterioration is evident. Most spark plug caps have an internal resistor designed to inhibit electrical interference with radio and television sets. On rare occasions the resistor may break down, thus preventing sparking. If this is suspected, fit new caps as a precaution.
- Spark plug caps loose. Check that the spark plug caps fit securely over the plugs and, where fitted, the screwed terminals on the plug ends are secure.
- Shorting due to moisture. Certain parts of the ignition system are susceptible to shorting when the machine is ridden or parked in wet weather. Check particularly the area from the spark plug caps back to the ignition coil. A water dispersant spray may be used to dry out waterlogged components. Recurrence of the problem can be prevented by using an ignition sealant spray after drying out and cleaning.
- Ignition or stop switch shorted. May be caused by water corrosion or wear. Water dispersant and contact cleaning sprays may be used. If this fails to overcome the problem dismantling and visual inspection of the switches will be required.
- Shorting or open circuit in wiring. Failure in any wire connecting any of the ignition components will cause ignition malfunction. Check also that all connections are clean, dry and tight.
- Ignition coil failure. Check the coil, referring to Chapter 4.
- Failure in CDI system. Refer to ignition Chapter for details of checks and tests on the system.

6 Weak spark at plugs

- Feeble sparking at the plug may be caused by any of the faults mentioned in the preceding Section other than those items in the first three paragraphs. Check first the spark plugs, these being the most likely culprits.

7 Compression low

- Spark plugs loose. This will be self-evident on inspection, and may be accompanied by a hissing noise when the engine is turned over. Remove the plugs and check that the threads in the cylinder head are not damaged. Check also that the plug sealing washers are in good condition.
- Cylinder head gasket leaking. This condition is often accompanied by a high pitched squeak from around the cylinder head and oil loss, and may be caused by insufficiently tightened cylinder head fasteners, a warped cylinder head or mechanical failure of the gasket material. Re-torqueing the fasteners to the correct specification may seal the leak in some instances but if damage has occurred this course of action will provide, at best, only a temporary cure.
- Low crankcase compression. This can be caused by worn main bearings and seals and will upset the incoming fuel/air mixture. A good seal in these areas is essential on any two-stroke engine.
- Piston rings sticking or broken. Sticking of the piston rings may be caused by seizure due to lack of lubrication or heating as a result of poor carburation or incorrect fuel type. Gumming of the rings may result from lack of use, or carbon deposits in the ring grooves. Broken rings result from over-revving, overheating or general wear. In either case a top-end overhaul will be required.

Engine stalls after starting

8 General causes

- Improper cold start mechanism operation. Check that the operating controls function smoothly and, where applicable, are correctly adjusted. A cold engine may not require application of an enriched mixture to start initially but may baulk without choke once firing. Likewise a hot engine may start with an enriched mixture but will stop almost immediately if the choke is inadvertently in operation.
- Ignition malfunction. See Section 9, Weak spark at plugs.
- Carburettors incorrectly adjusted. Maladjustment of the mixture strength or idle speed may cause the engine to stop immediately after starting. See Chapter 2.
- Fuel contamination. Check for filter blockage by debris or water which reduces, but does not completely stop, fuel flow or blockage of the slow speed circuit in the carburettor by the same agents. If water is present it can often be seen as droplets in the bottom of the float bowl. Clean the filter and, where water is in evidence, drain and flush the fuel tank and float bowl.
- Intake air leak. Check for security of the carburettor mounting and hose connections, and for cracks or splits in the hoses. Check also that the carburettor top is secure and that the vacuum gauge adaptor plug (where fitted) is tight.
- Air filter blocked or omitted. A blocked filter will cause an over-rich mixture; the omission of a filter will cause an excessively weak mixture. Both conditions will have a detrimental effect on carburation. Clean or renew the filter as necessary.
- Fuel filler cap air vent blocked. Usually caused by dirt or water. Clean the vent orifice.
- Choked exhaust system. Caused by excessive carbon build-up in the system, particularly around the silencer baffles. In many cases these can be detached for cleaning, though mopeds have one-piece systems which require a rather different approach. Refer to Chapter 2 for further information.
- Excessive carbon build-up in the engine. This can result from failure to decarbonise the engine at the specified interval or through excessive oil consumption. Check pump adjustment.

Poor running at idle and low speed

9 Weak spark at plug or erratic firing

- Battery voltage low. In certain conditions low battery charge, especially when coupled with a badly sulphated battery, may result in misfiring. If the battery is in good general condition it should be recharged; an old battery suffering from sulphated plates should be renewed.
- Spark plugs fouled, faulty or incorrectly adjusted. See Section 4 or refer to Chapter 3.
- Spark plug caps or high tension leads shorting. Check the condition of both these items ensuring that they are in good condition and dry and that the caps are fitted correctly.
- Spark plug type incorrect. Fit plugs of correct type and heat range as given in Specifications. In certain conditions plugs of hotter or colder type may be required for normal running.
- Faulty ignition coil. Partial failure of the coil internal insulation will diminish the performance of the coil. No repair is possible, a new component must be fitted.
- Defective flywheel generator ignition source. Refer to Chapter 4 for further details on test procedures.

10 Fuel/air mixture incorrect

- Intake air leak. Check carburettor mountings and air cleaner hoses for security and signs of splitting. Ensure that carburettor top is tight and that the vacuum gauge take-off plug (where fitted) is tight.
- Mixture strength incorrect. Adjust slow running mixture strength using pilot adjustment screw.
- Carburettor synchronisation.
- Pilot jet or slow running circuit blocked. The carburettors should be removed and dismantled for thorough cleaning. Blow through all jets and air passages with compressed air to clear obstructions.
- Air cleaner clogged or omitted. Clean or fit air cleaner element as necessary. Check also that the element and air filter cover are correctly seated.
- Cold start mechanism in operation. Check that the choke has not been left on inadvertently and the operation is correct. Where applicable check the operating cable free play.
- Fuel level too high or too low. Check the float height and adjust as necessary. See Section 3 or 4.
- Fuel tank air vent obstructed. Obstruction usually caused by dirt or water. Clean vent orifice.

11 Compression low

- See Section 7.

Acceleration poor

12 General causes

- All items as for previous Section.
- Choked air filter. Failure to keep the air filter element clean will allow the build-up of dirt with proportional loss of performance. In extreme cases of neglect acceleration will suffer.
- Choked exhaust system. This can result from failure to remove accumulations of carbon from the silencer baffles at the prescribed intervals. The increased back pressure will make the machine noticeably sluggish. Refer to Chapter 3 for further information on decarbonisation.
- Excessive carbon build-up in the engine. This can result from failure to decarbonise the engine at the specified intervals or through excessive oil consumption. Check pump adjustment.
- Ignition timing incorrect. Check the ignition timing as described in Chapter 4. Where no provision for adjustment exists, test the electronic ignition components and renew as required.
- Carburation fault. See Section 10.
- Mechanical resistance. Check that the brakes are not binding. On small machines in particular note that the increased rolling resistance caused by under-inflated tyres may impede acceleration.

Poor running or lack of power at high speeds

13 Weak spark at plugs or erratic firing

- All items as for Section 9.
- HT leads insulation failure. Insulation failure of the HT leads and spark plug caps due to old age or damage can cause shorting when the engine is driven hard. This condition may be less noticeable, or not noticeable at all at lower engine speeds.

14 Fuel/air mixture incorrect

- All items as for Section 10, with the exception of items relative exclusively to low speed running.
- Main jets blocked. Debris from contaminated fuel, or from the fuel tank, and water in the fuel can block the main jets. Clean the fuel filter, the float bowl area, and if water is present, flush and refill the fuel tank.
- Main jets are the wrong size. The standard carburettor jetting is for sea level atmospheric pressure. For high altitudes, usually above 5000 ft, smaller main jets will be required.
- Jet needles and needle jets worn. These can be renewed individually but should be renewed as a pair. Renewal of both items requires partial dismantling of the carburettors.
- Air bleed holes blocked. Dismantle carburettors and use compressed air to blow out all air passages.
- Reduced fuel flow. A reduction in the maximum fuel flow from the fuel tank to the carburettors will cause fuel starvation, proportionate to the engine speed. Check for blockages through debris or a kinked fuel line.

15 Compression low

- See Section 7.

Knocking or pinking

16 General causes

- Carbon build-up in combustion chamber. After high mileages have been covered large accumulations of carbon may occur. This may glow red hot and cause premature ignition of the fuel/air mixture, in advance of normal firing by the spark plug. Cylinder head removal will be required to allow inspection and cleaning.
- Fuel incorrect. A low grade fuel, or one of poor quality may result in compression induced detonation of the fuel resulting in knocking and pinking noises. Old fuel can cause similar problems. A too highly leaded fuel will reduce detonation but will accelerate deposit formation in the combustion chamber and may lead to early pre-ignition as described in item 1.
- Spark plug heat range incorrect. Uncontrolled pre-ignition can result from the use of spark plugs the heat range of which is too hot.
- Weak mixture. Overheating of the engine due to a weak mixture can result in pre-ignition occurring where it would not occur when engine temperature was within normal limits. Maladjustment, blocked jets or passages and air leaks can cause this condition.

Overheating

17 Firing incorrect

- Spark plugs fouled, defective or maladjusted. See Section 6.
- Spark plug type incorrect. Refer to the Specifications and ensure that the correct plug type is fitted.
- Incorrect ignition timing. Timing that is far too much advanced or far too much retarded will cause overheating. Check the ignition timing is correct.

18 Fuel/air mixture incorrect

- Slow speed mixture strength incorrect. Adjust pilot air screw.
- Main jets wrong size. The carburettors are jetted for sea level atmospheric conditions. For high altitudes, usually above 5000 ft, a smaller main jet will be required.
- Air filter badly fitted or omitted. Check that the filter element is in place and that it and the air filter box cover are sealing correctly. Any leaks will cause a weak mixture.
- Induction air leaks. Check the security of the carburettor mountings and hose connections, and for cracks and splits in the hoses. Check also that the carburettor top is secure and that the vacuum gauge adaptor plug (where fitted) is tight.
- Fuel level too low. See Section 4.
- Fuel tank filler cap air vent obstructed. Clear blockage.

19 Lubrication inadequate

- Oil pump settings incorrect. The oil pump settings are of great importance since the quantities of oil being injected are very small. Any variation in oil delivery will have a significant effect on the engine. Refer to Chapter 4 for further information.
- Oil tank empty or low. This will have disastrous consequences if left unnoticed. Check and replenish tank regularly.
- Transmission oil low or worn out. Check the level regularly and investigate any loss of oil. If the oil level drops with no sign of external leakage it is likely that the crankshaft main bearing oil seals are worn, allowing transmission oil to be drawn into the crankcase during induction.

20 Miscellaneous causes

- Radiator fins clogged. Accumulated debris in the radiator core will gradually reduce its ability to dissipate heat generated by the engine. It is worth noting that during the summer months dead insects can cause as many problems in this respect as road dirt and mud during the winter. Cleaning is best carried out by dislodging the debris with a high pressure hose from the back of the radiator. Once cleaned it is worth painting the matrix with a heat-dispersant matt black paint both to assist cooling and to prevent external corrocion. The fitting of some sort of mesh guard will help prevent the fins from becoming clogged, but make sure that this does not itself prevent adequate cooling.

Clutch operating problems

21 Clutch slip

- No clutch lever play. Adjust clutch lever end play according to the procedure in Chapter 1.
- Friction plates worn or warped. Overhaul clutch assembly, replacing plates out of specification.
- Steel plates worn or warped. Overhaul clutch assembly, replacing plates out of specification.
- Clutch spring broken or worn. Old or heat-damaged (from slipping clutch) springs should be replaced with new ones.
- Clutch release not adjusted properly. See the adjustments section of Chapter 1.
- Clutch inner cable snagging. Caused by a frayed cable or kinked outer cable. Replace the cable with a new one. Repair of a frayed cable is not advised.
- Clutch release mechanism defective. Worn or damaged parts in the clutch release mechanism could include the shaft, cam, actuating arm or pivot. Replace parts as necessary.
- Clutch hub and outer drum worn. Severe indentation by the clutch plate tangs of the channels in the hub and drum will cause snagging of the plates preventing correct engagement. If this damage occurs, renewal of the worn components is required.
- Lubricant incorrect. Use of a transmission lubricant other than that specified may allow the plates to slip.

22 Clutch drag

- Clutch lever play excessive. Adjust lever at bars or at cable end if necessary.
- Clutch plates warped or damaged. This will cause a drag on the clutch, causing the machine to creep. Overhaul clutch assembly.
- Clutch spring tension uneven. Usually caused by a sagged or broken spring. Check and replace springs.
- Transmission oil deteriorated. Badly contaminated transmission oil and a heavy deposit of oil sludge on the plates will cause plate sticking. The oil recommended for this machine is of the detergent type, therefore it is unlikely that this problem will arise unless regular oil changes are neglected.
- Transmission oil viscosity too high. Drag in the plates will result from the use of an oil with too high a viscosity. In very cold weather clutch drag may occur until the engine has reached operating temperature.
- Clutch hub and outer drum worn. Indentation by the clutch plate tangs of the channels in the hub and drum will prevent easy plate disengagement. If the damage is light the affected areas may be dressed with a fine file. More pronounced damage will necessitate renewal of the components.
- Clutch housing seized to shaft. Lack of lubrication, severe wear or damage can cause the housing to seize to the shaft. Overhaul of the clutch, and perhaps the transmission, may be necessary to repair damage.
- Clutch release mechanism defective. Worn or damaged release mechanism parts can stick and fail to provide leverage. Overhaul clutch cover components.
- Loose clutch hub nut. Causes drum and hub misalignment, putting a drag on the engine. Engagement adjustment continually varies. Overhaul clutch assembly.

Gear selection problems

23 Gear lever does not return

- Weak or broken centraliser spring. Renew the spring.
- Gearchange shaft bent or seized. Distortion of the gearchange shaft often occurs if the machine is dropped heavily on the gear lever. Provided that damage is not severe straightening of the shaft is permissible.

24 Gear selection difficult or impossible

- Clutch not disengaging fully. See Section 22.
- Gearshift shaft bent. This often occurs if the machine is dropped heavily on the gear lever. Straightening of the shaft is permissible if the damage is not too great.
- Gearchange arms, pawls or pins worn or damaged. Wear or breakage of any of these items may cause difficulty in selecting one or more gears. Overhaul the selector mechanism.
- Gearchange shaft centraliser spring maladjusted. This is often characterised by difficulties in changing up or down, but rarely in both directions. Adjust the centraliser anchor bolt as described in Chapter 1.
- Gearchange drum stopper cam or detent plunger damaged. Failure, rather than wear, of these items may jam the drum thereby preventing gearchanging or causing false selection at high speed.
- Selector forks bent or seized. This can be caused by dropping the machine heavily on the gearchange lever as a result of lack of lubrication. Though rare, bending of a shaft can result from a missed gearchange or false selection at high speed.
- Selector fork end and pin wear. Pronounced wear of these items and the grooves in the gearchange drum can lead to imprecise selection and, eventually, no selection. Renewal of the worn components will be required.
- Structural failure. Failure of any one component of the selector rod and change mechanism will result in improper or fouled gear selection.

25 Jumping out of gear

- Detent plunger assembly worn or damaged. Wear of the plunger and the cam with which it locates and breakage of the detent spring can cause imprecise gear selection resulting in jumping out of gear. Renew the damaged components.
- Gear pinion dogs worn or damaged. Rounding off the dog edges and the mating recesses in adjacent pinion can lead to jumping out of gear when under load. The gears should be inspected and renewed. Attempting to reprofile the dogs is not recommended.
- Selector forks, gearchange drum and pinion grooves worn. Extreme wear of these interconnected items can occur after high mileages especially when lubrication has been neglected. The worn components must be renewed.
- Gear pinions, bushes and shafts worn. Renew the worn components.
- Bent gearchange shaft. Often caused by dropping the machine on the gear lever.
- Gear pinion tooth broken. Chipped teeth are unlikely to cause jumping out of gear once the gear has been selected fully; a tooth which is completely broken off, however, may cause problems in this respect and in any event will cause transmission noise.

26 Overselection

- Pawl spring weak or broken. Renew the spring.
- Detent plunger worn or broken. Renew the damaged items.
- Stopper arm spring worn or broken. Renew the spring.
- Gearchange arm stop pads worn. Repairs can be made by welding and reprofiling with a file.
- Selector limiter claw components (where fitted) worn or damaged. Renew the damaged items.

Abnormal engine noise

27 Knocking or pinking

- See Section 16.

28 Piston slap or rattling from cylinder

- Cylinder bore/piston clearance excessive. Resulting from wear, partial seizure or improper boring during overhaul. This condition can often be heard as a high, rapid tapping noise when the engine is under little or no load, particularly when power is just beginning to be applied. Reboring to the next correct oversize should be carried out and a new oversize piston fitted.
- Connecting rod bent. This can be caused by over-revving, trying to start a very badly flooded engine (resulting in a hydraulic lock in the cylinder) or by earlier mechanical failure. Attempts at straightening a bent connecting rod from a high performance engine are not recommended. Careful inspection of the crankshaft should be made before renewing the damaged connecting rod.
- Gudgeon pin, piston boss bore or small-end bearing wear or seizure. Excess clearance or partial seizure between normal moving parts of these items can cause continuous or intermittent tapping noises. Rapid wear or seizure is caused by lubrication starvation.
- Piston rings worn, broken or sticking. Renew the rings after careful inspection of the piston and bore.

29 Other noises

- Big-end bearing wear. A pronounced knock from within the crankcase which worsens rapidly is indicative of big-end bearing failure as a result of extreme normal wear or lubrication failure. Remedial action in the form of a bottom end overhaul should be taken;

continuing to run the engine will lead to further damage including the possibility of connecting rod breakage.

● Main bearing failure. Extreme normal wear or failure of the main bearings is characteristically accompanied by a rumble from the crankcase and vibration felt through the frame and footrests. Renew the worn bearings and carry out a very careful examination of the crankshaft.

● Crankshaft excessively out of true. A bent crank may result from over-revving or damage from an upper cylinder component or gearbox failure. Damage can also result from dropping the machine on either crankshaft end. Straightening of the crankshaft is not possible in normal circumstances; a replacement item should be fitted.

● Engine mounting loose. Tighten all the engine mounting nuts and bolts.

● Cylinder head gasket leaking. The noise most often associated with a leaking head gasket is a high pitched squeaking, although any other noise consistent with gas being forced out under pressure from a small orifice can also be emitted. Gasket leakage is often accompanied by oil seepage from around the mating joint or from the cylinder head holding down bolts and nuts. Leakage results from insufficient or uneven tightening of the cylinder head fasteners, or from random mechanical failure. Retightening to the correct torque figure will, at best, only provide a temporary cure. The gasket should be renewed at the earliest opportunity.

● Exhaust system leakage. Popping or crackling in the exhaust system, particularly when it occurs with the engine on the overrun, indicates a poor joint either at the cylinder port or at the exhaust pipe/silencer connection. Failure of the gasket or looseness of the clamp should be looked for.

Abnormal transmission noise

30 Clutch noise

- Clutch outer drum/friction plate tang clearance excessive.
- Clutch outer drum/spacer clearance excessive.
- Clutch outer drum/thrust washer clearance excessive.
- Primary drive gear teeth worn or damaged.
- Clutch shock absorber assembly worn or damaged.

31 Transmission noise

● Bearing or bushes worn or damaged. Renew the affected components.

● Gear pinions worn or chipped. Renew the gear pinions.

● Metal chips jammed in gear teeth.This can occur when pieces of metal from any failed component are picked up by a meshing pinion. The condition will lead to rapid bearing wear or early gear failure.

● Gearbox oil level too low. Top up immediately to prevent damage to gearbox and engine.

● Gearchange mechanism worn or damaged. Wear or failure of certain items in the selection and change components can induce mis-selection of gears (see Section 24) where incipient engagement of more than one gear set is promoted. Remedial action, by the overhaul of the gearbox, should be taken without delay.

● Chain snagging on cases or cycle parts. A badly worn chain or one that is excessively loose may snag or smack against adjacent components.

Exhaust smokes excessively

32 White/blue smoke (caused by oil burning)

● Oil pump settings incorrect. Check and reset the oil pump as described in Chapter 3.

● Crankshaft main bearing oil seals worn. Wear in the main bearing oil seals, often in conjunction with wear in the bearings themselves, can allow transmission oil to find its way into the crankcase and thence to the combustion chamber. This condition is often indicated by a mysterious drop in the transmission oil level with no sign of external leakage.

● Accumulated oil deposits in exhaust system. If the machine is used for short journeys only it is possible for the oil residue in the exhaust gases to condense in the relatively cool silencer. If the machine is then taken for a longer run in hot weather, the accumulated oil will burn off producing ominous smoke from the exhaust.

33 Black smoke (caused by over-rich mixture)

● Air filter element clogged. Clean or renew the element.

● Main jets lose or too large. Remove the float chambers to check for tightness of the jets. If the machine is used at high altitudes rejetting will be required to compensate for the lower atmospheric pressure.

● Cold start mechanism jammed on. Check that the mechanism works smoothly and correctly and that, where fitted, the operating cable is lubricated and not snagged.

● Fuel level too high. The fuel level is controlled by the float height which can increase as a result of wear or damage. Remove the float bowls and check the float height. Check also that floats have not punctured; a punctured float will lose buoyancy and allow an increased fuel level.

● Float valve needles stuck open. Caused by dirt or worn valves. Clean the float chambers or renew the needle and, if necessary, the valve seats.

Poor handling or roadholding

34 Directional instability

● Steering head bearing adjustment too tight. This will cause rolling or weaving at low speeds. Re-adjust the bearings.

● Steering head bearing worn or damaged. Correct adjustment of the bearing will prove impossible to achieve if wear or damage has occurred. Inconsistent handling will occur including rolling or weaving at low speed and poor directional control at indeterminate higher speeds. The steering head bearing should be dismantled for inspection and renewed if required. Lubrication should also be carried out.

● Bearing races pitted or dented. Impact damage caused, perhaps, by an accident or riding over a pot-hole can cause indentation of the bearing, usually in one position. This should be noted as notchiness when the handlebars are turned. Renew and lubricate the bearings.

● Steering stem bent. This will occur only if the machine is subjected to a high impact such as hitting a curb or a pot-hole. The lower yoke/stem should be renewed; do not attempt to straighten the stem.

● Front or rear tyre pressures too low.

● Front or rear tyre worn. General instability, high speed wobbles and skipping over white lines indicates that tyre renewal may be required. Tyre induced problems, in some machine/tyre combinations, can occur even when the tyre in question is by no means fully worn.

● Swinging arm bearings worn. Difficulties in holding line, particularly when cornering or when changing power settings indicates wear in the swinging arm bearings. The swinging arm should be removed from the machine and the bearings renewed.

● Swinging arm flexing. The symptoms given in the preceding paragraph will also occur if the swinging arm fork flexes badly. This can be caused by structural weakness as a result of corrosion, fatigue or impact damage, or because the rear wheel spindle is slack.

● Wheel bearings worn. Renew the worn bearings.

● Tyres unsuitable for machine. Not all available tyres will suit the characteristics of the frame and suspension, indeed, some tyres or tyre combinations may cause a transformation in the handling characteristics. If handling problems occur immediately after changing to a new tyre type or make, revert to the original tyres to see whether an improvement can be noted. In some instances a change to what are, in fact, suitable tyres may give rise to handling deficiences. In this case a thorough check should be made of all frame and suspension items which affect stability.

35 Steering bias to left or right

- Rear wheel out of alignment. Caused by uneven adjustment of chain tensioner adusters allowing the wheel to be askew in the fork ends. A bent rear wheel spindle will also misalign the wheel in the swinging arm.
- Wheels out of alignment. This can be caused by impact damage to the frame, swinging arm, wheel spindles or front forks. Although occasionally a result of material failure or corrosion it is usually as a result of a crash.
- Front forks twisted in the steering yokes. A light impact, for instance with a pot-hole or low curb, can twist the fork legs in the steering yokes without causing structural damage to the fork legs or the yokes themselves. Re-alignment can be made by loosening the yoke pinch bolts, wheel spindle and mudguard bolts. Re-align the wheel with the handlebars and tighten the bolts working upwards from the wheel spindle. This action should be carried out only when there is no chance that structural damage has occurred.

36 Handlebar vibrates or oscillates

- Tyres worn or out of balance. Either condition, particularly in the front tyre, will promote shaking of the fork assembly and thus the handlebars. A sudden onset of shaking can result if a balance weight is displaced during use.
- Tyres badly positioned on the wheel rims. A moulded line on each wall of a tyre is provided to allow visual verification that the tyre is correctly positioned on the rim. A check can be made by rotating the tyre; any misalignment will be immediately obvious.
- Wheel rims warped or damaged. Inspect the wheels for runout as described in Chapter 6.
- Swinging arm bearings worn. Renew the bearings.
- Wheel bearings worn. Renew the bearings.
- Steering head bearings incorrectly adjusted. Vibration is more likely to result from bearings which are too loose rather than too tight. Re-adjust the bearings.
- Loose fork component fasteners. Loose nuts and bolts holding the fork legs, wheel spindle, mudguards or steering stem can promote shaking at the handlebars. Fasteners on running gear such as the forks and suspension should be check tightened occasionally to prevent dangerous looseness of components occurring.
- Engine mounting bolts loose. Tighten all fasteners.

37 Poor front fork performance

- Damping fluid level incorrect. If the fluid level is too low poor suspension control will occur resulting in a general impairment of roadholding and early loss of tyre adhesion when cornering and braking. Too much oil is unlikely to change the fork characteristics unless severe overfilling occurs when the fork action will become stiffer and oil seal failure may occur.
- Damping oil viscosity incorrect. The damping action of the fork is directly related to the viscosity of the damping oil. The lighter the oil used, the less will be the damping action imparted. For general use, use only one of the recommended types of oil, changing to a slightly higher or heavier oil only when a change in damping characteristic is required. Overworked oil, or oil contaminated with water which has found its way past the seals, should be renewed to restore the correct damping performance and to prevent bottoming of the forks.
- Damping components worn or corroded. Advanced normal wear of the fork internals is unlikely to occur until a very high mileage has been covered. Continual use of the machine with damaged oil seals which allows the ingress of water, or neglect, will lead to rapid corrosion and wear. Dismantle the forks for inspection and overhaul.
- Weak fork springs. Progressive fatigue of the fork springs, resulting in a reduced spring free length, will occur after extensive use. This condition will promote excessive fork dive under braking, and in its advanced form will reduce the at-rest extended length of the forks and thus the fork geometry. Renewal of the springs as a pair is the only satisfactory course of action.
- Bent stanchions or corroded stanchions. Both conditions will prevent correct telescoping of the fork legs, and in an advanced state can cause sticking of the fork in one position. In a mild form corrosion will cause stiction of the fork thereby increasing the time the suspension takes to react to an uneven road surface. Bent fork stanchions should be attended to immediately because they indicate that impact damage has occurred, and there is a danger that the forks will fail with disastrous consequences.
- Leaking fork oil seals. If the seals wear sufficiently, the fork will be unable to hold air pressure. If this occurs in one leg only the instability will be worsened. Check that air pressure settings are being maintained and renew the seals if leakage is evident.

38 Front fork judder when braking (see also Section 46)

- Wear between the fork stanchions and the fork legs. Renewal of the affected components is required.
- Slack steering head bearings. Re-adjust the bearings.
- Warped brake disc or drum. If irregular braking action occurs fork judder can be induced in what are normally serviceable forks. Renew the damaged brake components.

39 Poor rear suspension performance

- Rear suspension unit damper worn out or leaking. The damping performance of most rear suspension units falls off with age. This is a gradual process, and thus may not be immediately obvious. Indications of poor damping include hopping of the rear end when cornering or braking, and a general loss of positive stability.
- Weak rear spring. If the suspension unit spring fatigues, it will promote excessive pitching of the machine and reduce the ground clearance when cornering. Although replacement springs are available separately from the rear suspension damper unit it is probable that if spring fatigue has occurred the damper unit will also require renewal.
- Swinging arm flexing or bearings worn. See Sections 34 and 36.
- Bent suspension unit damper rod. This is likely to occur only if the machine is dropped or if seizure of the piston occurs. If either happens the suspension unit should be renewed.

Abnormal frame and suspension noise

40 Front end noise

- Oil level low or too thin. This can cause a 'spurting' sound and is usually accompanied by irregular fork action.
- Spring weak or broken. Makes a clicking or scraping sound. Fork oil will have a lot of metal particles in it.
- Steering head bearings loose or damaged. Clicks when braking. Check, adjust or replace.
- Fork clamps loose. Make sure all fork clamp pinch bolts are tight.
- Fork stanchion bent. Good possibility if machine has been dropped. Repair or replace tube.
- Excessive play in damper assembly. See Chapter 6.

41 Rear suspension noise

- Fluid level too low. Leakage of a suspension unit, usually evident by oil on the outer surfaces, can cause a spurting noise. The suspension unit should be renewed.
- Defective rear suspension unit with internal damage.

Brake problems

42 Brakes are spongy or ineffective

- Air in brake circuit. This is only likely to happen in service due to neglect in checking the fluid level or because a leak has developed. The problem should be identified and the brake system bled of air.
- Pad worn. Check the pad wear against the wear lines provided or measure the friction material thickness and renew the pads if necessary.
- Contaminated pads. Cleaning pads which have been contaminated with oil, grease or brake fluid is unlikely to prove successful; the pads should be renewed.
- Pads glazed. This is usually caused by overheating. The surface of the pads may be roughened using glass-paper or a fine file.
- Brake fluid deterioration. A brake which on initial operation is firm but rapidly becomes spongy in use may be failing due to water contamination of the fluid. The fluid should be drained and then the system refilled and bled.
- Master cylinder seal failure. Wear or damage of master cylinder internal parts will prevent pressurisation of the brake fluid. Overhaul the master cylinder unit.
- Caliper seal failure. This will almost certainly be obvious by loss of fluid, a lowering of fluid in the master cylinder reservoir and contamination of the brake pads and caliper. Overhaul the caliper assembly.
- Brake lever or pedal improperly adjusted. Adjust the clearance between the lever end and master cylinder plunger to take up lost motion, as recommended in Routine Maintenance.

43 Brakes drag

- Disc warped. The disc must be renewed.
- Caliper piston, caliper or pads corroded. The brake caliper assembly is vulnerable to corrosion due to water and dirt, and unless cleaned at regular intervals and lubricated in the recommended manner, will become sticky in operation.
- Piston seal deteriorated. The seal is designed to return the piston in the caliper to the retracted position when the brake is released. Wear or old age can affect this function. The caliper should be overhauled if this occurs.
- Brake pad damaged. Pad material separating from the backing plate due to wear or faulty manufacture. Renew the pads. Faulty installation of a pad also will cause dragging.
- Wheel spindle bent. The spindle may be straightened if no structural damage has occurred.
- Brake lever or pedal not returning. Check that the lever or pedal works smoothly throughout its operating range and does not snag on any adjacent cycle parts. Lubricate the pivot if necessary.
- Twisted caliper support bracket. This is likely to occur only after impact in an accident. No attempt should be made to re-align the caliper; the bracket should be renewed.

44 Brake lever or pedal pulsates in operation

- Disc warped or irregularly worn. The disc must be renewed.
- Wheel spindle bent. The spindle may be straightened provided no structural damage has occurred.

45 Disc brake noise

- Brake squeal. This can be caused by the omission or incorrect installation of the anti-squeal shim fitted to the rear of one pad. The arrow on the shim should face the direction of wheel normal rotation. Squealing can also be caused by dust on the pads, usually in combination with glazed pads, or other contamination from oil, grease brake fluid or corrosion. Persistent squealing which cannot be traced to any of the normal causes can often be cured by applying a thin layer of high temperature silicone grease to the rear of the pads. Make absolutely certain that no grease is allowed to contaminate the braking surface of the pads.
- Glazed pads. This is usually caused by high temperatures or contamination. The pad surfaces may be roughened using glass-paper or a fine file. If this approach does not effect a cure the pads should be renewed.
- Disc warped. This can cause a chattering, clicking or intermittent squeal and is usually accompanied by a pulsating brake lever or pedal or uneven braking. The disc must be renewed.
- Brake pads fitted incorrectly or undersize. Longitudinal play in the pads due to omission of the locating springs (where fitted) or because pads of the wrong size have been fitted will cause a single tapping noise every time the brake is operated. Inspect the pads for correct installation and security.

46 Brake induced fork judder

- Worn front fork stanchions and legs, or worn or badly adjusted steering head bearings. These conditions, combined with uneven or pulsating braking as described in Section 44 will induce more or less judder when the brakes are applied, dependent on the degree of wear and poor brake operation. Attention should be given to both areas of malfunction. See the relevant Sections.

Electrical problems

47 Battery dead or weak

- Battery faulty. Battery life should not be expected to exceed 3 to 4 years, particularly where a starter motor is used regularly. Gradual sulphation of the plates and sediment deposits will reduce the battery performance. Plate and insulator damage can often occur as a result of vibration. Complete power failure, or intermittent failure, may be due to a broken battery terminal. Lack of electrolyte will prevent the battery maintaining charge.
- Battery leads making poor contact. Remove the battery leads and clean them and the terminals, removing all traces of corrosion and tarnish. Reconnect the leads and apply a coating of petroleum jelly to the terminals.
- Load excessive. If additional items such as spot lamps, are fitted, which increase the total electrical load above the maximum alternator output, the battery will fail to maintain full charge. Reduce the electrical load to suit the electrical capacity.
- Rectifier or ballast resistor failure.
- Alternator generating coils open-circuit or shorted.
- Charging circuit shorting or open circuit. This may be caused by frayed or broken wiring, dirty connectors or a faulty ignition switch. The system should be tested in a logical manner. See Section 50.
- Battery short circuited – battery positive terminal in contact with fuel tank. Remove fuel tank and check for marks on underside, near battery. If plaster cover for battery positive terminal is missing, obtain a new one or make up a suitable alternative.

48 Battery overcharged

- Regulator or ballast resistor faulty. Overcharging is indicated if the battery becomes hot or it is noticed that the electrolyte level falls repeatedly between checks. In extreme cases the battery will boil causing corrosive gases and electrolyte to be emitted through the vent pipes.
- Battery wrongly matched to the electrical circuit. Ensure that the specified battery is fitted to the machine.

49 Total electrical failure

- Fuse blown. Check the main fuse. If a fault has occurred, it must be rectified before a new fuse is fitted.
- Battery faulty. See Section 47.

● Earth failure. Check that the main earth strap from the battery is securely affixed to the gearbox and is making a good contact.
● Ignition switch or power circuit failure. Check for current flow through the battery positive lead (red) to the ignition switch. Check the ignition switch for continuity.

50 Circuit failure

● Cable failure. Refer to the machine's wiring diagram and check the circuit for continuity. Open circuits are a result of loose or corroded connections, either at terminals or in-line connectors, or because of broken wires. Occasionally, the core of a wire will break without there being any apparent damage to the outer plastic cover.
● Switch failure. All switches may be checked for continuity in each switch position, after referring to the switch position boxes incorporated in the wiring diagram for the machine. Switch failure may be a result of mechanical breakage, corrosion or water.
● Fuse blown. Refer to the wiring diagram to check whether or not a circuit fuse is fitted. Replace the fuse, if blown, only after the fault has been identified and rectified.

51 Bulbs blowing repeatedly

● Vibration failure. This is often an inherent fault related to the natural vibration characteristics of the engine and frame and is, thus, difficult to resolve. Modifications of the lamp mounting, to change the damping characteristics may help.
● Intermittent earth. Repeated failure of one bulb, particularly where the bulb is fed directly from the generator, indicates that a poor earth exists somewhere in the circuit. Check that a good contact is available at each earthing point in the circuit.
● Reduced voltage. Where a quartz-halogen bulb is fitted the voltage to the bulb should be maintained or early failure of the bulb will occur. Do not overload the system with additional electrical equipment in excess of the system's power capacity and ensure that all circuit connections are maintained clean and tight.

Routine maintenance

For information relating to the RD350 F II, N II and R models, refer to Chapter 8

Specifications

Engine/transmission		
Spark plug gap	0.7 – 0.8 mm (0.028 – 0.031 in)	
Idle speed	1150 – 1250 rpm	
Throttle cable free play	3 – 7 mm (0.12 – 0.28 in)	
Oil pump minimum stroke setting	0.10 – 0.15 mm (0.004 – 0.006 in)	
Clutch cable free play at lever end	10 – 15 mm (0.4 – 0.6 in)	
Cycle parts		
Front brake lever free play	5 – 8 mm (0.2 – 0.3 in)	
Rear brake pedal free play	35 – 40 mm (1.38 – 1.57 in)	
Tyre pressures (cold):	**Front**	**Rear**
Up to 90 kg (198 lb) load	26 psi (1.8 kg/cm²)	28 psi (1.8 kg/cm²)
90 – 211 kg (198 – 428) load	32 psi (2.2 kg/cm²)	40 psi (2.8 kg/cm²)
High speed riding	28 psi (1.9 kg/cm²)	32 psi (2.2 kg/cm²)
Recommended lubricants		
Engine:		
Capacity (tank)	1.6 litre (2.8 Imp pt)	
Oil grade	Air-cooled 2-stroke engine oil	
Transmission:		
Capacity at oil change	1.5 litres (2.6 Imp pt)	
Oil grade	SAE 10W30 type SE motor oil	
Cooling system:		
Capacity:		
Overall	1.5 litres (2.6 Imp pt)	
From low to full mark	185 cc (0.32 Imp pt)	
Reservoir tank	215 cc (0.38 Imp pt)	
Coolant type	50% water, 50% high quality ethylene glycol antifreeze mixture with aluminium engine corrosion inhibitors	
Front forks:		
Capacity:		
RD350 LC II	253 cc (8.92 Imp fl oz)	
RD350 F and N	297 cc (10.48 Imp fl oz)	
Oil grade:		
RD350 LC II	SAE 10W30 SE motor oil	
RD350 F and N	SAE 10W fork oil	
Oil level:		
RD350 LC II	120 mm (4.72 in)	
RD350 F and N	106.1 mm (4.18 in)	
Final drive chain	Aerosol chain lubricant	
Wheel bearings	High melting-point grease	
Steering head bearings	General purpose grease	
Rear suspension and swinging arm bearings	General purpose grease	
Hydraulic brake fluid	SAE J1703 or DOT 4 or 3 hydraulic fluid	
Pivot points	Motor oil or WD40	
Control cables	Light machine oil	

Introduction

Periodic routine maintenance is a continuous process which should commence immediately the machine is used. The object is to maintain all adjustments and to diagnose and rectify minor defects before they develop into more extensive, and often more expensive, problems.

It follows that if the machine is maintained properly, it will both run and perform with optimum efficiency, and be less prone to unexpected breakdowns. Regular inspection of the machine will show up any parts which are wearing, and with a little experience, it is possible to obtain the maximum life from any one component, renewing it when it becomes so worn that it is liable to fail.

Regular cleaning can be considered as important as mechanical maintenance. This will ensure that all the cycle parts are inspected regularly and are kept free from accumulations of road dirt and grime.

Cleaning is especially important during the winter months, despite its appearance of being a thankless task which very soon seems pointless. On the contrary, it is during these months that the paintwork, chromium plating, and the alloy casings suffer the ravages of abrasive grit, rain and road salt. A couple of hours spent weekly on cleaning the machine will maintain its appearance and value, and highlight small points, like chipped paint, before they become a serious problem.

The various maintenance tasks are described under their respective mileage and calendar headings, and are accompanied by diagrams and photographs where pertinent.

It should be noted that the intervals between each maintenance task serve only as a guide. As the machine gets older, or if it is used under particularly arduous conditions, it is advisable to reduce the period between each check.

For ease of reference, most service operations are described in detail under the relevant heading. However, if further general information is required, this can be found under the pertinent Section heading and Chapter in the main text.

Although no special tools are required for routine maintenance, a good selection of general workshop tools is essential. Included in the tools must be a range of metric ring or combination spanners, a selection of crosshead screwdrivers, and two pairs of circlip pliers, one external opening and the other internal opening. Additionally, owning to the extreme tightness of most casing screws, an impact screwdriver, together with a choice of large or small cross-head screw bits, is absolutely indispensable. This is particularly so if the engine has not been dismantled since leaving the factory.

Daily

The checklist shown below should be carried out prior to riding the machine each day. The procedure should take only a few moments, and will significantly reduce the risk of unexpected failure in use.

a) Check brake operation, fluid levels and lever/pedal adjustment
b) Check clutch and throttle operations
c) Check oil tank level
d) Check fuel tank level
e) Check coolant level
f) Check tyre pressures and condition of tread and sidewalls
g) Check the electrical system, particularly the horn, lights, brake lamp and turn signals

Weekly, or every 100 miles (160 km)

1 Topping up the engine oil tank

Unlock and open the dualseat to gain access to the oil filler cap; this is located just to the rear of the fuel tank. Top up with any good quality **air-cooled** (not water-cooled as might be expected) two-stroke engine oil to within about one inch of the filler neck. It is important that the tank level is maintained at all times. If it is suspected that the level has fallen too low, or if the system has been disconnected at any point, it should be bled to remove any air which may have entered. Note that air can cause a lubrication failure which in turn can lead to engine seizure, so do not put off bleeding in the hope that it will correct itself. For details refer to Chapter 3.

2 Checking the tyre condition and pressures

The importance of maintaining the correct tyre pressures cannot be stressed too highly, the safety of the rider and other road users being at risk if regular checks are ignored or postponed. It is a good idea to keep an accurate pocket pressure gauge with the machine's toolkit and to have access to a simple footpump at home. It should be noted that not every filling station gauge is accurate, and that tyre pressure checks should only be made when the tyres are cold to ensure consistent readings. The tyre pressures shown below are for original equipment tyres, and non-standard fitments may require modified settings. A reputable tyre supplier will be able to advise on this point when the tyres are fitted.

Tyre pressures (cold)	Front	Rear
Up to 90 kg (198 lb) load	26 psi	28 psi
90 – 211 kg (198 – 428 lb) load	32 psi	40 psi
High speed riding	28 psi	32 psi

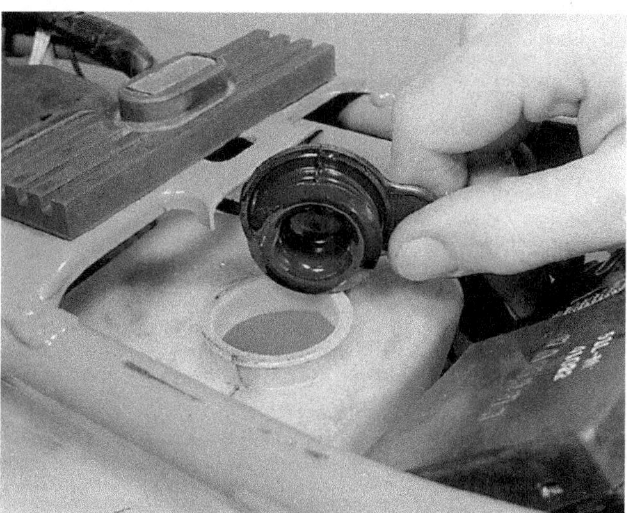

Engine oil tank filler can be reached after seat has been removed

Carry a pocket pressure gauge to check tyre and fork air pressures

When checking the tyre pressures, examine the tyre treads and sidewalls for signs of damage or excessive wear. If cracks or splitting are noted, renew the tyre or have it examined by a specialist. Remove any small stones which may have become embedded in the treads, ensuring that no damage to the fabric carcass has occurred. Check the tread depth around the tyre's circumference. The minimum permissible depth in the UK is 1 mm; this should be considered an absolute lower limit, 2 mm being a safer allowance.

3 Checking the battery electrolyte level

The level of the battery electrolyte can be checked via the translucent case, making the operation quick and simple. It is unlikely that any topping up will be required; this is usually necessary only after several months of normal use. If a sudden drop in electrolyte level is noted, find out the cause of the problem. If confined to one cell, the fault may be a cracked or split casing. If this is left unattended, the leaking electrolyte will damage the machine and the battery will fail soon afterwards. If all cells are abnormally low there may be an electrical fault causing overcharging. Again, prompt attention is called for if the battery is to be saved. Top up using only distilled water to the "MAX" level line on the battery. For full details, including precautions when dealing with battery acid, refer to Chapter 7.

4 Checking the coolant level

Check the level of coolant in the expansion tank, located below the left-hand side of the fuel tank, near its front mounting. On fully-faired models, it is advisable to remove the left-hand fairing panel for improved access.

Note that the level of coolant will vary according to the engine temperature and is no cause for concern. When checking the level always ensure the engine is cold. With the motorcycle upright, the level should lie between the FULL and LOW lines on the expansion tank body.

If topping up is required, pry off the filler cap at the top of the expansion tank and add the specified coolant (see Specifications) to restore the level. Tap water can be used in an emergency, but only if it is known to be soft; hard tap water can lead to scaling of the cooling system and impaired cooling. It is preferred to use only distilled or de-ionised water to avoid this problem. When handling coolant, note the following precautions:

a) Take care to avoid splashing coolant in the eyes or on the skin or clothing. If coolant enters the eyes, wash it out with copious quantities of water and seek medical advice at once.

b) Never leave antifreeze lying around in an open container or in puddles on the floor; children and pets are attracted by its sweet smell and may drink it.

c) If swallowed, induce vomiting and seek immediate medical advice.

d) If coolant contacts the painted surfaces of the motorcycle, wash it off immediately.

If the coolant level falls steadily, check the system for leaks. If no leaks are found and the level continues to fall, it is recommended that the motorcycle be taken to a Yamaha dealer who will pressure test the system.

Periodically check the radiator pressure cap, located below the nose of the fuel tank. Remove the fuel tank for access (see Chapter 3). With the engine cold, turn the radiator cap anticlockwise until it reaches a stop. If you hear a hissing sound (indicating there is still pressure in the system), wait until it stops. Now press down on the cap with the palm of your hand and continue turning the cap anticlockwise until it can be removed.

Check the condition of the coolant in the system. If it is rust coloured or if accumulations of scale are visible, drain, flush and refill the system with new coolant (see Chapter 2). Check the radiator cap gaskets for cracks and other damage, renewing it if necessary. Install the radiator cap by turning it clockwise until it reaches its first stop, then push down on the cap and continue turning until it is fully secure.

5 General maintenance and inspection

It is recommended that one month be considered the maximum interval for cleaning the machine, but that if possible, cleaning should be carried out on a weekly basis, especially during the winter months. This will make the cleaning job much easier and will usually bring to the owner's attention faults such as loose fasteners which might otherwise be missed. Although a less pleasant task in winter, remember that cleaning is even more important, if only to remove potentially corrosive road salt. If appearance is not of paramount importance, spray the cleaned machine's cycle parts (avoiding the brakes, seat and tyres!) with a silicone-based maintenance spray, such as WD-40. This will inhibit corrosion and prevent electrical problems, and cleaning will be easier on the next occasion.

Monthly, or every 300 miles (500 km)

Lubricating and adjusting the final drive chain

The monthly interval prescribed for chain adjustment should be regarded as an absolute maximum. It is preferable to reduce this to about once each week, particularly during the winter months when the rate of chain wear is higher due to adverse weather conditions.

Clean off any accumulated road dirt using a stiff brush soaked in petrol or paraffin. Examine the general condition of the chain, paying particular attention to indications of impending failure such as loose or rattling rollers or cracks in rollers or side plates. If damage of this nature is found, the chain will almost certainly be in need of renewal.

Radiator pressure cap is located below nose of fuel tank. **Do not** attempt to remove it when the engine is hot

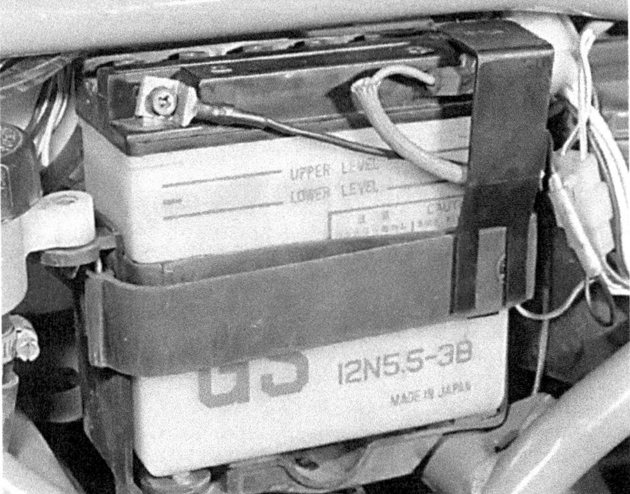

Check that battery electrolyte level lies between upper and lower level lines. Do not omit to fit black plastic cover over positive terminal

Lubricate the chain using one of the proprietary aerosol chain lubricants, making sure that it finds its way onto both sides of the side plates and onto the rollers. Engine oil may be used in an emergency, but note that this will tend to be flung off the chain quite quickly.

Most chains have a "tight spot", and it is here that the free play check should be made. Place the machine on its centre stand and check the amount of up-and-down movement in the lower run of the chain. Turn the wheel and repeat this check until the tightest spot is found. The machine should now be placed on its wheels and the amount of free play measured about four inches forward of the rear wheel sprocket. The correct clearance is 30 – 40 mm (1.2 – 1.6 in).

If adjustment is required, slacken the rear brake torque arm nut, remove the split pin from the rear wheel nut and slacken it. Turn the adjusters on each side of the swinging arm by a similar amount until chain free play is set correctly. Check that the alignment marks on each side are at the same position, then tighten the rear wheel spindle nut to 10.0 kgf m (72 lbf ft) in the case of the RD350 LC II model, and 10.5 kgf m (75 lbf ft) on the later models. Recheck the adjustment, then secure the nut with a new split pin and tighten the torque arm nut.

Adjust drive chain tension, noting alignment marks on swinging arm. Tighten wheel spindle nut and secure using a new split pin

6 Monthly, or every 4000 miles (6000 km)

1 Checking, cleaning and re-gapping the spark plugs

Remove each spark plug cap in turn and unscrew the plugs using a proper plug spanner to avoid damage. The appearance of the plugs can be used to assess the general condition of the engine,

Examine the condition of the plug electrodes. If they are worn or badly contaminated the plug should be renewed. Plugs are relatively inexpensive, and attempting to reuse an old or worn plug is a false economy.

A sound plug can be cleaned using a brass wire brush of the type sold for this purpose in motor accessory shops, or by abrasive cleaning. Many garages offer this service, and inexpensive home units are available. If the abrasive method is chosen, make sure that any residual particles are removed from the plug before it is refitted.

Check the electrode gap using feeler gauges. The recommended gap is 0.7 – 0.8 mm (0.028 – 0.031 in). If adjustment is required, bend the outer, earth, electrode to give the specified gap. On no account try to bend the centre electrode; the porcelain insulator will invariably be broken.

The specified spark plug fitment is NGK BR8ES. If NGK plugs cannot be obtained, an equivalent can be used, but be sure that these are of the correct reach and grade. Do not fit plugs of a different value or engine damage may result. The plug threads should be greased lightly prior to installation. Tighten the plugs firmly by hand, then tighten them by a further quarter turn with the plug spanner. This will ensure that the plugs seal correctly without risk of damage to the plug threads in the cylinder head. Alternatively, tighten the plugs using a torque wrench in conjunction with a spark plug socket to 2.0 kgf m(14 lbf ft). Remember to keep new plugs of the correct type and gap setting in the toolkit.

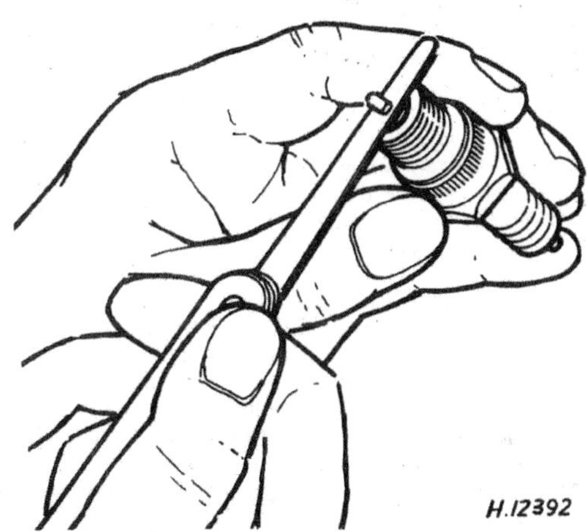

Measuring the spark plug gap

2 Cleaning the air filter element

Remove the seat and both side panels. Check that the fuel tap is turned to the "ON" position (not "PRI"), then disconnect the fuel and vacuum pipes. Remove the single retaining bolt at the rear of the tank and remove the tank by grasping it and pulling it rearwards until the front of the tank comes free of the two mounting rubbers.

Release the three screws which retain the flat air filter cover and lift it away. Lift away the element and wash it in methylated spirit or in clean petrol. If using petrol, take care to avoid any possible risk of fire. When the element is dry, apply SAE 30W motor oil to it, squeezing out any excess to leave the element moist but not dripping. Refit the element, making sure that it locates properly and that the cover seals correctly.

Every 1600 miles, or every fourth cleaning, the element should be renewed. If used after this time its filtering properites will have been impaired. On no account run the machine with the filter missing or damaged; dust in the unfiltered air will enter the engine, causing accelerated wear. Before refitting the fuel tank, check the condition of the fuel and vacuum pipes. If these show signs of splitting or leakage they should be renewed.

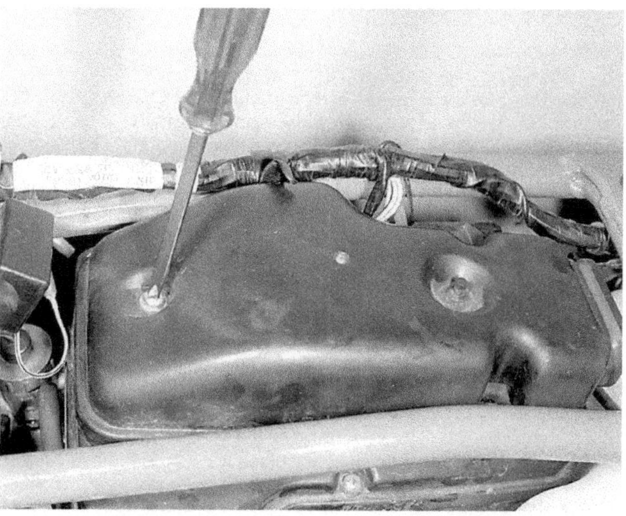

Air filter is housed below the seat, the cover being retained by three screws

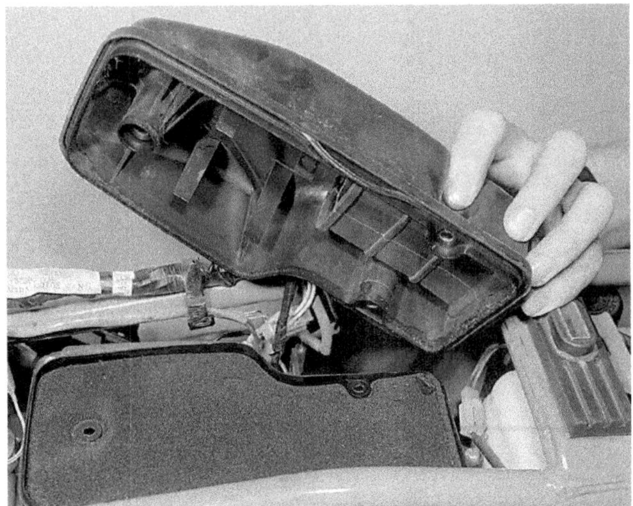
Lift away the cover to reveal the oiled foam element ...

... which can be carefully lifted out for cleaning and inspection

3 *Throttle cable adjustment*

The throttle cable should be adjusted to give 3 – 7 mm (0.12 – 0.28 in) free play measured at the twistgrip flange. To effect adjustment, slacken the adjuster locknut and turn the adjuster in or out as required, then secure the locknut. The adjuster is located at the upper end of the cable, at the right-angle bend from the twistgrip control. After adjustment, start the engine and check that the idle speed does not alter when the handlebars are turned from lock to lock. If necessary, readjust or re-route the cable to prevent this.

4 *Carburettor synchronisation*

If the engine is to run smoothly and efficiently it is essential that the carburettor throttle valves are in perfect synchronisation. If one throttle valve lags behind the other the engine will run erratically, power output will be less than normal and fuel consumption will suffer.

On each carburettor body there is a small inspection plug about one inch (25 mm) down from the carburettor top. Remove the plugs, then open the throttle twistgrip fully and check the position of the synchronisation marks through the inspection holes. If necessary, slacken the adjuster locknuts on the top of the carburettors and move the adjusters to align both marks correctly. Open and close the throttle twistgrip a few times, then recheck the settings. Finally, secure the locknuts and refit the inspection plugs.

5 *Checking the engine idle speed*

Set the pilot air screw of each carburettor to $1^1/_4$ turns out. The setting can be found by turning the screw clockwise until it seats lightly, then unscrewing it by the prescribed amount. The pilot air screw is located on the right-hand side of each carburettor, near the inlet mounting rubber.

Start the engine and allow it to reach full operating temperature. Check the idle speed which should be between 1150 and 1250 rpm. If adjustment is required, turn the knurled adjuster on each instrument by an equal amount until the idle speed falls within the above range.

6 *Checking the transmission oil level*

Run the engine for a few minutes to allow the oil to reach its normal working level and temperature. Stop the engine and place the motorcycle on its centre stand. Wait a further few minutes to allow the oil level to stabilize, then remove the combined filler plug and dipstick. Wipe the dipstick and then place it back in the filler hole so that the plug just rests on the edge of the hole; do not screw it home. Check that the level of the oil is between the upper and lower level marks. If topping up is required, use SAE 10W30 type SE motor oil.

Note that at every fourth check (every 16 000 miles) the oil should be changed as described later in Routine Maintenance.

7 *Checking the oil pump cable adjustment*

RD350 LC II

Note that the oil pump cable adjustment should be checked after any carburettor or throttle cable adjustments have been made. Remove the oil pump cover on the right-hand side of the engine. Open the throttle twistgrip fully and hold it in this position. Check that the plunger pin aligns with the reference mark on the pump pulley. If necessary, adjust the cable adjuster until the pin and mark align. When adjusted correctly, secure the adjuster locknut and grease the cable and the pulley.

RD350 F and RD350 N

The pump alignment on the later models is checked with the throttle closed. Refer to the accompanying line drawing for details of the relevant alignment work.

Transmission oil filler plug incorporates a dipstick. Level is measured with the plug resting on its threads – not screwed home

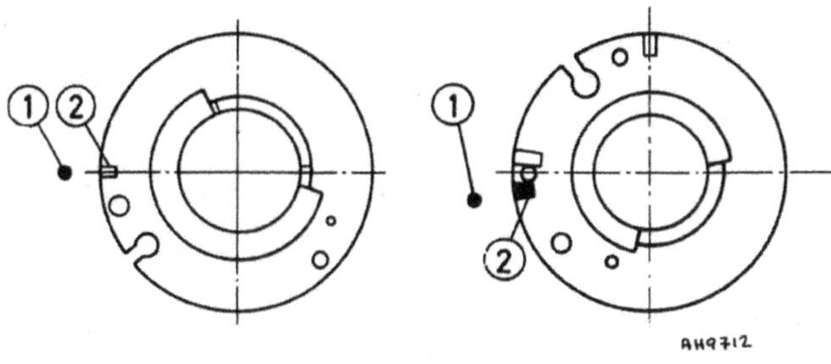

Oil pump pulley alignment marks

1 Plunger pin 2 Pulley alignment mark

8 Checking the oil pump minimum stroke setting

With the pump cover removed, start the engine and allow it to idle while observing the movement of the pump plunger. The pump is at its minimum stroke when the plunger and adjuster plate have moved fully outwards. Stop the engine at the minimum stroke position, and measure the gap between the raised boss on the pulley and the adjuster plate. Take care not to force the feeler gauge into the gap and thus get a false reading. Note the clearance, then start the engine and repeat the check several times.

Using the largest gap reading, check that this is between 0.10 mm and 0.15 mm (0.004 – 0.006 in). If the clearance is incorrect, remove the locknut and adjuster plate and lift away the shims behind it. Add or subtract shims to obtain the correct clearance. Finally, refit the shims, adjuster plate and locknut and re-check the clearance.

9 Bleeding the oil pump

This operation is necessary whenever the oil tank has run dry or if any part of the oil pump system or pipes have been disconnected. Note that air in the system can lead to lubrication failure and possible engine seizure, so do not take any chances. If the system has not been disturbed and the oil tank level has been carefully maintained, this operation can be ignored.

Check that the oil tank is full (topping up as required with any good quality 2-stroke oil for air-cooled engines), then remove the oil pump bleed screw. Allow the oil to flow out for a few minutes, until any air has been expelled. Refit the bleed screw, using a new sealing washer if required.

To bleed the oil delivery pipe, start the engine and allow it to run at about 2000 rpm or so for two minutes. With the engine running at this speed, pull the pump cable so that it is at maximum stroke. Any air will be expelled during this time. Note that there will be a tendency for the exhaust to smoke for a while after bleeding, but this should subside after excess oil in the exhaust system has burnt off.

10 Clutch adjustment

Check the operation of the clutch, and check that there is about 10 – 15 mm (0.4 – 0.6 in) free play measured at the lever end. If adjustment is required, this is carried out using the adjuster at the upper end of the cable which threads into the clutch lever bracket. Slacken the adjuster lockwheel and turn the adjuster out (anti-clockwise) to decrease freeplay or turn it in (clockwise) to increase it; tighten the lockwheel on completion.

Some models also have an in-line adjuster situated mid-way along the cable's length, enabling further adjustment should all free play have been taken up by the adjuster at the handlebar end of the cable.

If the cable has stretched beyond the range of the adjuster(s) it must be renewed.

The clutch release mechanism can be adjusted to compensate for wear of the clutch assembly, ie plate wear. It should not require frequent attention, but should be carried out if clutch slip or drag are experienced.

a) Slacken off the cable adjuster at the handlebar, and where fitted, the in-line cable adjuster.
b) Remove the oil pump cover and disconnect the oil pump cable
c) Drain the transmission oil

Set the oil pump cable so that the marks align correctly. These vary with model (see text)

d) Drain the cooling system
e) Disconnect the radiator hose from the side cover
f) Remove the kickstart lever
g) Remove the right-hand side cover

In view of the amount of work involved it is probably best to time any adjustment to coincide with a normal transmission oil or engine coolant change. For details of the removal of the various components and assemblies detailed above, refer to the appropriate sections of Chapters 1 and 2.

Slacken the clutch centre adjuster locknut, then set the clutch cable in-line adjuster so that the clutch actuating lever aligns with the index mark on the casing. Turn the clutch centre adjuster screw inwards until slight resistance is noted, then back it off by a quarter turn. Tighten the locknut to 0.8 kgf m (5.8 lbf ft). When reassembly is complete, check the cable adjustment as described above.

11 Checking the braking system

Check the operations of the front brake, noting any signs of leakage or sponginess in the system. If air has entered the system, or if there are any signs of leakage around the hose unions, master cylinder or brake calipers, it is imperative that the cause of the problem is traced and rectified immediately. Refer to Chapter 6 for further information.

Check the level of the hydraulic fluid via the inspection window in

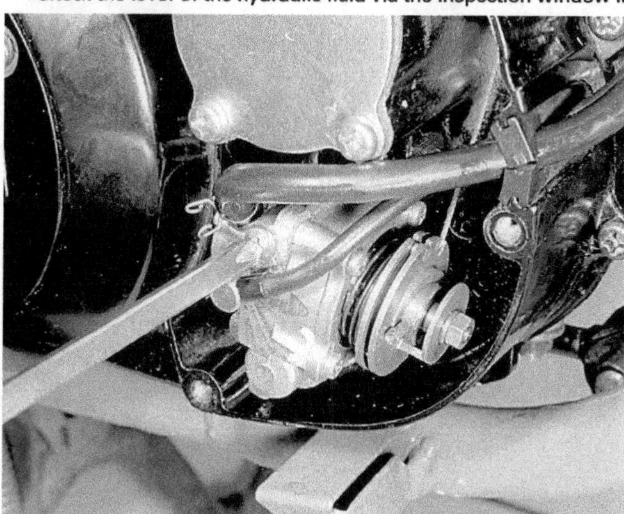

Remove bleed screw and allow oil to flow from tank until free from air bubbles

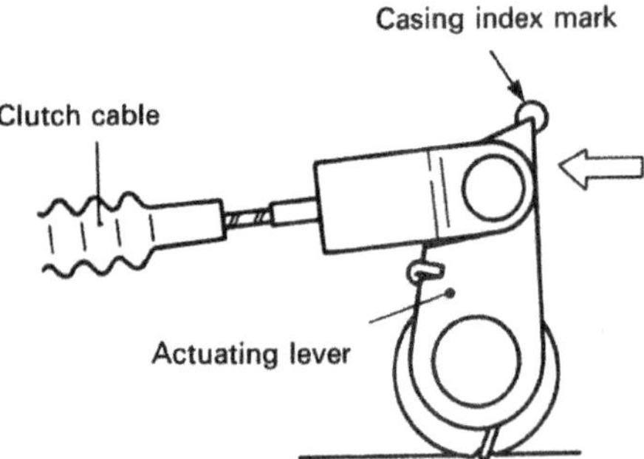

Clutch adjustment

the master cylinder reservoir. The fluid level will fall very gradually due to wear in the brake pads, but any sudden or drastic drop in level is indicative of a leak or seal failure and should be treated as serious.

If routine topping up is required, carefully clean the reservoir to remove any dirt or moisture. Remove the screws which retain the reservoir cap and lift away the cap and the diaphragm below it, taking great care to avoid spilling fluid on painted or plastic parts. Any drips or splashes must be washed off promptly before the surfaces become damaged. Top up the reservoir using only new DOT 4 or 3 or SAE J1703 specification hydraulic fluid, then refit the diaphragm and cap ensuring that a good seal is made.

Check the condition of the brake pads noting that if either is worn down to the wear limit indicators they should be renewed as a set. Details of the pad renewal procedure will be found in Chapter 6.

The rear brake system can be checked in much the same way as described above for the front brake, noting that the rear brake reservoir is mounted remotely from the master cylinder to which it is connected by a short hose. The reservoir cap screws onto the body. Again, check the pads for wear, referring to Chapter 6 if renewal is required.

Finally, check the free play of the front brake lever and rear brake pedal. In the case of the former, there should be 5 – 8 mm (0.2 – 0.3 in) lever movement, measured at the lever end, before the brake begins to operate. This is adjusted using the adjuster screw and locknut near the lever pivot. The rear brake pedal should be set so that it is 35 – 40 mm (1.38 – 1.57 in) below the top of the footrest. To adjust the pedal height, slacken the locknut on the pushrod between the pedal and the master cylinder, and set the pushrod to give the specified clearance. When set correctly, the distance between the pushrod hexagon and the round shank of the fork end should be 11 – 13 mm (0.43 – 0.51 in).

If adjustment was made, check that the brake light comes on just before the brake begins to operate. If this is not the case, adjust the height of the brake light switch by holding the switch body and turning the adjuster nut.

12 Checking the wheels

The conditon of each wheel should be checked, looking for indications of corrosion or damage. It follows that this is best carried out after the machine has been cleaned thoroughly. Check for free play or wear in the wheel bearings and for excessive runout at the wheel rim. For further details, refer to Chapter 6.

13 Checking the steering and front suspension

Check the operation of the front forks by pushing and pulling on the handlebars with the front brake applied. The forks should move smoothly, with no signs of stiffness or excessive play. Any knocking in the steering head area may be indicative of worn or badly adjusted steering head bearings.

Raise the front wheel clear of the ground. Grasp the lower ends of the fork legs and push and pull them to check for play. It will be possible to feel any free movement in the steering head bearings. Turn the handlebars from lock to lock to check for roughness. If the above checks indicate wear in the steering head area, refer to Chapter 5 for details on adjustment and renewal of the steering head bearings.

Examine the surface of the fork stanchions for wear or scoring. Scoring can lead to leakage of the fork seals and consequent loss of fork air pressure and oil. Slide the fork dust seals up and clear of the lower leg to check for leaks. If oil leakage is noted, or if the forks lose air pressure quickly, the forks should be overhauled and new seals fitted. Details of this operation will be found in Chapter 5.

14 Checking the rear suspension

Raise the rear wheel clear of the ground and check for play in the swinging arm pivot by pushing it from side to side. If any play is found, it will be necessary to dismantle and overhaul the swinging arm pivot assembly and to check for wear in the suspension linkage. This is described in detail in Chapter 5.

Check the suspension unit for signs of leakage. If necessary, remove the unit as described in Chapter 5 for closer examination. Leakage will necessitate renewal of the unit; it is of sealed construction and cannot be overhauled.

15 Checking the fuel pipes and coolant hoses

Check the fuel and vacuum pipes for damage or deterioration, and check for fuel leakage; refer to Chapter 3, Section 4 if attention is required.

Similarly check the cooling system hoses, referring to Chapter 2, Section 6 for further details.

Clutch centre incorporates a screw and locknut adjuster to set the initial clearance

Check that hydraulic fluid level lies between minimum and maximum

Yearly, or every 8000 miles (12 000 km)

Carrry out the operations listed under the previous headings, then complete the following:

1 Cleaning the fuel tap

Remove the left-hand side panel to gain access to the fuel tap assembly. Ensure that the tap lever is turned to either the 'On' or 'Res' positions and unscrew the sediment bowl situated at the base of the tap. If traces of water or dirt are found in the bowl the tap should be removed from the tank for cleaning of the fuel strainer and flushing of the tank, for details refer to Chapter 3, Section 3. Before refitting the sediment bowl check that its sealing O-ring has not become compressed or damaged, renewing it if necessary. Do not overtighten the bowl on refitting.

2 Lubricate the rear suspension pivots

Once each year the swinging arm and linkage assembly should be removed from the machine, cleaned and checked for wear, and the pivots repacked with wheel bearing grease. Preventative maintenance will greatly extend the life of the various bushes and should not be ignored. This is especially true if the machine is used through the winter months when the exposed pivots are subjected to considerable attack from salt, road dirt and water. For further details, see Chapter 5.

Two yearly, or every 16 000 miles (24 000 km)

1 Checking and greasing the steering head bearings

Remove the fork legs from the machine, then dismantle and check the steering head bearings. Renew the bearings if worn, otherwise repack with wheel bearing grease and readjust after assembly. See Chapter 5 for more information. Before the fork legs are refitted it is a good idea to change the damping oil as described below.

2 Changing the front fork oil

The front fork oil should be drained and renewed at the specified intervals to ensure consistent fork operation and to prevent internal corrosion which might occur if condensation were allowed to build up in the fork legs. It is important to make sure that the fork oil level is correct and equal in each leg, and to this end it is preferable to remove the fork legs from the machine so that this can be measured accurately with the fork leg held vertical. It should be noted that if the oil level is higher or lower than that specified, the fork spring rate will be affected. This is because the internal air volume of the fork determines the effect of fork air pressure on the overall spring rate. It follows that a difference in oil level between legs must be avoided at all costs. For a full description of the removal and refitting sequence, refer to Chapter 5, Section 3.

3 Changing the coolant

The coolant should be drained, the system flushed and new coolant added every two years. This will ensure adequate frost protection and will minimise corrosion in the system, prolonging the working life of the various parts. The procedure is described in detail in Chapter 2.

4 Changing the transmission oil

The transmission oil should be changed at the specified intervals to ensure full lubrication of the transmission components and to avoid premature wear. The operation should be undertaken with the engine hot, preferably after a run, to ensure that the oil drains fully. Place the machine on its centre stand and slide a bowl or drain tray of at least 2.0 litre (4.5 Imp pint) capacity beneath the drain plug on the underside of the crankcase. Remove the plug and wait until the old oil drains fully.

Clean the drain plug hole and the plug itself, then check that the sealing washer is in good condition; if in doubt, renew it. Refit the plug and tighten to 2.0 kgf m (14.0 lbf ft). Remove the combined filler plug and dipstick and add SAE 10W30 motor oil until it reaches the full mark on the dipstick. Note that the dipstick reading is taken with the plug just resting in the filler hole, and **not** screwed fully home. The transmission casing will take approximately 1.5 litres (2.6 Imp pint) of oil, but note that the oil level, rather than the quantity, is important. Finally, run the engine for a few minutes to distribute the new oil, then re-check the level.

5 Renewing the brake caliper seals

The manufacturer recommends that the piston fluid and dust seals be renewed every two years to preserve brake performance. This operation will require the dismantling of the caliper assembly as described in Chapter 6.

Additionally, every four years the hydraulic brake hoses should be renewed, irrespective of their apparent condition.

General maintenance work

Apart from the specific tasks listed in the foregoing text there are numerous items of general maintenance requiring attention. These are mostly concerned with cleaning and lubrication of the various exposed pivots, such as stands and control levers, and also checking, lubrication and renewal of the control cables.

It is not easy to give realistic mileage or calendar headings for this type of work, and much will depend on the type of use to which the machine is put and the conditions under which it is operated. As a rough guide, work to a maximum of one monthly intervals, though if the machine is used for frequent short trips during the winter, this interval can be reduced to weekly or two-weekly intervals.

Start by cleaning the machine thoroughly. This will make subsequent examination more pleasant and will highlight areas which might otherwise be overlooked. Check areas such as stand pivots carefully and tighten or renew pivot bolts. Note that only the correct hardened pivots bolts must be used. Lubrication can be by engine oil or one of the many maintenance aerosols such as WD40 or similar.

Check all control cables for signs of damage, looking in particular for fraying around the exposed ends of the inner cables. If damaged in any way, renew the cable. Cables can be lubricated overnight using the arrangement shown in the accompanying line drawing. This works well enough, but for quicker results one of the various hydraulic or aerosol type cable oilers can be used. These can be obtained from most motorcycle dealers.

Other jobs which should be undertaken after cleaning are checking for possible sources of future breakdowns; loose fasteners should be tightened and any worn or damaged electrical leads renewed or repaired. Switches can be kept clean and free from corrosion by regular applications of WD40.

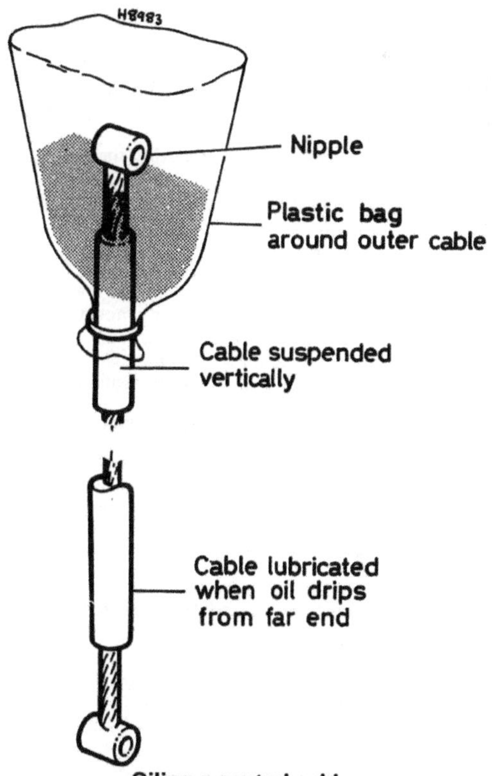

Oiling a control cable

Chapter 1 Engine, clutch and gearbox

For information relating to the RD350 F II, N II and R models, refer to Chapter 8

Contents

Specifications

Note: *Specifications relate to all models unless shown differently*

Engine

Type	Water-cooled, parallel twin cylinder two-stroke
Bore	64.0 mm (2.520 in)
Stroke	54.0 mm (2.126 in)
Compression ratio	6.0:1
Capacity	347 cc (21.2 cu in)

Cylinder head

Type	Cast aluminium alloy
Maximum warpage	0.03 mm (0.0012 in)
Head gasket thickness:	
RD350 LC II	0.7 mm (0.026 in)
Other models	Not available
Combustion chamber volume	21.3 – 21.9 cc (1.29 – 1.34 cu in)

Cylinder barrel

Type	Aluminium alloy, cast-in iron sleeve
Bore size	64.00 – 64.02 mm (2.5197 – 2.5205 in)
Service limit	64.1 mm (2.5236 in)
Taper limit	0.05 mm (0.0020 in)
Maximum ovality	0.01 mm (0.0004 in)

Pistons

Piston diameter	64.0 mm (2.520 in)
Piston/bore clearance	0.060 – 0.065 mm (0.0024 – 0.0026 in)
Service limit	0.1 mm (0.004 in)
Piston oversizes	+0.25, +0.50 mm (0.010, 0.020 in)

Piston rings

Type:	
Top	Keystone
Height	1.2 mm (0.047 in)
Width	2.6 mm (0.102 in)
2nd	Plain, with expander
Height	1.5 mm (0.059 in)
Width	2.15 mm (0.085 in)
End gap (installed):	
Top	0.30 – 0.45 mm (0.012 – 0.018 in)
2nd	0.35 – 0.50 mm (0.014 – 0.020 in)
Ring/groove clearance:	
Top	0.02 – 0.06 mm (0.0008 – 0.0024 in)
2nd	0.03 – 0.07 mm (0.0012 – 0.0028 in)

Crankshaft assembly

Width:	
(F)	54.00 – 54.05 mm (2.130 – 2.132 in)
(A)	155.90 – 156.05 mm (6.136 – 6.142 in)
Maximum deflection (S)	0.05 mm (0.002 in)
Big-end side clearance (D)	0.25 – 0.75 mm (0.01 – 0.03 in)
Connecting rod small-end deflection (P)	0.36 – 0.98 mm (0.0142 – 0.0386 in)
Service limit	2.0 mm (0.08 in)

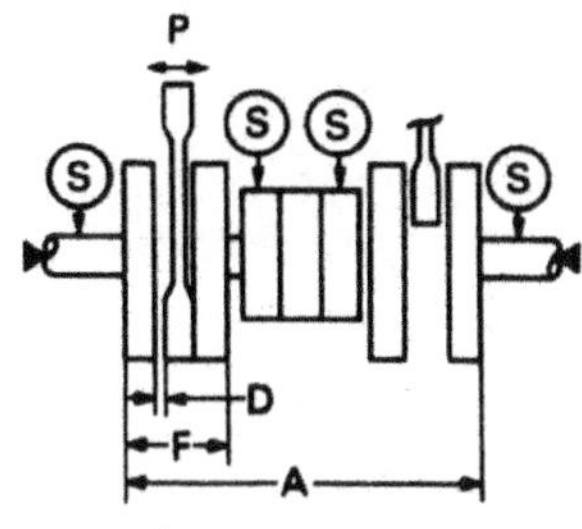

Clutch

Type	Wet, multiplate
Friction plates:	
Quantity	7
Thickness	3.0 mm (0.12 in)
Service limit	2.7 mm (0.106 in)
Plain plates:	
Quantity	6
Thickness	1.2 mm (0.047 in)
Warpage limit	0.05 mm (0.002 in)
Clutch springs:	
Quantity	6
Free length	36.4 mm (1.43 in)
Clutch drum:	
Thrust clearance	0.07 – 0.12 mm (0.003 – 0.005 in)
Radial clearance	0.011 – 0.048 mm (0.0004 – 0.0019 in)
Pushrod warpage limit	0.2 mm (0.008 in)

Primary drive

Type	Helical gear
Reduction ratio	66/23T (2.870:1)
Backlash tolerance	154 – 156 (sum of gear tolerance numbers)
Drive gear backlash number	90 – 98
Driven gear backlash number	57 – 65

Secondary drive

Type	Chain and sprockets
Reduction ratio	39/17T (2.294:1)

Gear selector mechanism

Type	Cam drum
Guide bar bend limit	0.025 mm (0.001 in)

Kickstart mechanism

Friction clip force	0.8 – 1.2 kg (1.8 – 2.9 lb)

Gearbox

Type	6-speed, constant mesh
Ratios:	
1st	36/14 (2.571:1)
2nd	32/18 (1.778:1)
3rd	29/22 (1.318:1)
4th	26/24 (1.083:1)
5th	25/26 (0.962:1)
Top	24/27 (0.889:1)

Torque wrench settings

Component	kgf m	lbf ft
Cylinder head	2.8	20.0
Cylinder barrel	2.5	18.0
Spark plug	2.0	14.0
YPVS valve Allen screw:		
RD350 LC II	0.7	5.1
Other models	0.6	4.3
YPVS pulley:		
RD350 LC II	Not available	
Other models	1.0	7.2
Primary drive gear	6.5	47.0
Clutch centre nut	9.0	65.0
Clutch spring retaining bolts	1.0	7.2
Gearbox sprocket	8.0	58.0
Kickstart lever	2.5	18.0
Gearchange pedal:		
RD350 LC II	1.5	10.0
Other models	1.6	11.0
Reed valve assembly	1.5	11.0
Reed valve petals:		
RD350 LC II	Not available	
Other models	0.1	0.7
Flywheel generator rotor:		
RD350 LC II	8.0	58.0
Other models	8.5	61.0
Exhaust pipe	1.8	13.0
Silencer bracket bolt	6.5	47.0
Thermostat cover	1.2	8.7
Housing cover:		
RD350 LC II	1.0	7.2
Other models	0.8	5.8
Radiator cover	0.3	2.0
Hose union (cylinder head)	1.2	8.7
Oil pump:		
RD350 LC II	0.4	3.0
Other models	0.5	3.6
Coolant drain plug	1.4	10.0
Transmission drain plug	2.0	14.0
Crankcase cover screws:		
RD350 LC II	1.0	7.2
Other models (R)	1.0	7.2
Other models (L)	0.7	5.1
Crankcase bolts (see text):		
Lower	2.5	18.0
Upper	1.0	7.2

Component	kgf m	lbf ft
Bearing cover plate	1.0	7.2
Tachometer gear stopper plate:		
RD350 LC II	0.4	2.9
Other models	0.5	3.6
Gear selector cam stopper plate:		
RD350 LC II	1.4	10.0
Other models	1.0	7.2
Stopper lever	1.4	10.0
Neutral switch	0.4	2.9
Gearchange lever adjuster screw	3.0	22.0
Thermosenser (temperature gauge sender unit):		
RD350 LC II	1.2	8.7
Other models	1.4	10.0
Engine mounting bracket bolts	2.4	17.0
Engine mounting bolts	6.5	47.0

1 General description

The Yamaha RD350 YPVS models employ a water-cooled twin cylinder two-stroke engine built in unit with the primary drive, clutch and gearbox. The engine features a light alloy one-piece cylinder head incorporating cast-in passages for the coolant. Separate light alloy cylinders are fitted, each having an integral cast iron liner.

Induction is controlled by a combination of conventional piston porting, reed valves and the Yamaha power valve system (YPVS). The YPVS system consists of a spool-type valve unit mounted transversely across the two exhaust ports. The valve is able to rotate in the port, thus altering its shape. This allows the exhaust port timing to be varied to suit any given engine speed. The YPVS valve is controlled via two Bowden cables from a servomotor unit mounted below the fuel tank. A microprocessor in the servomotor unit senses engine speed and adjusts the YPVS valve to the necessary setting. In this way, the engine is able to produce high torque at low engine speeds, and has unrestricted performance at high engine speeds.

A pressed-up crankshaft is used, carried on four caged ball main bearings. Both the big-end and small-end bearings are of the needle roller type.

Primary drive is by gears to the wet multi-plate clutch mounted on the end of the gearbox input shaft. The gearbox is of the six-speed constant mesh type. Gearbox lubrication is by oil bath, whilst the engine is lubricated by direct injection via a metered pump driven off the crankshaft.

2 Operations with the engine/gearbox unit in the frame

The items listed below can be overhauled with the engine/gearbox unit in place. Where a number of these operations need to be undertaken simultaneously it may prove advantageous to remove the unit to gain better access and more comfortable working. Engine removal is fairly straightforward, and will take about one hour.

a) Cylinder head, barrels and pistons
b) YPVS valves
c) Clutch assembly and primary drive pinion
d) Oil pump
e) Water pump
f) Ignition pickup
g) Alternator assembly
h) Gear selector mechanism (external components only)
i) Kickstart mechanism
j) Final drive sprocket

3 Operations with the engine/gearbox unit removed from the frame

To gain access to the following items it is first necessary to remove the unit from the frame and to separate the crankcase halves:

a) Crankshaft assembly
b) Gearbox components
c) Gear selector drum and forks

4 Removing the engine/gearbox unit from the frame

1 To allow access to the engine area for removal purposes, it will be necessary to remove the fairing belly pan (RD350 LC II) or the fairing lowers (RD350 F). This operation is described in Chapter 5. Once removed, place the fairing sections away from the work area to avoid accidental damage.

2 Place the machine on its centre stand, leaving working space on both sides and at the front. Start the engine and allow it to reach normal operating temperature. Place a drain tray or bowl of at least 2.0 litre (0.44 Imp gal) capacity below the engine unit, remove the transmission drain plug and allow the oil to drain. When draining is complete, refit the plug for safekeeping.

3 Remove the seat and place it to one side. Check that the fuel tap is turned to the "ON" position (vacuum tap) or "OFF" position (gravity-fed tap), then remove the left and right-hand side panels. Disconnect the fuel and vacuum pipes (where fitted) at the fuel tap. Remove the single holding bolt at the rear of the tank, lift it slightly and pull it rearwards until it comes free of the mounting rubbers on each side of the steering head.

4 When the engine has cooled fully, and not before, the cooling system should be drained. Note that care must be taken to avoid removing the radiator cap or removing drain plugs while there is residual pressure in the system. If the coolant was changed recently it can be retained for re-use, otherwise discard it and fill with a fresh solution during reassembly. The procedure for draining the system, including precautions to be taken when handling the coolant mixture, will be found in Chapter 2. Disconnect the radiator top and bottom hose at the engine end, and also the smaller bypass hose between the engine and radiator filler neck.

5 Free the exhaust pipe retaining nuts at the exhaust ports, then release the silencer to footrest bracket mounting bolts. Lift each half of the system clear of the machine and place them to one side. Disconnect and remove the gearchange pedal and linkage. Disconnect both spark plug caps and the temperature gauge sender lead. Lodge them clear of the engine around the frame top tubes.

6 Remove the oil pump cover. Prise off the pipe from the oil tank at the pump stub, plugging its open end to prevent oil leakage. Rotate the pump pulley to its fully open position to allow the cable to be disengaged. Free the cable and return spring and lodge the cable clear of the engine.

7 Release the pulley cover on the left-hand end of the YPVS valve. Slacken the two cable adjuster locknuts, then screw the adjusters fully inwards. Unscrew the central retaining bolt to free the pulley, and disconnect the cables. Remove the two cross-head screws to free the cables, together with the pulley housing.

8 Moving to the carburettors, make a simple sketch of the drain and vent pipe routing so that they can be refitted correctly. Remove the pipes and place them to one side. Unscrew the carburettor tops and withdraw the throttle valve assemblies, taking care not to bend the needles. It is not necessary to disconnect the cables from the valves. Slacken the clips which retain the carburettor stubs to the inlet and air filter rubbers. Disconnect the rubbers, then manoeuvre the carburettors clear of the engine and remove them.

9 Free the tachometer cable at the crankcase by unscrewing the knurled retaining ring (RD350 LC II model). Slacken the adjuster at the upper end of the clutch cable. Disconnect the cable at the

handlebar lever, then at the lower end, lodging it clear of the engine unit.

10 Trace the wiring from the left-hand side of the engine unit (alternator, ignition pickup and neutral switch) and disconnect it at the two multi-pin connectors. Remove the left-hand outer cover to gain access to the gearbox sprocket. Knock back the locking tab which secures the sprocket retaining nut. Apply the rear brake, holding it on while the retaining nut is slackened. Slide the sprocket off its splines and disengage it from the chain. The latter can be left to rest against the frame tube.

11 Remove the two bolts which pass up through the ends of the engine stabiliser bars and into the underside of the crankcase. The bars can be either removed or just pivoted down and clear of the engine. Remove the engine mounting bolts and plates as shown in the accompanying photographs. As these are removed, the engine will tend to drop and trap the remaining bolts. To remove these, carefully lift or lever the engine upwards until they can be displaced and removed. The engine is now free and will be left sitting on the frame cradle.

12 Make a final careful check around the engine to ensure that nothing remains which might impede removal. Pay particular attention to cables or wiring which might get caught on the unit as it is lifted out. In an emergency, it is just feasible for one strong person to remove the engine unit unaided, but this approach is not recommended. It is much better to involve at least two people at this stage to allow the unit to be manoeuvred out of the frame without damage. Take the weight of the unit, then carefully remove it from the right-hand side, placing it on a bench to await further attention.

5 Dismantling the engine/gearbox unit: general

1 Before commencing work on the engine unit, the external surfaces must be cleaned thoroughly. A motorcycle engine has very little protection from road grit and other foreign matter, which will sooner or later find its way into the dismantled engine if this simple precaution is not observed.

2 One of the proprietary engine cleaning compounds such as 'Gunk' or 'Jizer' can be used to good effect, especially if the compound is worked into the film of oil and grease before it is washed away. When washing down, make sure that water cannot enter the inlet or exhaust ports or the electrical system, particularly if these parts are now more exposed.

3 Never use force to remove any stubborn part, unless mention is made of this requirement in the text. There is invariably good reason why a part is difficult to remove, often because the dismantling operation has been tackled in the wrong sequence.

4.3a On the vacuum type fuel tap fitted to early models, turn the tap to ON and disconnect the fuel and vacuum pipes

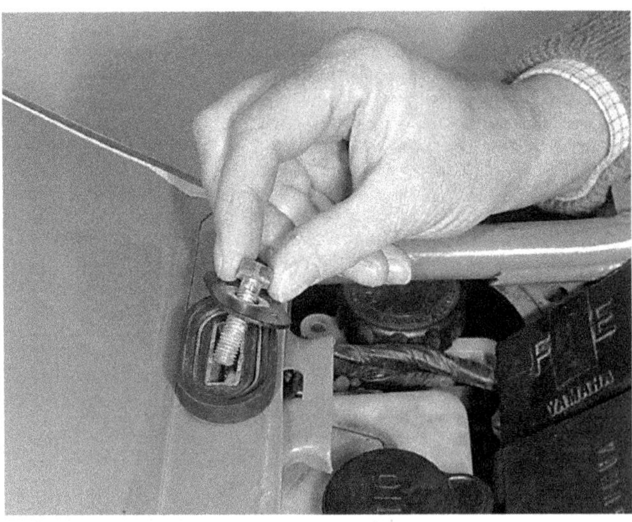

4.3b Tank is secured at the rear by a single bolt

4.4 Disconnect coolant hoses between engine and radiator ...

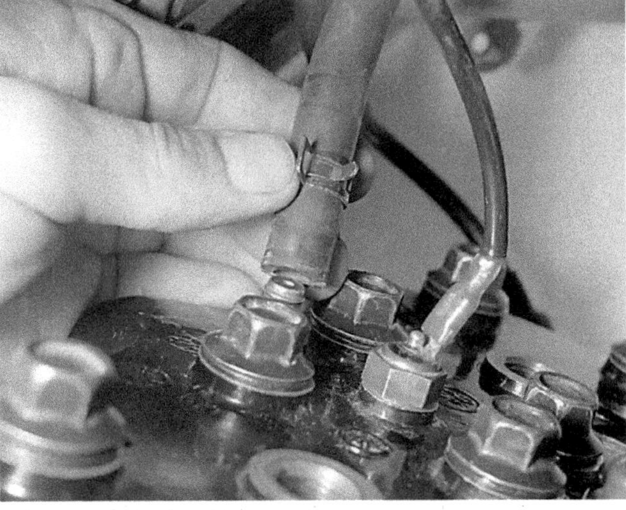

4.5a ... not forgetting the bypass hose

4.5b Disconnect the water temperature sender lead

4.5c It is a good idea to remove the sender to avoid damage during engine removal

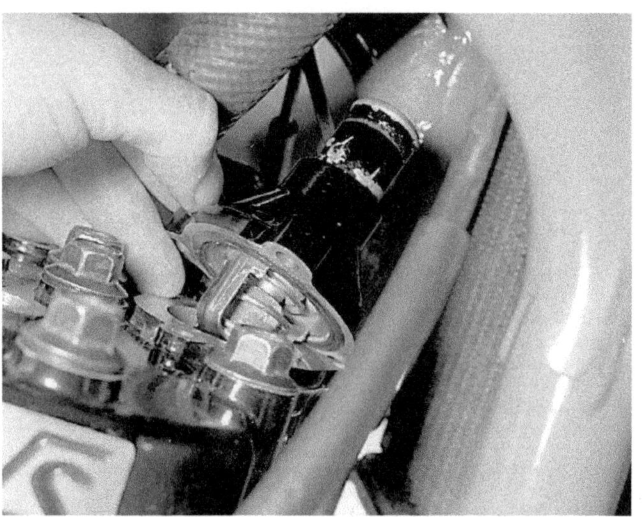
4.5d The thermostat housing and unit can also be removed to give improved clearance

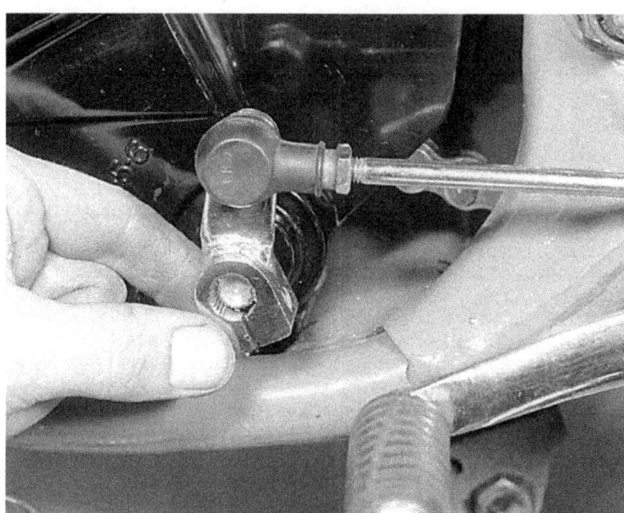
4.5e Disconnect the gearchange linkage as shown

4.10a Release the retaining screws and lift away the left-hand cover

4.10b Knock back the tab washer and remove nut to free the gearbox sprocket

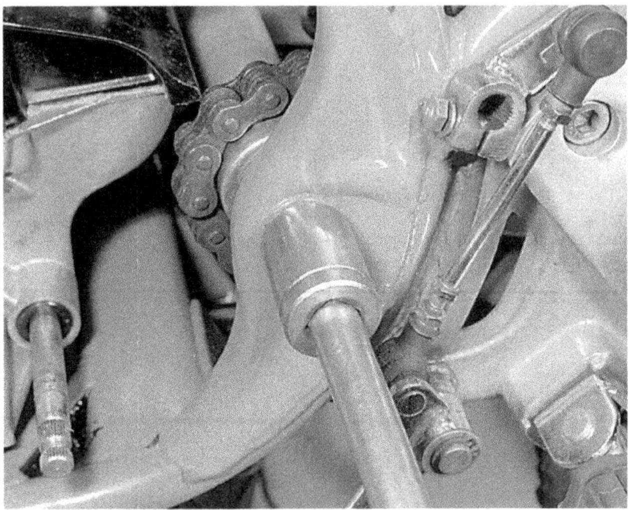

4.11a Swinging arm pivot bolt doubles as engine rear mounting bolt

4.11b Upper rear engine mounting bracket arrangement

6 Dismantling the engine/gearbox unit: removing the cylinder head, barrels and pistons

1 Free the balance pipe which connects the two inlet rubbers. The pipe is secured by wire clips, the ends of which should be squeezed together to allow the pipe end to be pulled clear. Take care not to strain the inlet rubber; it is easily torn. Remove the four Allen bolts which retain each reed valve unit to its cylinder barrel and remove it.

2 Slacken the clamps holding the coolant hose to the underside of the cylinder head and to the crankcase stub. Disengage the hose and remove it. Unscrew the temperature gauge sender from the cylinder head, taking care not to lose the sealing washer. Release the Allen screws which retain the thermostat housing to the cylinder head. Lift away the housing and remove the thermostat.

3 Remove or slacken the spark plugs before the head is removed; it is easier to do so at this stage than with the head detached. To free the head, slacken the cylinder head sleeve bolts by following the numbers cast into the head material in descending order. Each bolt should be slackened by a half turn until all are loose. Then remove the bolts and lift the head away. In the unlikely event that the head is stuck to the gasket, tap around the joint using a hide mallet, or a hammer and a hardwood block.

4 Before the barrels can be removed, it is necessary to disconnect the two power valves at the centre connection joint by releasing the two screws. Remove the four retaining nuts from the flange of each barrel. Lift each barrel slightly and pack some clean rag into the crankcase mouth to catch any debris or residual coolant. Carefully lift each barrel away, supporting its piston as it emerges from the bore. Place the inverted barrels and head on some rag to allow any remaining coolant to drain away.

5 The power valve should now be removed from each barrel. Although this is not an essential operation, it is normal practice to remove and clean the valve and to inspect it for damage before reassembly. Start by releasing the single Allen screw which secures the retainer plate at the inner end of the valve. Hold the outer end of the valve with a pair of pliers, using a strip of thin card to protect the valve from damage from the plier jaws. Slacken and remove the long through bolt which retains the two halves of the valve. If the through bolt is unusually stiff, a hardwood strip can be introduced through the exhaust port to wedge the valve.

6 Using the retainer groove in the inner end of the valve, lever the valve half out with a screwdriver. Take care not to damage the valve or barrel material. The outer half of the valve can now be displaced and removed. Check that the two small dowel pins are in place, then fit the two halves of the valve together and refit the through bolt to hold them. Place the valve with the barrel to which it belongs; do not interchange them.

7 Prise out the end of one gudgeon pin (piston pin) circlip using a pair of pointed-nose pliers. Push out the gudgeon pin from the opposite end and remove it, lifting the piston clear of the connecting rod. If the gudgeon pin proves tight, warm the piston with a rag soaked in hot water to expand the alloy, taking care to avoid burnt fingers. If available, a hot air gun can be used instead. Repeat for the other piston. As each piston is removed, mark the inside of the skirt to indicate the bore to which it belongs. Displace the small-end bearing from the connecting rod and place it with its piston and gudgeon pin.

6.1a Remove the balance pipe from between the two inlet rubbers

6.1b Note that pipe is held by spring clips which can be slid off after squeezing the 'ears' together as shown

6.2 Remove the two Allen bolts to free the coolant hose adaptor

6.3 Slacken and remove the cylinder head sleeve bolts and lift the head away

6.4a Disconnect the joint piece between the two power valves

6.4b Remove the cylinder base nuts and lift away the cylinder barrels

6.5a Each power valve is secured at its inner end by a retainer plate

6.5b Hold the valve with pliers and remove the long Allen bolt

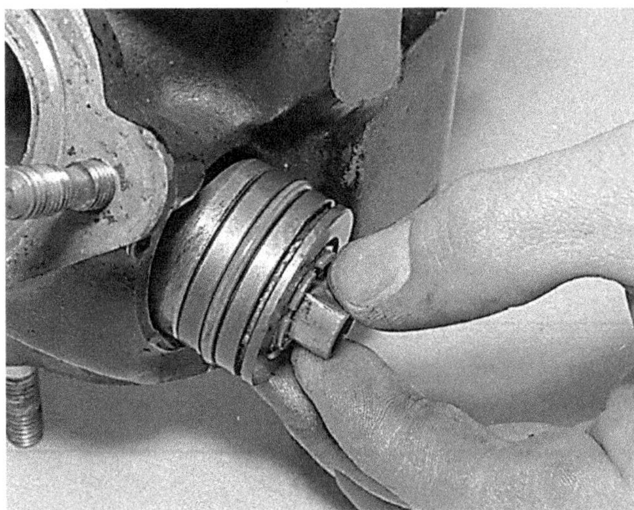

6.6a The two valve halves can now be removed. Lever out the inner half using the retainer groove ...

6.6b ... then displace and remove the outer half, taking care to retain the two small dowel pins

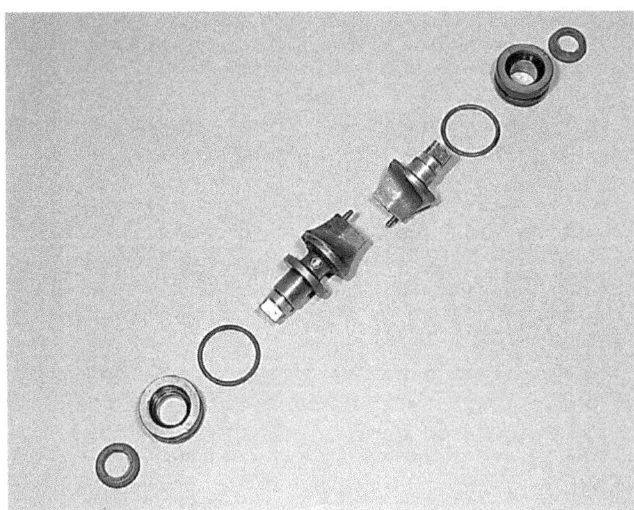

6.6c A complete power valve assembly shown here in its component parts

6.7a Disengage the gudgeon pin circlips using snipe-nosed pliers

6.7b Gudgeon pins can now be displaced to allow the pistons to be lifted away

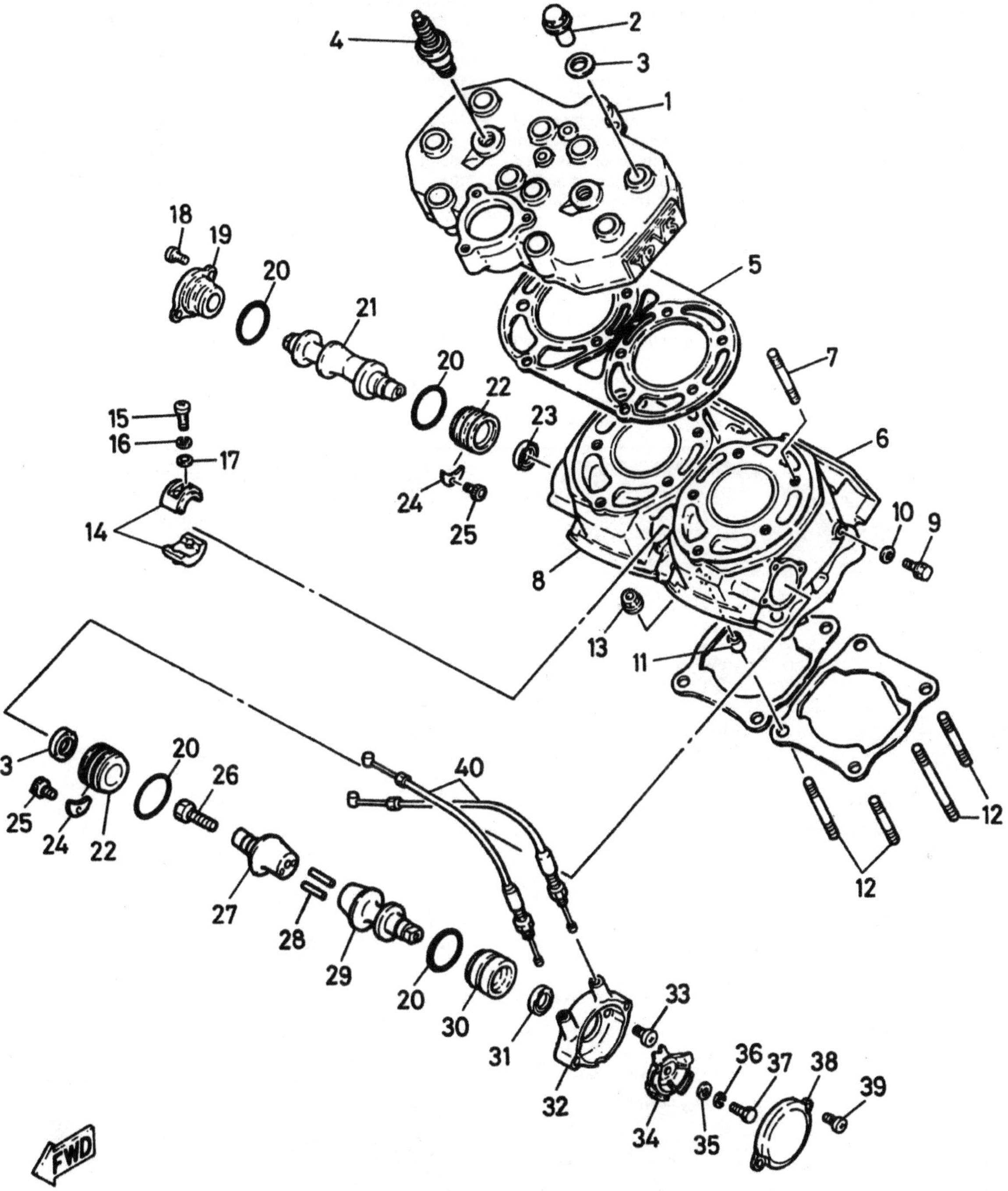

Fig. 1.1 Cylinder head, barrels and power valve

1 Cylinder head
2 Sleeve bolt – 8 off
3 Washer – 8 off
4 Spark plug – 2 off
5 Cylinder head gasket
6 Left-hand barrel
7 Stud – 8 off
8 Right-hand barrel
9 Drain plug
10 Sealing washer
11 Dowel
12 Stud – 8 off
13 Nut – 8 off
14 Connection joint
15 Screw – 2 off
16 Spring washer – 2 off
17 Washer – 2 off
18 Screw – 2 off
19 End cap
20 O-ring
21 Right-hand cylinder valve complete
22 Valve holder
23 Oil seal
24 Retaining plate
25 Allen screw
26 Bolt
27 Inner valve half
28 Dowel pin
29 Outer valve half
30 Valve holder
31 Oil seal
32 Pulley housing
33 Screw – 2 off
34 Pulley
35 Washer
36 Spring washer
37 Bolt
38 Pulley cover
39 Screw – 2 off
40 Operating cables

7 Dismantling the engine/gearbox unit: removing the alternator

1 The alternator asembly can be removed with the engine unit in or out of the frame. In the former case it will first be necessary to remove the gearchange linkage and the left-hand outer cover, and to trace and disconnect the alternator wiring if the stator is to be removed. A rotor extractor tool will be needed to draw the rotor off its taper safely; do not attempt to remove it by levering. The tool can be obtained as Part Number 90890-01189 from Yamaha dealers, but many motorcycle dealers stock a range of suitable extractors from other suppliers.
2 Slacken and remove the rotor holding nut whilst holding the rotor to prevent it from turning. There are a number of ways to hold the rotor; the best method being the use of a home-made tool like that shown in the accompanying photograph. This was made up with some steel strip and a few bolts and need not be very elaborate. If you prefer, you can buy a made up tool of this type from Yamaha dealers as Part Number 90890-01235.
3 In the absence of a holding tool, and with the engine unit in the frame, select top gear and apply the rear brake to lock the crankshaft. If the engine is on the bench, postpone the operation until the cylinder barrels have been removed, then pass a smooth round bar through one of the connecting rod eyes. Support the ends of the bar on hardwood blocks placed against the crankcase, thus locking the crankshaft.
4 With the rotor immobilised by one of the above methods, remove the nut and washer. Fit the extractor into the thread in the centre of the rotor. Holding the body of the extractor with a spanner, tighten the centre bolt to draw the rotor off its taper. If it proves stubborn, try tapping the head of the centre bolt; this will usually jar the rotor free. If necessary, tighten the bolt further and repeat the process.
5 Once the rotor has been removed, unscrew the three bolts which secure the stator assembly to the crankcase and lift it away. The wiring and block connector can be fed through the hole in the crankcase once the wiring grommet has been displaced. Note that the neutral switch lead must also be freed.

7.2 Use fabricated holding tool to allow the rotor nut to be removed

7.4 A proprietary extractor can be used to draw the rotor off its taper, but do not attempt to use a legged puller

7.5a Neutral switch lead must be disconnected before stator can be removed

7.5b Remove the stator mounting bolts and lift the assembly clear

8 Dismantling the engine/gearbox unit: removing the clutch

1 The above mentioned parts may be removed with the engine in or out of the frame. In the former case it will be necessary to remove first the transmission drain plug and allow the oil to drain, and to release the kickstart lever and oil pump cable and pipes. Note that the water pump is housed in the right-hand engine casing and thus the cooling system must be drained before it can be removed, though it is not necessary to disturb either the water pump or the oil pump.
2 Release the screws around the outer edge of the engine casing. These are invariably tight and will require the use of an impact driver to loosen them without damaging the screw heads. The cover can now be lifted away complete with the pumps and placed to one side. If the work is being undertaken with the engine in the frame, the nut which secures the primary drive pinion must be slackened at this stage if it is wished to remove the pinion. To prevent crankshaft rotation as the nut is removed, select top gear and apply the rear brake. Once the nut has been loosened, the clutch can be dismantled. Where the engine is being stripped on the workbench it is easier to lock the crankshaft by passing a round metal bar through one of the connecting rod small-end eyes, its ends being supported on small wooden blocks placed against the crankcase mouth to protect the gasket face.
3 Slacken and remove the six bolts which secure the clutch pressure plate, releasing them evenly by about one turn at a time until they are no longer under spring tension. Lift the pressure plate clear together with the six clutch springs. Displace and remove the clutch plain and friction plates.
4 Before the clutch centre nut can be removed, some method of holding the centre must be devised. Yamaha produce a special holding tool, Part Number 90890-04086, and this can be used if available. Alternatively, the home-made equivalent shown in the accompanying photograph will prove equally effective. The tool was made up from 1 in x 1/8 in mild steel strip, the edges of the angled jaws being ground to fit snugly in the clutch centre splines. An assistant will be required to hold the clutch centre with the improvised tool while the nut is removed. Take care not to allow the tool to slip or the soft alloy splines will be damaged.

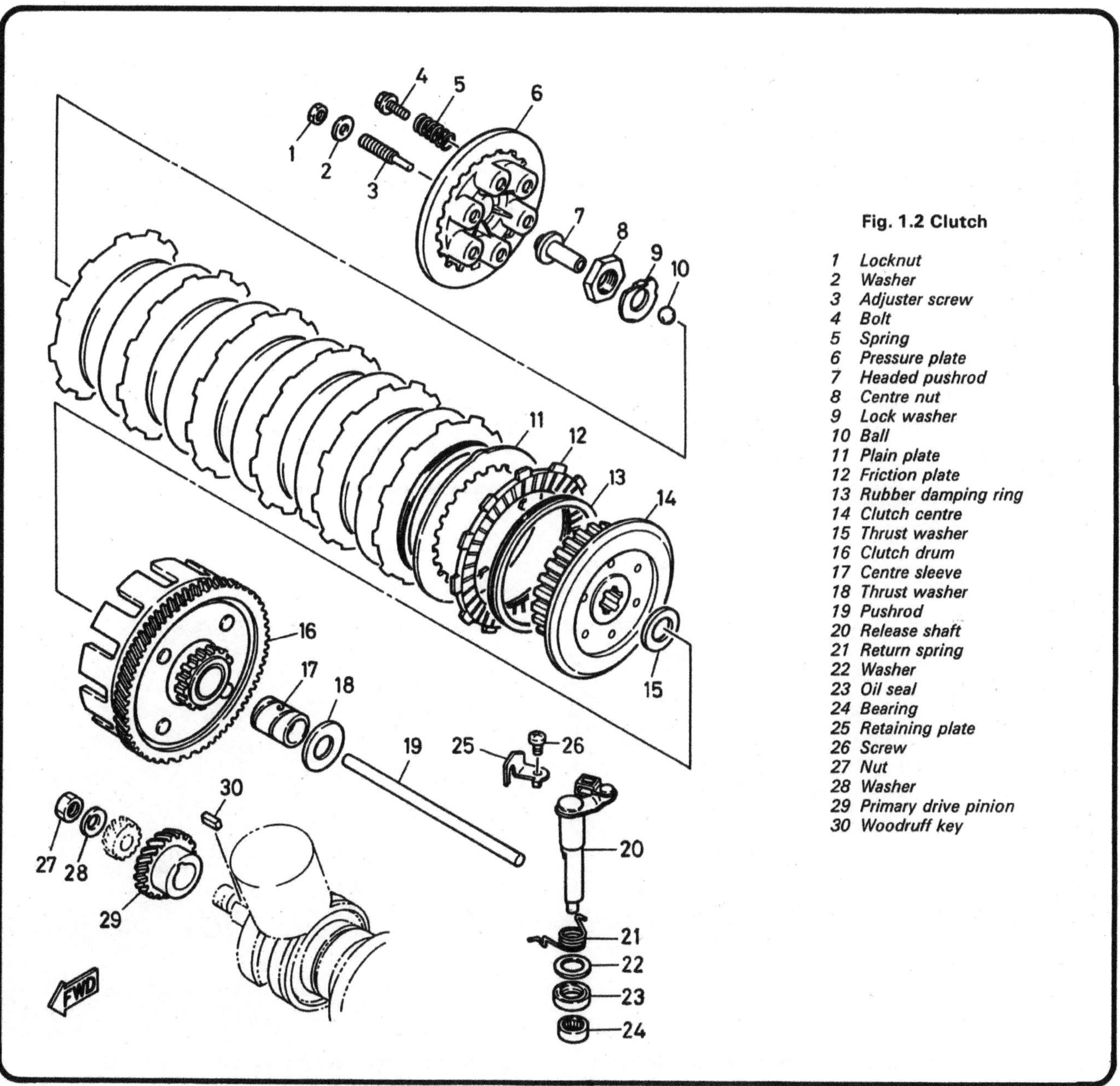

Fig. 1.2 Clutch

1 *Locknut*
2 *Washer*
3 *Adjuster screw*
4 *Bolt*
5 *Spring*
6 *Pressure plate*
7 *Headed pushrod*
8 *Centre nut*
9 *Lock washer*
10 *Ball*
11 *Plain plate*
12 *Friction plate*
13 *Rubber damping ring*
14 *Clutch centre*
15 *Thrust washer*
16 *Clutch drum*
17 *Centre sleeve*
18 *Thrust washer*
19 *Pushrod*
20 *Release shaft*
21 *Return spring*
22 *Washer*
23 *Oil seal*
24 *Bearing*
25 *Retaining plate*
26 *Screw*
27 *Nut*
28 *Washer*
29 *Primary drive pinion*
30 *Woodruff key*

8.2a Remove the outer cover and gasket to gain access to the clutch

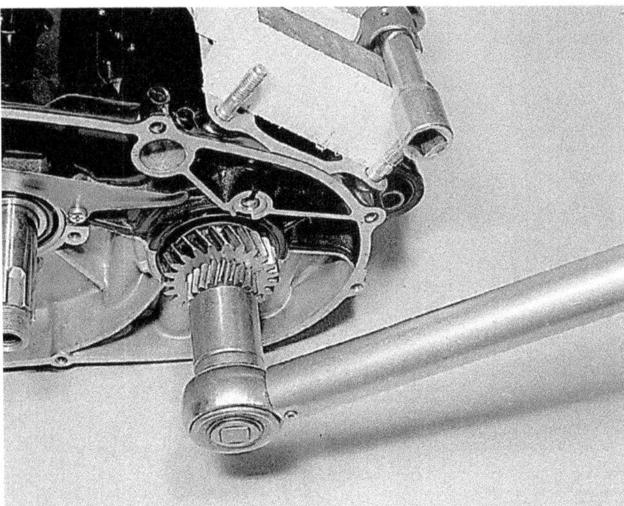
8.2b Method of locking the crankshaft to permit the removal of the primary drive pinion nut

9 Dismantling the engine/gearbox unit: removing the primary drive and pump drive pinions

1 The pinions referred to in the heading are mounted on the right-hand end of the crankshaft and are secured by a large nut. The crankshaft (primary drive) pinion is located by a Woodruff key, whilst the outer pinion, which drives the oil and water pumps, is located by pressure from the securing nut. It follows that the nut is very tight and will require the use of a stout socket and lever bar to facilitate loosening. A ring spanner can be used with good effect but on no account attempt removal with an open-ended spanner.
2 As mentioned in the previous Section, a secure method of holding the crankshaft is essential. If the engine is installed in the frame, it is recommended that the securing nut is slackened before the clutch is removed. This will allow the crankshaft to be locked through the transmission by selecting top gear and applying the rear brake.
3 If, on the other hand, the unit is to be dismantled on the workbench, wait until the cylinder head, barrels and pistons have been removed. A close-fitting round metal bar can now be passed through one of the connecting rod small-end eyes and its free ends supported by wooden blocks placed against the crankcase mouth. This will provide positive restraint for the crankshaft without risk of damage to any component. Slacken and remove the nut, followed by the Belville washer. The pump pinion can now be removed together with the primary drive pinion and its Woodruff key. If the clutch is to be dismantled, refer to Section 8.

10 Dismantling the engine/gearbox unit: removing the kickstart mechanism

1 The kickstart mechanism can be removed with the engine in or out of the frame after the engine right-hand casing has been detached. Note that if it is wished to remove the idler pinion which conveys drive to the clutch drum, and thus to the crankshaft, the clutch must be removed first. The idler pinion does not impair crankcase separation and may be left in position unless specific attention to it or the input shaft components is required.
2 Release the kickstart return spring by grasping its outer end with a pair of pliers and disengaging the end from its anchor pin. Allow the spring to unwind in a controlled manner, then pull the kickstart shaft assembly from its casing hole. If it is wished to remove the idler gear pinion after the clutch has been removed, release the circlip which retains it to the end of the gearbox output shaft.

8.4 Make up the holding tool shown to allow the clutch centre nut to be slackened.

10.2 Unhook the return spring and withdraw the kickstart mechanism from the crankcase

11 Dismantling the engine/gearbox unit: separating the crankcase halves

1 The remaining external engine components can be left in position and do not impair crankcase separation. Their removal is described in Section 12, and it should be noted that if required they can be removed without complete dismantling of the unit. Bear in mind, however, that any internal components, such as the tachometer drive shaft cannot be dealt with unless the crankcase halves have been parted. The only remaining item to be removed at this stage is the input shaft right-hand bearing retainer, which bridges the crankcase halves. It is retained by two cross-head screws which are invariably stubborn and will require the use of an impact driver to effect safe removal.
2 The crankcase bolts are numbered to indicate the correct tightening sequence and should be released in reverse order, starting at the highest number and working backwards. Each bolt should be slackened initially by about $^1/_4$ turn, then removed completely. There are eight bolts on the underside of the unit and a further eight on the upper face of the crankcase. Once all of the bolts have been removed, separate the joint by striking the front and rear edges of the upper crankcase half with a soft-faced mallet.
3 When the joint has been broken the upper crankcase half can be lifted away. Note that the connecting rods will tend to fall against the crankcase edge and they should be supported to prevent this. The gearbox shafts and the crankshaft should remain in the lower crankcase half.

12 Dismantling the engine/gearbox unit: final dismantling

1 Grasp the ends of the crankshaft assembly and lift it away from the lower crankcase half. Note the half-ring which locates the right-hand main bearing. This will probably be displaced as the crankshaft is removed and should be placed in a safe place to avoid its loss. The gearbox input shaft and output shaft assemblies should be removed in a similar manner, again noting the locating half-rings.
2 Disengage the selector claw from the end of the selector drum and remove the gearchange shaft assembly by displacing the selector shaft on the opposite side of the crankcase. Note that the seal through which the shaft must pass is easily damaged and if it is necessary to re-use it, protect the seal lip by wrapping some pvc tape around the shaft splines.
3 Release the selector drum stopper arm by removing its single retaining bolt, then remove the selector drum retainer which is held by two countersunk crosshead screws. Note that the retainer also locates one of the selector fork shafts. The remaining shaft is retained by the selector mechanism centralising spring anchor pin which should be removed together with the retainer plate.
4 Working from inside the crankcase, use a pair of pointed-nose pliers to displace the circlips on the inner ends of the selector fork shafts whilst the shafts are pushed through the casing. Support the selector forks and withdraw the shafts completely, then slide the shafts back through the forks to keep them in the correct relative positions as a guide during reassembly. The selector drum and its bearing can now be pushed out of the casing and removed.
5 The tachometer drive (where fitted) need not be disturbed unless it requires specific attention, but if removal proves necessary proceed as follows. Remove the circlip and plain washer which retain the white plastic drive pinion, then remove the pinion from the shaft end. Displace the drive pin and place it with the pinion for safe keeping. Release the single bolt which retains the tachometer drive body to the crankcase and remove it complete with the driven shaft. Remove the single screw which retains the drive shaft locating plate to allow the shaft to be displaced and removed. The drive gear should be slid off the shaft as the latter is pulled clear of the crankcase, having first released the circlips which retain it.
6 The clutch release arm is held in position by a retainer plate, and this will have been removed together with the crankcase bolt which retains it. The arm can be withdrawn from the crankcase upper half by lifting it upwards.

11.1 This bearing retainer bridges the crankcase halves and must be removed to allow separation

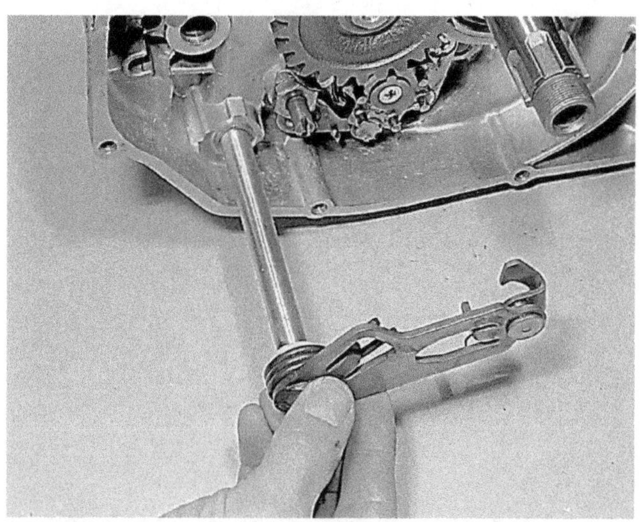

12.2 Disengage and withdraw the selector claw assembly

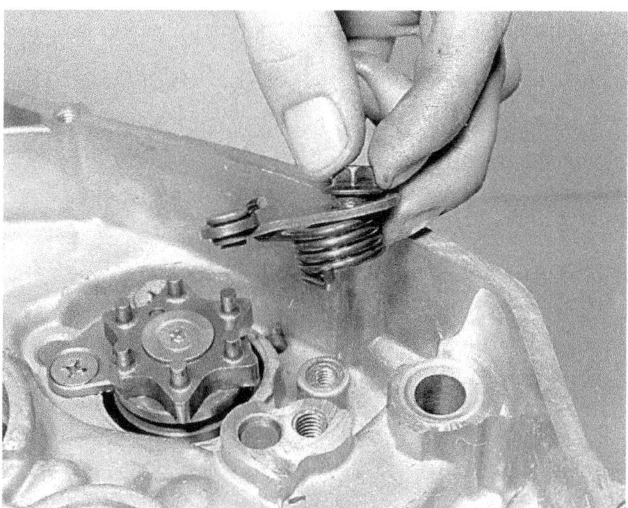

12.3a Stopper arm assembly is held by a single pivot bolt

12.3b Selector drum retainer is secured by two countersunk screws

12.3c Knock back the locking tab and unscrew the anchor pin

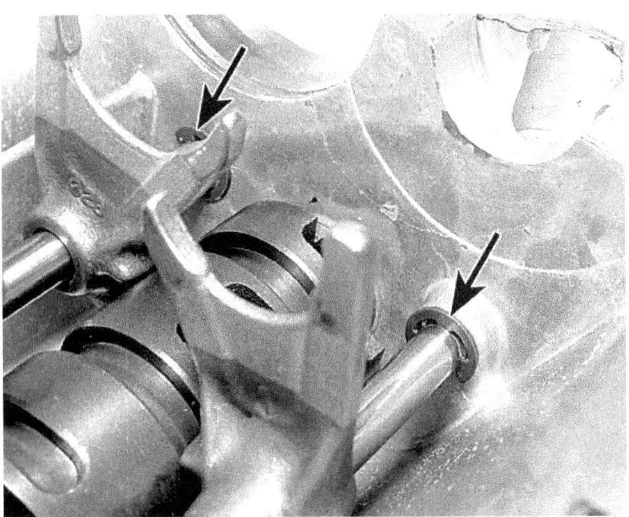
12.4a Release the circlips which retain the selector fork shafts ...

12.4b .. then slide shafts out of casing while supporting the selector forks

12.4c The gear selector drum can now be withdrawn from the crankcase

12.6a The clutch release arm is normally held in place by this retainer and one of the crankcase bolts (arrowed) ...

12.6b ... and can be lifted out of the casing bore for examination

13 Examination and renovation: general

1 Before examining the parts of the dismantled engine unit for wear, it is essential that they should be cleaned thoroughly. Use a paraffin/petrol mix to remove all traces of old oil and sludge that may have accumulated within the engine.
2 Examine the crankcase castings for cracks or other signs of damage. If a crack is discovered, it will require professional repair.
3 Examine carefully each part to determine the extent of wear, checking with the tolerance figures listed in the main text or in the Specifications section of this Chapter. If there is any question of doubt, play safe and renew.
4 Use a clean, lint-free rag for cleaning and drying the various components. This will obviate the risk of small particles obstructing the internal oilways, causing the lubrication system to fail.

14 Gearbox input and output shafts: dismantling and reassembly

1 The gearbox clusters should not be disturbed needlessly, and need only be stripped where careful examination of the whole assembly fails to resolve the source of a problem, or where obvious damage, such as stripped or chipped teeth is discovered.
2 The input and output shaft components should be kept separate to avoid confusion during reassembly. Using circlip pliers, remove the

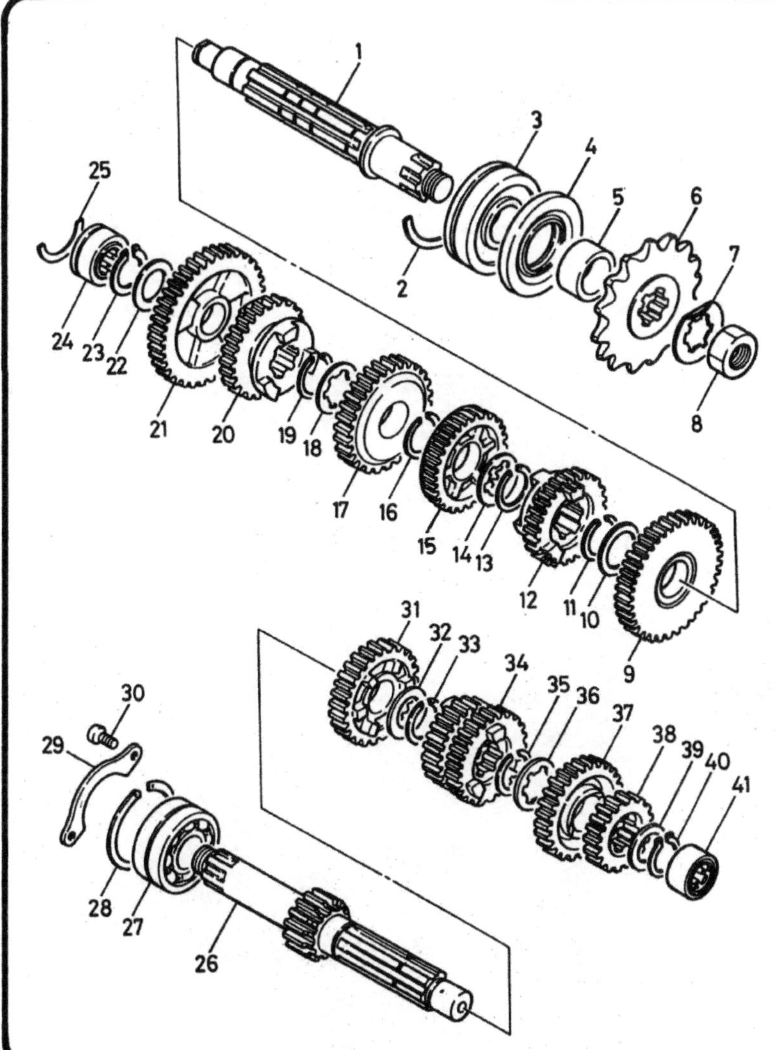

Fig. 1.3 Gearbox components

1 Output shaft
2 Bearing locating ring
3 Output shaft left-hand bearing
4 Oil seal
5 Spacer
6 Final drive sprocket
7 Tab washer
8 Nut
9 Output shaft 2nd gear pinion
10 Thrust washer
11 Circlip
12 Output shaft 6th gear pinion
13 Circlip
14 Splined thrust washer
15 Output shaft 4th gear pinion
16 Circlip
17 Output shaft 3rd gear pinion
18 Splined thrust washer
19 Circlip
20 Output shaft 5th gear pinion
21 Output shaft 1st gear pinion
22 Thrust washer
23 Circlip
24 Needle roller bearing
25 Bearing locating ring
26 Input shaft and 1st gear pinion
27 Input shaft right-hand bearing
28 Circlip
29 Bearing retainer
30 Screw
31 Input shaft 5th gear pinion
32 Thrust washer
33 Circlip
34 Input shaft 3rd and 4th gear pinion
35 Circlip
36 Splined thrust washer
37 Input shaft 6th gear pinion
38 Input shaft 2nd gear pinion
39 Thrust washer
40 Circlip
41 Needle roller bearing

circlip and plain washer which retain each part. As each item is removed, place it in order on a clean surface so that the reassembly sequence is self evident and the risk of parts being fitted the wrong way round or in the wrong sequence is avoided. Care should be exercised when removing circlips to avoid straining or bending them excessively. The clips must be opened just sufficiently to allow them to be slid off the shaft. Note that a loose or distorted circlip might fail in service, and any dubious items must be renewed as a precautionary measure. The same applies to worn or distorted thrust washers.

3 Having checked and renewed the gearbox components as required (see Section 20) reassemble each shaft, referring to the accompanying line drawing and photographs for guidance. The correct assembly sequence is detailed below.

Input shaft (mainshaft)

4 Note that the input shaft is readily identified by its integral 1st gear pinion. Slide the 5th gear into position with the dogs facing away from the 1st gear.

5 Fit the plain thrust washer and secure the 5th gear pinion with its circlip. The double 3rd/4th gear pinion is fitted next, with the smaller, 22 tooth, gear towards the 5th gear pinion. Fit a circlip to the next exposed groove, followed by a splined thrust washer. This retains the 3rd/4th gear pinion but allows it to move along the shaft to effect gear changes.

6 Slide the 6th gear pinion into place, noting that the engagement dogs face inwards, towards the 3rd/4th gear. The 2nd gear pinion is fitted next and is retained by a plain thrust washer and a circlip. The needle roller bearing should now be lubricated and slid into place to complete assembly. If it has been removed, fit the large caged ball bearing and large thrust washer to the right-hand end of the shaft.

Output shaft (layshaft)

7 Slide the 2nd gear pinion up against the shouldered portion of the output shaft, noting that it must be fitted from the right-hand end, with the engagement webs away from the shoulder. Fit a plain thrust washer and retain the pinion with a circlip.

8 Slide the 6th gear pinion into position with the selector groove away from the previous gear. Fit a circlip to limit the 6th gear pinion's movement then slide a splined thrust washer into place.

9 The 4th gear pinion is fitted next, noting that the heavily chamfered teeth face outwards, towards the right-hand end of the shaft. Secure it with a circlip, then fit the 3rd gear pinion, plain face inwards, and retain it with a splined thrust washer and a circlip.

10 The 5th gear pinion can now be slid into place with the selector groove inwards, followed by the large 1st gear pinion with its plain face outwards. Fit a plain thrust washer and a circlip to retain the above components.

11 Place the caged needle roller bearing over the right-hand end of the output shaft and the large ball bearing, seal and spacer over the left-hand end. The idler gear which runs on the right-hand end of the output shaft can be fitted at this stage noting that a thrust washer is fitted on each side of the pinion and that the assembly is retained by a circlip.

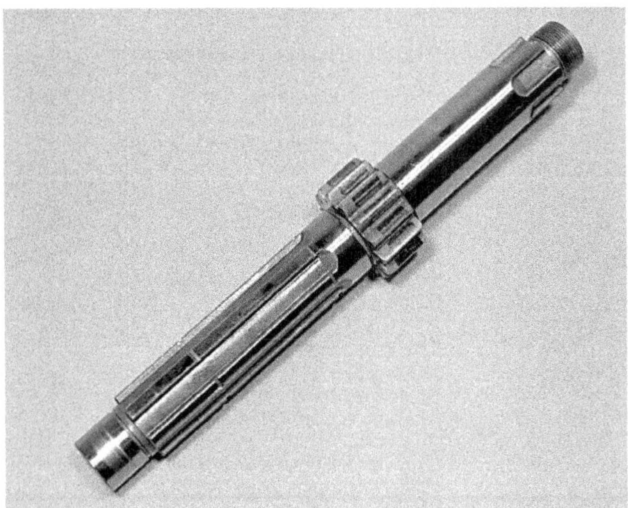

14.4a Input shaft can be identified by its integral 1st gear pinion

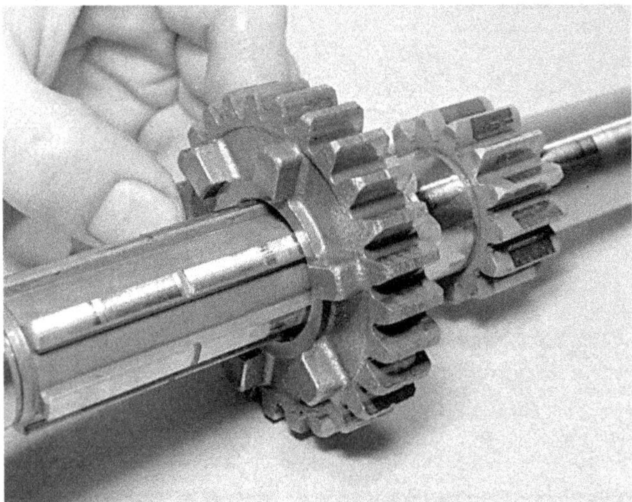

14.4b Fit the 5th gear pinion with the dogs facing away from the 1st gear pinion as shown

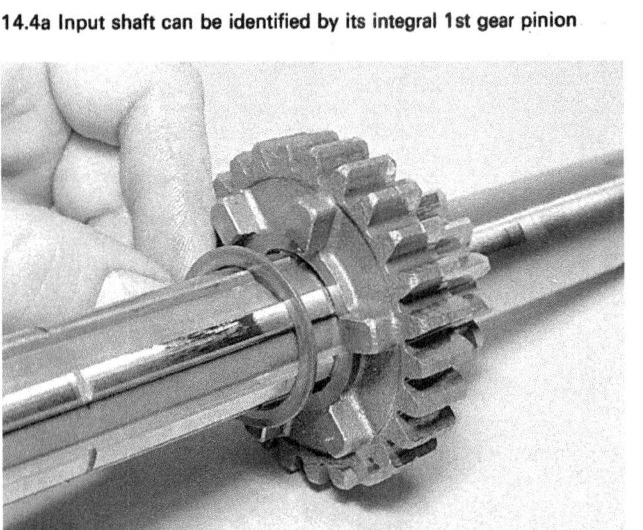

14.5a Slide a plain thrust washer up against the pinion ...

14.5b ... and secure it with a circlip

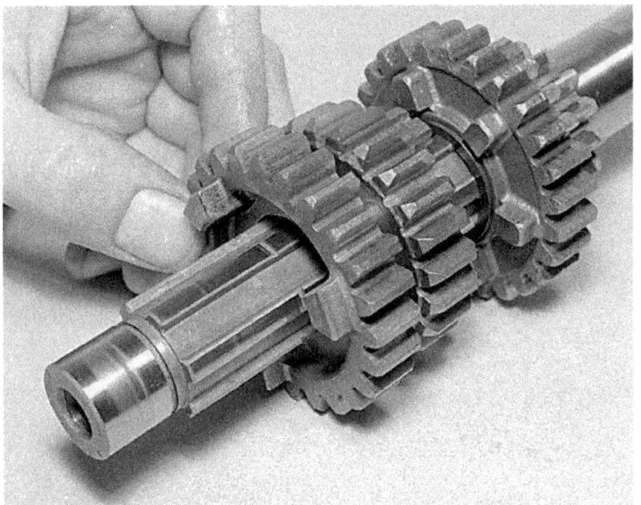

14.5c The double 3rd/4th gear pinion is fitted next, facing in the direction shown

14.5d Fit the circlip in the position shown, leaving the gear free to slide on the shaft, then fit the splined thrust washer

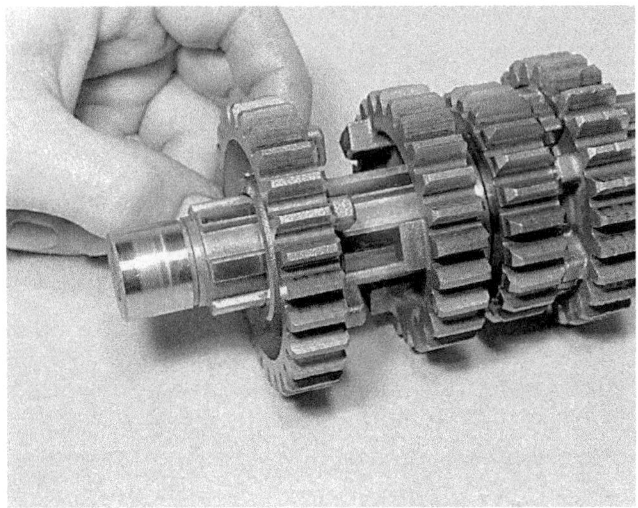

14.6a The 6th gear pinion can now be fitted with the engagement dogs innermost

14.6b The 2nd gear pinion is fitted, directly against the 6th gear ...

14.6c ... and is retained by a thrust washer and circlip

14.6d Complete input shaft assembly by fitting the needle roller bearing on the left-hand end of the shaft ...

14.6e ... and the larger caged ball bearing to the right-hand end of the shaft

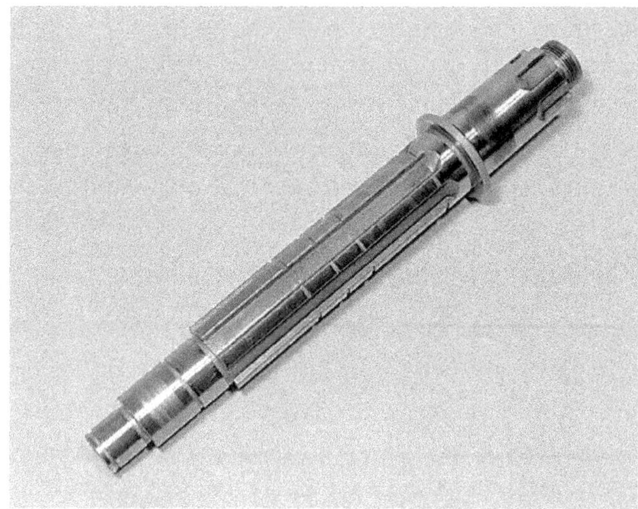
14.7a The output shaft has a plain shoulder at the left-hand end

14.7b Slide the 2nd gear pinion into place, facing in the direction shown ...

14.7c ... and retain with a thrust washer and circlip

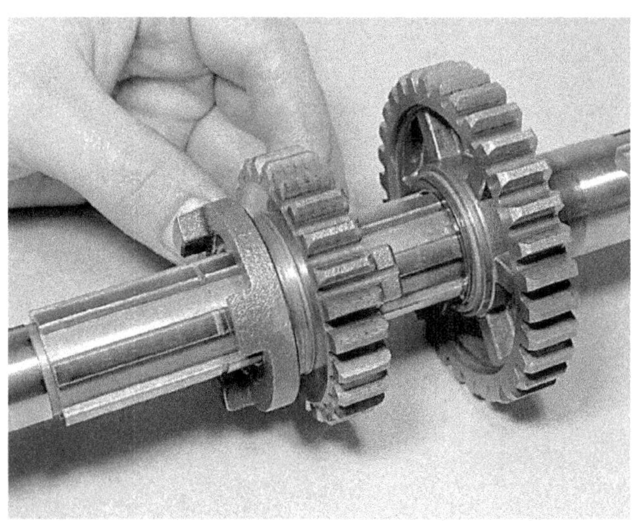
14.8a Slide the 6th gear pinion onto the shaft with the selector groove outwards as shown ...

14.8b ... and fit the circlip and splined washer to limit its movement

14.9a The 4th gear pinion is fitted next with its chamfered teeth outermost

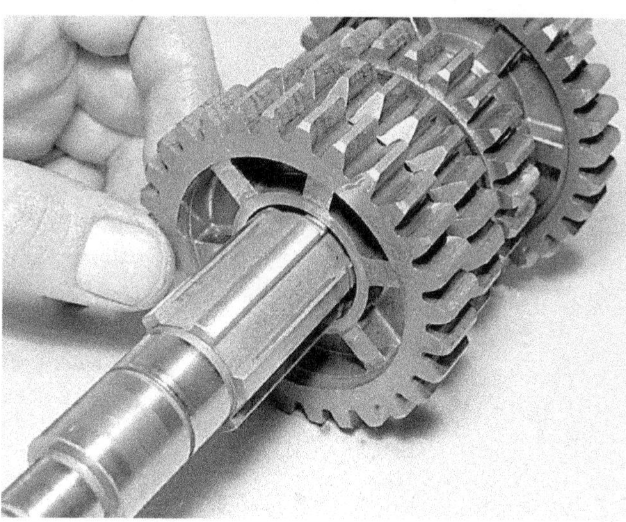
14.9b Now fit the 3rd gear pinion ...

14.9c ... splined washer and circlip

14.10a The 5th gear pinion is fitted with the selector groove inwards ...

14.10b ... followed by the large 1st gear pinion, plain face outermost

14.10c Secure the last two gears with a plain thrust washer and a circlip

14.11a Fit the caged needle roller bearing over the shaft right-hand end

14.11b Fit the thrust washer as shown ...

14.11c ... followed by the idler gear ...

14.11d ... and its special thrust washer (note flat in its internal bore)

14.11e Secure the idler gear with the circlip

14.11f Fit the large caged ball bearing to the left-hand end of the shaft

15 Big-end and main bearings: examination and renovation

1 Failure of the big-end bearing is invariably accompanied by a knock within the crankcase that progressively becomes worse. Some vibration will also be experienced.

2 There should be no vertical play whatsoever in the big-end bearings, after the oil has been washed out. If even a small amount of vertical play is evident, the bearings are due for replacement. (A small amount of endfloat is both necessary and acceptable.) Do not continue to run the machine with worn big-end bearings, for there is risk of breaking the connecting rods or crankshaft.

3 If a dial gauge and V-blocks are available check the amount of radial clearance in the big-end bearings, measuring this as lateral deflection at the small-end of each connecting rod to magnify the clearance in the big-end bearings. A serviceable bearing will allow between 0.36 and 1.0 mm (0.0142 – 0.0394 in) free play, whilst 2.0 mm (0.0787 in) of movement is indicative of the need for renewal of the bearing concerned.

4 Check the connecting rod side clearance by measuring the gap between it and the adjacent flywheel boss with feeler gauges. Clearance should be between 0.25 mm (0.0098 in) minimum and 0.75 mm (0.0295 in) maximum.

5 The crankshft main bearings are of the ball journal type. If wear is evident in the form of play, or if the bearings feel rough as they are rotated, replacement is necessary. Always check after the old oil has been washed out of the bearings. Whilst it is possible to remove the outer bearings at each end of the crankshaft, it is probable that the centre bearing will also require attention.

6 In the event that the big-end or main bearings prove to be in need of renewal it will be necessary to have the work done by an authorised Yamaha dealer. It is not practicable to attempt to overhaul the crankshaft without the necessary press and trueing equipment. The Yamaha dealer will also be able to check and correct runout in a crankshaft that has become twisted or distorted for any reason. If the owner possesses a dial gauge and stand, runout may be checked by supporting the crankshaft on its centre bearings and then measuring deflection as the crank is rotated with the dial gauge needle resting on the end of each mainshaft and on both outer main bearings. No one reading should exceed 0.05 mm (0.0020 in). Correction of excessive runout requires a large degree of skill and experience if the problem is not to be made worse by the operator misunderstanding the cause of the problem. For this reason, professional help should be enlisted.

7 Failure of both the big-end bearings and the main bearings may not necessarily occur as the result of high mileage covered. If the machine is used only infrequently, it is possible that condensation within the engine may cause premature bearing failure. The condition of the flywheels is usually the best guide. When condensation troubles have occurred, the flywheels will rust and become discoloured. Note too that lack of care when disturbing the cylinder head or barrels can allow coolant to find its way into the crankcase. This will soon corrode and destroy the bearings and should be avoided for obvious reasons.

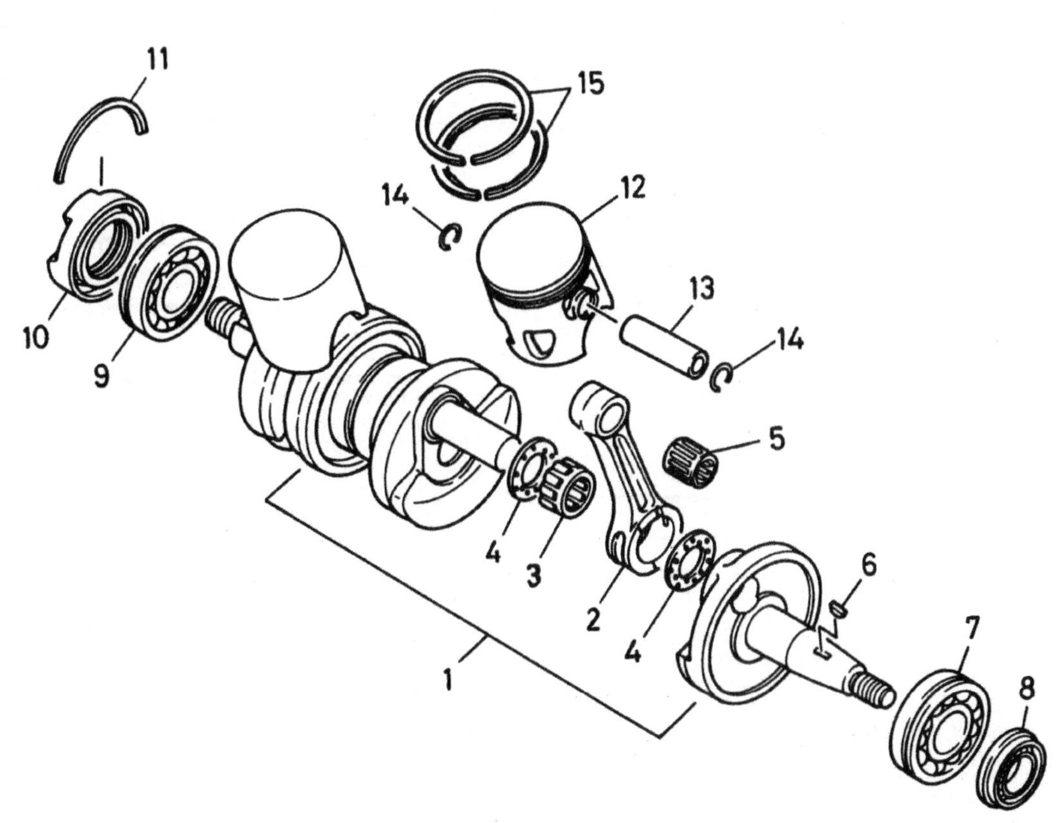

Fig. 1.4 Crankshaft

1 Crankshaft
2 Connecting rod
3 Big-end bearing
4 Thrust washer
5 Small-end bearing
6 Woodruff key
7 Left-hand main bearing
8 Oil seal
9 Right-hand main bearing
10 Oil seal
11 Half-ring
12 Piston
13 Gudgeon pin
14 Circlip
15 Piston rings

15.4a Big-end bearing side clearance can be checked using feeler gauges

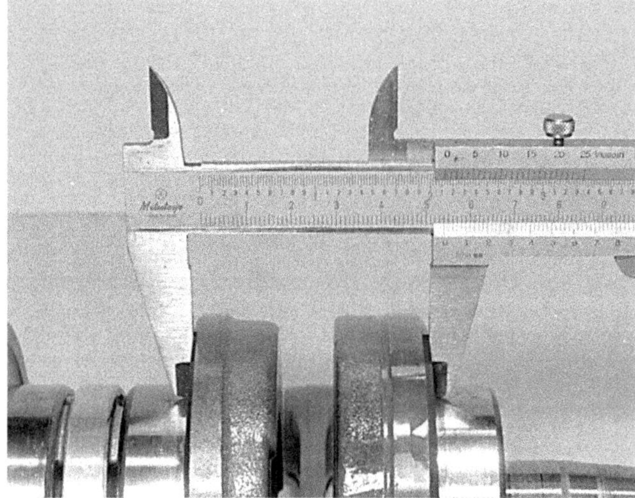

15.4b Check the dimensions of the crankshaft assembly (see specifications)

16 Oil seals: examination and renovation

1 The crankshaft oil seals form one of the most critical parts in any two-stroke engine because they perform the dual function of preventing oil from leaking along the crankshaft and preventing air from leaking into the crankcase when the incoming mixture is under crankcase vacuum during induction.
2 Oil seal failure is difficult to define precisely, although in most cases the machine will become difficult to start, particularly when warm. The engine will also tend to run unevenly and there will be a marked fall-off in performance, especially in the higher gears. This is caused by the intake of air into the crankcases which dilutes the mixture, giving an exceptionally weak mixture for ignition.
3 It is unusual for the crankshaft seals to become damaged during normal service, but instances have occurred when particles of broken piston rings have fallen into the crankcases and lacerated the seals. A defect of this nature will immediately be obvious.
4 In view of the foregoing remarks it is recommended that the two crankshaft oil seals are renewed as a matter of course during engine overhaul.

17 Cylinder barrels: examination and renovation

1 The usual indication of badly worn cylinder barrels and pistons is piston slap, a metallic rattle that occurs when there is little or no load on the engine. If the top of the bore of the cylinder barrels is examined carefully, it will be found that there is a ridge on the thrust side, the depth of which will vary according to the amount of wear that has taken place. This marks the limit of travel of the uppermost piston ring.
2 Measure the bore diameter just below the ridge, using an internal micrometer. Compare this reading with the diameter at the bottom of the cylinder bore, which has not been subjected to wear. If the difference in readings exceeds 0.10 mm (0.004 in) the cylinder should be rebored and fitted with an oversize piston and rings.
3 Bore ovality should also be checked, the maximum allowable being 0.01 mm (0.0004 in). Given that the bores are within the above limits and that the pistons are in serviceable condition (see Section 18) the parts may be re-used. Ovality may be corrected to some extent by honing, provided that this does not cause the maximum piston to bore clearance to be exceeded. A Yamaha dealer or a reputable engineering company will be able to assist with honing work should this prove necessary.
4 If scoring of the cylinder walls is evident it will normally prove necessary to have it re-bored to the next oversize, though light scratching may sometimes be removed by careful honing or by judicious use of abrasive paper. If the latter approach is adopted be careful to avoid removing more than the absolute minimum of material. The paper should be applied with a rotary motion **never** up and down the bore, which would cause more problems than it solves. One of the proprietary 'glaze busting' attachments for use in electric drills can be used to good effect for this operation. Even where the bore is in good condition, the glaze busting operation should be undertaken prior to reassembly. The light scratch marks around the bore surface assist in bedding in the rings and help initial lubrication by holding a certain amount of oil.
5 If reboring is necessary, obtain the pistons first, then have the boring done to suit the new pistons. Most Yamaha dealers have an arrangement with a local engineering company and will be able to get the reboring work carried out promptly.
6 Carefully remove any accumulated carbon deposits from the cylinder bore and ports, taking care not to damage the bore surface. It is recommended that the ports are cleaned completely but carefully, taking great care to avoid burring the edges of the ports where they enter the bore. To prevent the rings from becoming chipped or broken dress any burrs with fine emery paper.
7 It is inadvisable to attempt modification of the port sizes or profiles to obtain more power from the engine. Such modifications are feasible but should only be considered for racing purposes. Generally speaking, the changed characteristics of the engine would make it unwieldy for road use, and it should be noted that the machine's warranty would be invalidated.
8 Check the water passages for rust and scale. These may have built up, especially where the correct coolant has not been used. If necessary, scrape the passages clean using wire or an old screwdriver, taking care to flush out any debris. Bear in mind that any residual debris may clog the radiator or pump if it is not removed.

18 Pistons and piston rings: examination and renovation

1 If a rebore is necessary, the existing pistons and piston rings can be disregarded because they will have to be replaced with their new oversize equivalents as a matter of course.
2 Remove all traces of carbon from the piston crowns, using a blunt-ended scraper to avoid scratching the surface. Finish off by polishing the crowns with metal polish, so that carbon will not adhere so readily in the future. Never use emery cloth on the soft aluminium.
3 Piston wear usually occurs at the skirt or lower end of the piston and takes the form of vertical streaks or score marks on the thrust face. There may also be some variation in the thickness of the skirt, in an extreme case.
4 The piston ring grooves may have become enlarged in use, allowing the rings to have greater side float. If the clearances exceed those given, the rings, and possibly the pistons, must be renewed.

Piston to ring clearances

Top .... 0.02 – 0.06 mm (0.0008 – 0.0024 in)
2nd .... 0.03 – 0.07 mm (0.0012 – 0.0028 in)

5 Piston ring wear is measured by removing the rings from the piston and inserting them in the cylinder, using the crown of a piston to locate them about 20 mm from the bottom of the bore. Make sure they rest squarely in the bore. Measure the end gap with a feeler gauge; if the gap exceeds that given below, the rings must be replaced.

Piston ring end gap (installed)

Top .... 0.30 – 0.45 mm (0.0118 – 0.0177 in)
2nd .... 0.35 – 0.50 mm (0.0138 – 0.0197 in)

19 Cylinder head: examination and renovation

1 Remove all traces of carbon from the cylinder head, using a blunt-ended scraper. Finish by polishing with metal polish, to give a smooth, shiny surface. This will aid gas flow and will also prevent carbon from adhering so firmly in the future.
2 Check the condition of the threads in the sparking plug holes. If the threads are worn or stretched as the result of overtightening the plugs, they can be reclaimed by a 'Helicoil' thread insert. Most dealers have the means of providing this cheap but effective repair.
3 Inspect the water passages cast into the cylinder head, and where necessary remove any accumulated corrosion or scale. As mentioned previously, this can result from failure to use the recommended coolant mixture. Be sure to remove any debris from the passages by flushing them through with clean water.
4 Lay the cylinder head on a sheet of plate glass to check for distortion. Aluminium alloy cylinder heads will distort very easily, especially if the cylinder head bolts are tightened down unevenly. If the amount of distortion is only slight, it is permissible to rub the head down until it is flat once again by wrapping a sheet of very fine emery cloth around the plate glass sheet and rubbing with a rotary motion.
5 If the cylinder head is distorted badly, it is advisable to fit a new replacement. Although the head joint can be restored by skimming, this will raise the compression ratio of the engine and may adversely affect performance.

20 Gearbox components: examination and renovation

1 Give the gearbox components a close visual inspection for signs of wear or damage such as broken or chipped teeth, worn dogs, damaged or worn splines and bent selectors. Replace any parts found unserviceable because they cannot be reclaimed in a satisfactory manner.
2 The gearbox shafts are unlikely to sustain damage unless the lubricating oil has been run low or the engine has seized and placed an unusually high loading on the gearbox. Check the surfaces of the shaft, especially where a pinion turns on it, and renew the shaft if it is scored or has picked up. The shafts can be checked for trueness by setting them up in V-blocks and measuring any bending with a dial gauge.
3 Examine the gear selector claw assembly noting that worn or rounded ends of the claw can lead to imprecise gear selection. The springs in the selector mechanism and the detent or stopper arm should be unbroken and not distorted or bent in any way.
4 The gearbox bearings must be free from play and show no signs of roughness when they are rotated. Each shaft has a ball journal bearing at one end and a caged needle-roller bearing at the other.
5 It is advisable to renew the gearbox oil seals irrespective of their condition. Should a re-used oil seal fail at a later date, a considerable amount of dismantling is necessary to gain access and renew it.
6 Check the gear selector rods for straightness by rolling them on a sheet of plate glass. A bent rod will cause difficulty in selecting gears and will make the gear change action particularly heavy.
7 The selector forks should be examined closely, to ensure that they are not bent or badly worn. Wear is unlikely to occur unless the gearbox has been run for a period with a particularly low oil content.
8 The tracks in the gear selector drum, with which the selector forks engage, should not show any undue signs of wear unless neglect has led to under lubrication of the gearbox.

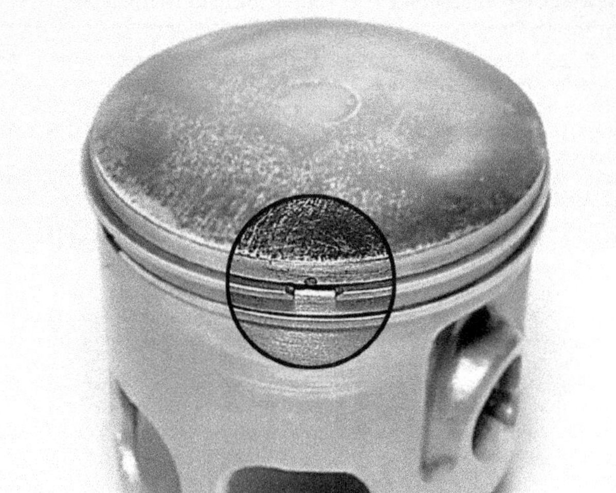

18.5a Note how ring ends locate over peg in ring grooves

18.5b Scraper ring is backed by a thin steel expander

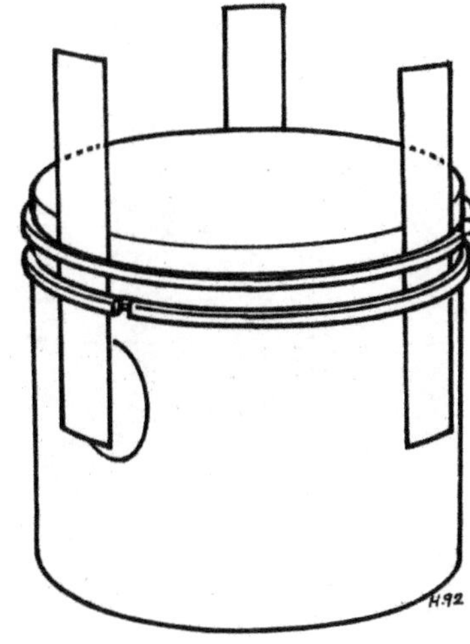

Fig. 1.5 Method of removing gummed piston rings

21 Kickstart mechanism: examination and renovation

1 The kickstart mechanism is a robust assembly and should not normally require attention. Apart from obvious defects such as a broken return spring, the friction clip is the only component likley to cause problems if it becomes worn or weakened. The clip is intended to apply a known amount of drag on the kickstart pinion, causing the latter to run up its quick thread and into engagement when the kickstart lever is operated.

2 The clip can be checked using a spring balance. Hook one end of the balance onto the looped end of the friction clip. Pull on the free end of the balance and note the reading at the point where pressure overcomes the clip's resistance. This should normally be 0.8 – 1.2 kg (1.8 – 2.9 lb). If the reading is higher or lower than this and the mechanism has been malfunctioning, renew the clip as a precaution. Do not attempt to adjust a worn clip by bending it.

3 Examine the kickstart pinion for wear or damage, remembering to check it in conjunction with the output shaft-mounted idler pinion. In view of the fact that these components are not subject to continuous use a significant amount of wear or damage is unlikely to be found.

22 Primary drive: examination and renovation

1 The primary drive consists of a crankshaft pinion which engages a large gear mounted on the inner face of the clutch drum. Both components are relatively lightly loaded and will not normally wear until very high mileages have been covered.

2 If wear or damage is discovered it will be necessary to renew the component concerned. In the case of the large driven gear it will be necessary to purchase a complete clutch drum because the two items form an integral unit and cannot be obtained separately.

3 When obtaining new primary drive parts note that the two components are matched to give a prescribed amount of backlash. To this end, ensure that the match marks marked on the inner face of each are similar to avoid excessive or insufficient clearance.

4 To check the backlash of the two components, examine the back face of each and make a note of the number etched into the metal. These two numbers, when added together, give the backlash figure. The number on the drive (crankshaft) pinion will be between 90 and 98, whilst that of the driven gear (clutch) will be between 57 and 65. The prescribed backlash tolerance is 154-156.

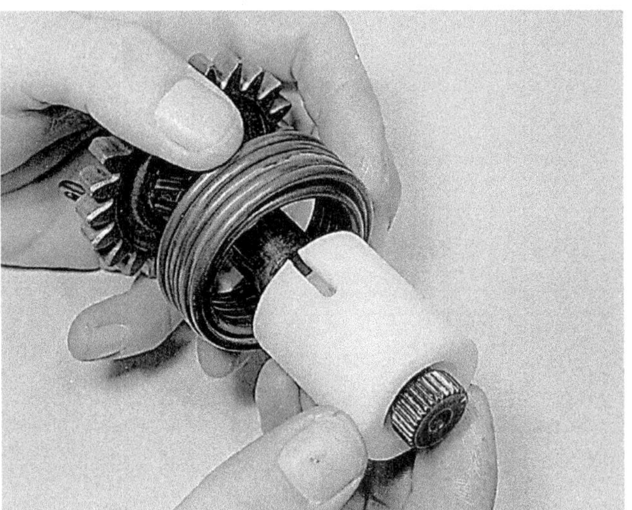

21.1a Kickstart return spring is held central by a white plastic spacer

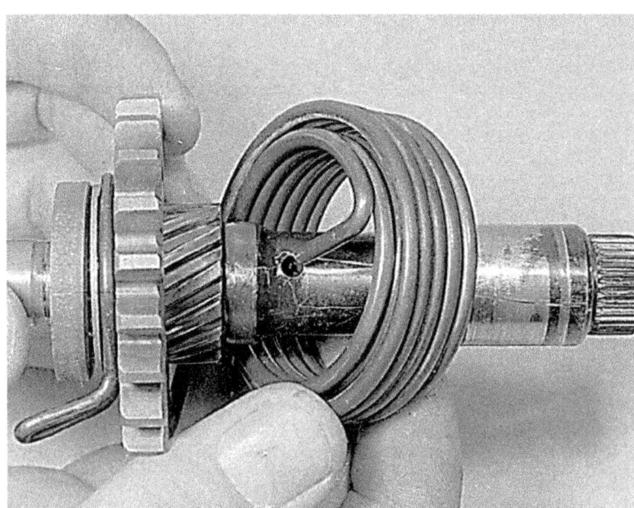

21.1b Spring end hooks into hole in the kickstart shaft

21.1c Kickstart pinion is removed as shown

21.1d Use spring balance to check force needed to move the friction clip

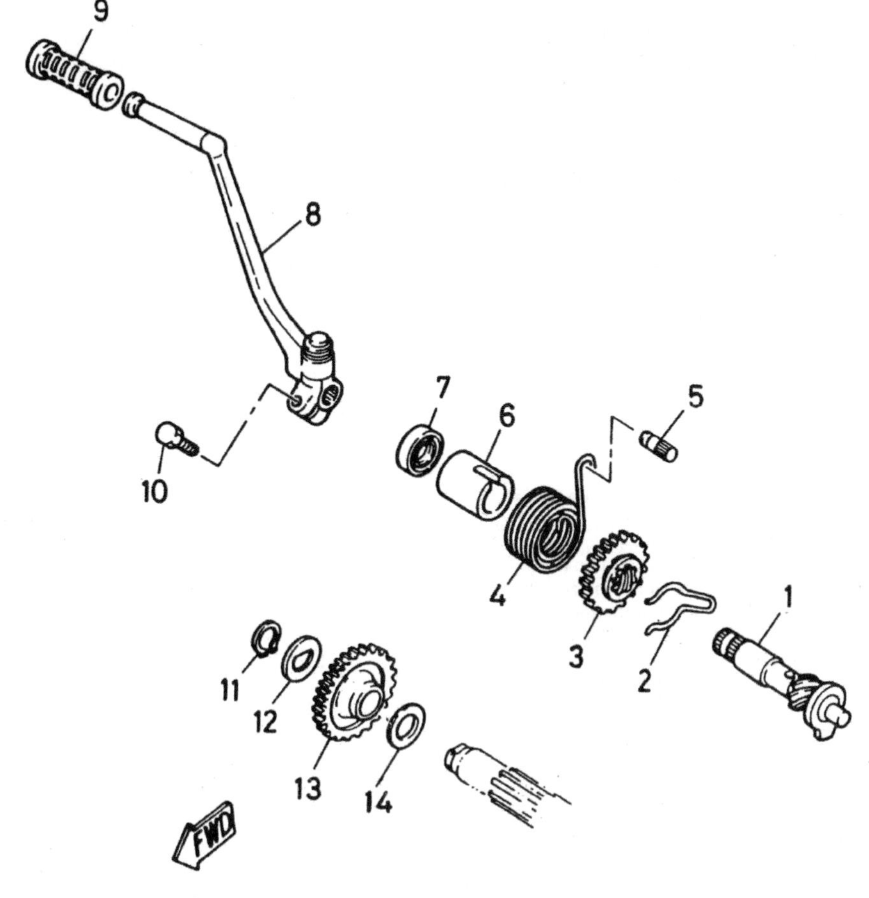

Fig. 1.6 Kickstart mechanism

1 Kickstart shaft
2 Friction clip
3 Kickstart pinion
4 Return spring
5 Spring anchor
6 Spring guide
7 Oil seal
8 Operating lever
9 Rubber
10 Pinch bolt
11 Circlip
12 Special washer
13 Idler pinion
14 Thrust washer

22.4a The backlash figure of the primary gear shown is 94 (circled) ...

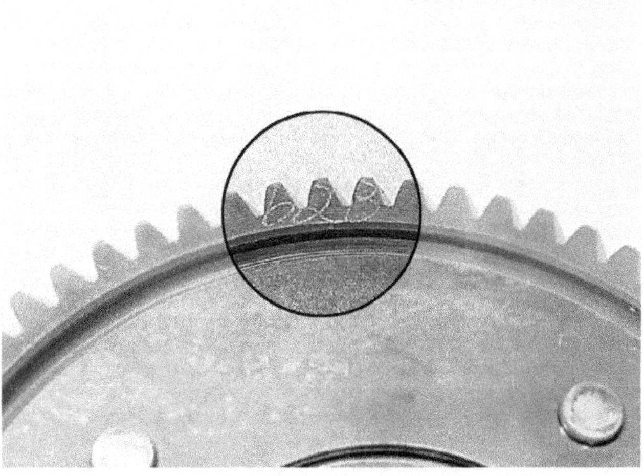

22.4b ... while that of the clutch drum is 62 (ignore the 'B'), giving a backlash total of 156 (see text)

23 Clutch assembly: examination and renovation

1 After an extended period of service, the friction plates will have become worn sufficiently to warrant renewal, to avoid subsequent problems with clutch slip. The lining thickness is measured across the friction plate using a vernier caliper. When new, each plate measures 3.0 mm (0.118 in). If any plate is worn to 2.7 mm (0.106 in) or less the friction plates must be renewed.
2 The plain plates should be free from any signs of blueing, which would indicate that the clutch had overheated in the past. Check each plate for distortion by laying it on a flat surface, such as a sheet of plate glass or similar, and measuring any detectable gap using feeler gauges. The plates must be less than 0.05 mm (0.002 in) out of true.
3 The clutch springs may, after a considerable mileage, require renewal, and their free length should be checked as a precautionary measure. When new, each spring measures 36.4 mm (1.43 in) and the set should be renewed if they have compressed to 34.4 mm (1.35 in) or less.
4 Check the condition of the slots in the outer surface of the clutch centre and the inner surfaces of the outer drum. In an extreme case, clutch chatter may have caused the tongues of the inserted plates to make indentations in the slots of the outer drum, or the tongues of the plain plates to indent the slots of the clutch centre. These indentations will trap the clutch plates as they are freed and impair clutch action. If the damage is only slight the indentations can be removed by careful work with a file and the burrs removed from the tongues of the clutch plates in similar fashion. More extensive damage will necessitate renewal of the parts concerned.
5 Check the clutch pushrod for bending or signs of wear or overheating. The limit for bending is 0.2 mm (0.008 in) and the trueness of the pushrod is easily checked by rolling it on a sheet of glass or a surface plate. In rare cases, the ends of the rod may have become overheated due to the machine being run with insufficient free play in the cable or release mechanism. This can cause the hardening of the rod ends to break down, allowing it to wear away. If blued, or if the clutch needs frequent adjustment, renew the rod. Finally, check the clutch release shaft and arm. Minor wear damage can be dressed out with abrasive paper, but if it is badly worn it should be renewed.

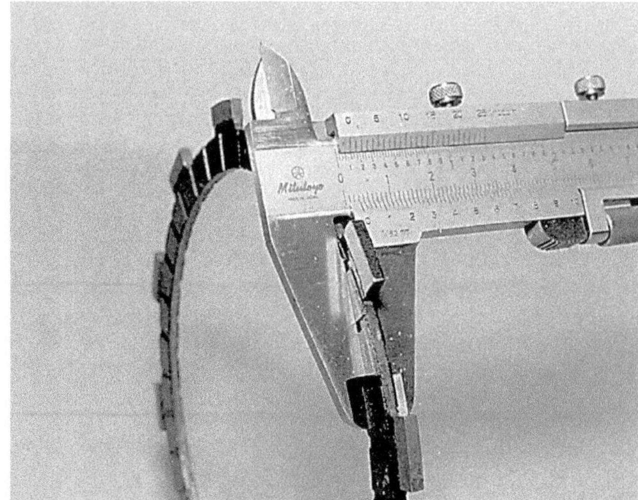

23.1 Measure the thickness of the clutch friction plates to assess wear

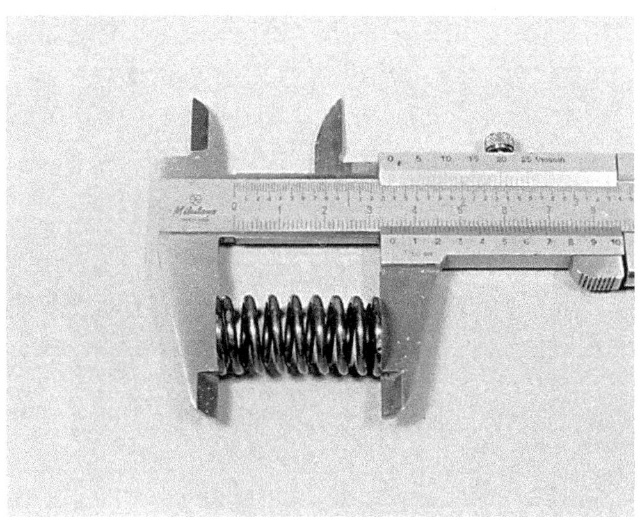

23.3 Check the free length of the clutch springs, renew if compressed

24 Engine reassembly: general

1 Before reassembly of the engine/gearbox unit is commenced, the various component parts should be cleaned thoroughly and placed on a sheet of clean paper, close to the working area.
2 Make sure all traces of old gaskets have been removed and that the mating surfaces are clean and undamaged. One of the best ways to remove old gasket cement is to apply a rag soaked in methylated spirit. This acts as a solvent and will ensure that the cement is removed without resort to scraping and the consequent risk of damage. If the gasket cement proves particularly stubborn it may be necessary to resort to using an aluminium or brass scraper. Do not use a screwdriver or a steel scraper because this will almost invariably damage the gasket face. One safe method is to use a brass wire brush such as those sold for cleaning suede shoes. This will usually prove very effective and will not damage the alloy.
3 Gather together all the necessary tools and have available an oil can filled with clean engine oil, Make sure all new gaskets and oil seals are to hand, also all replacement parts required. Nothing is more frustrating that having to stop in the middle of a reassembly sequence because a vital gasket or replacement has been overlooked.
4 Make sure that the reassembly area is clean and that there is adequate working space. Refer to the torque and clearance settings wherever they are given. Many of the smaller bolts are easily sheared if over-tightened. Always use the correct sized screwdriver bit for the cross-head screws and never an ordinary screwdriver. If the existing screws show evidence of maltreatment in the past it is advisable to renew them as a complete set. It is strongly recommended that a set of Allen screws are used instead of the original cross-head screws. Allen screw sets can be obtained through most good accessory retailers and are an inexpensive but thoroughly practical improvement to most Japanese machines.

25 Engine reassembly: refitting the tachometer drive – RD350 LC II

1 Slide the tachometer drive shaft part way into the upper half of the crankcase and fit the innermost of the drive pinion retaining circlips. Fit the drive pinion over its splines and secure it with the remaining circlip. The shaft can now be pushed fully home having lubricated it and the gear with clean engine oil.
2 On the outside of the casing, fit the retaining plate and screws to hold the shaft in position. Note that Loctite or a similar thread locking compound should be used on the two screws. Slide the drive gear locating pin through the shaft end, and place the gear over the end, securing it with its plain washer and circlip.
3 Assemble the tachometer driven gear and holder, having fitted a new O-ring to the latter where necessary. Slide the assembly into place and secure the single retaining bolt.

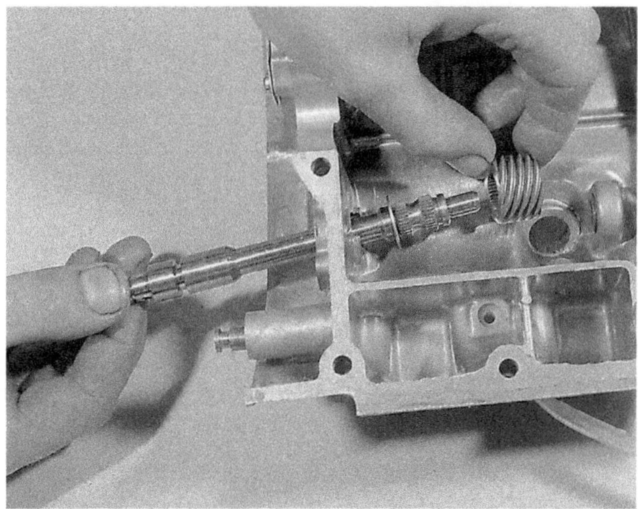

25.2a Where fitted, slide tachometer driveshaft into casing, fitting the inner circlip and drive gear

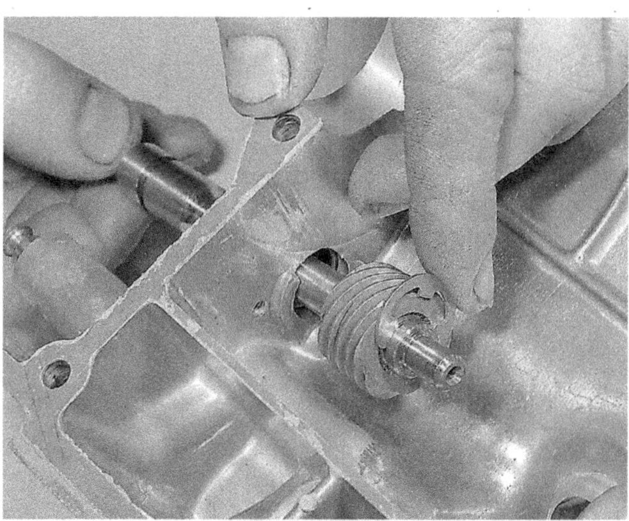

25.2b The gear is retained on the shaft as shown

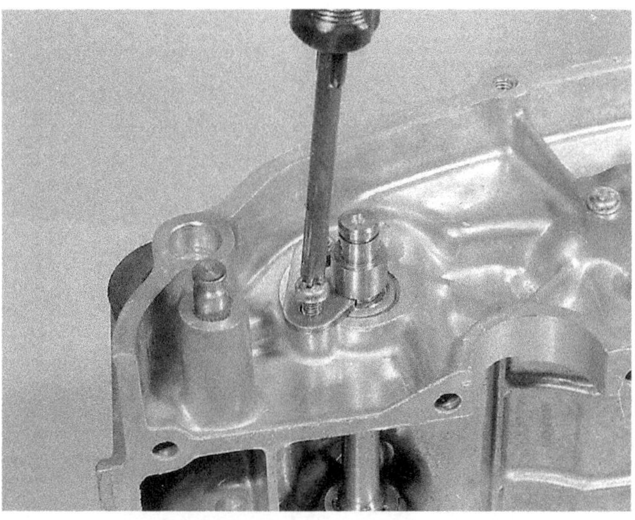

25.2c Fit the retainer plate and its two screws

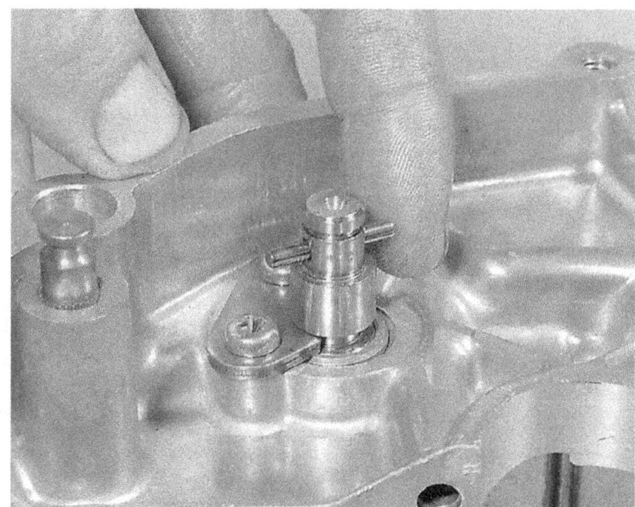

25.2d Place drive pin through the hole in the shaft end ...

25.2e ... and fit the pinion over it

25.2f Retain the assembly with the plain washer and circlip

25.2g Fit the tachometer drive body using a new O-ring ...

25.2h ... and secure it using its single bolt

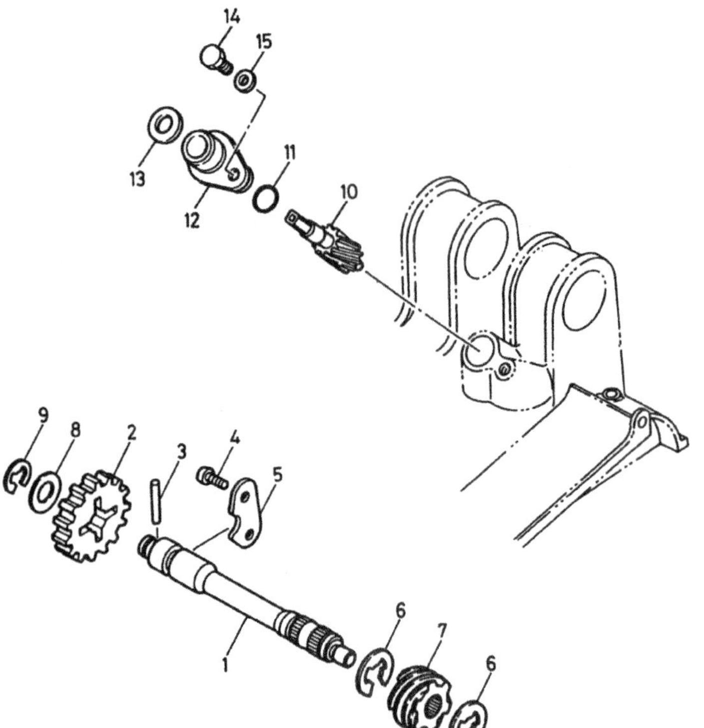

1 Tachometer driveshaft
2 Plastic drive pinion
3 Pin
4 Screw – 2 off
5 Locating plate
6 Circlip – 2 off
7 Drive gear
8 Washer
9 Circlip
10 Driven gear
11 O-ring
12 Drive body
13 Washer
14 Bolt
15 Washer

Fig. 1.7 Tachometer drive assembly – RD350 LC II only

26 Engine reassembly: refitting the selector mechanism

1 Lubricate and fit the large needle roller bearing to the right-hand end of the selector drum and retain it with its circlip. Assemble the cam plate on the end of the drum, noting the small pin and the flat which locates it. Position the special cam retaining washer and secure the screw.
2 Moving to the left-hand end of the drum, fit the neutral switch plate assembly and retain it with its single screw. Make sure that the spring and contact are properly located.
3 Lubricate the plain (left-hand) end of the selector drum and slide it into position in the lower casing half. Do not fit the retaining plate at this stage. The selector forks and shafts should be fitted next. Note that two of the forks are identical, one being fitted to the front shaft and the other to the right-hand side of the rear shaft. The remaining fork is fitted to the left-hand end of the rear shaft.
4 Slide each shaft part way into the casing and fit the appropriate forks over it. The location pins should be arranged so that they engage with the selector drum tracks. Once the shafts are through the forks fit the circlips to the grooves on their inner end so that when they are pushed fully home the circlips serve to locate the shaft ends.
5 Fit the selector drum retainer, noting that it also retains the front selector fork shaft. It is advisable to use Loctite on the two securing screws. Next, fit the selector mechanism centralising spring anchor pin, noting that the tab washer beneath it has an extended section which retains the rear selector fork shaft. The selector drum detent arm (stopper arm) can be fitted next and its pivot bolt tightened firmly. The selector shaft and claw assembly should be left off for the time being.

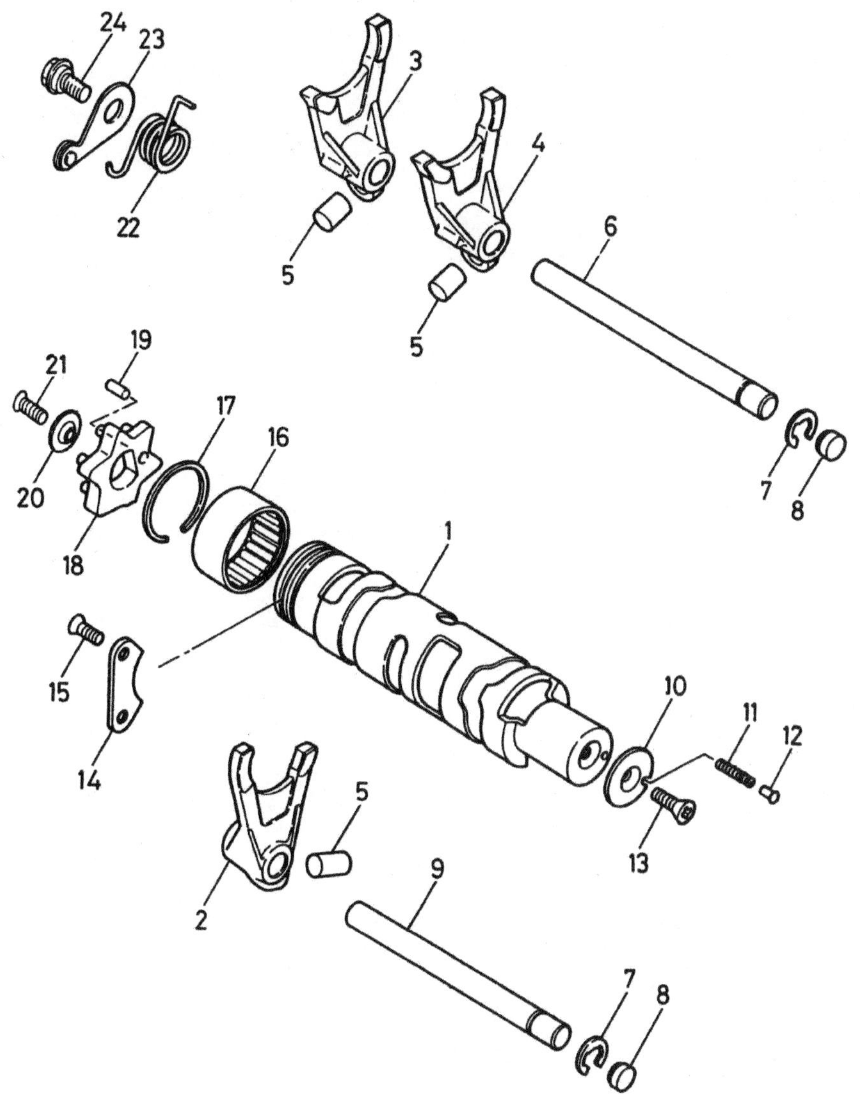

Fig. 1.8 Selector mechanism

1 Selector drum
2 Front selector fork
3 Rear selector fork
4 Rear selector fork
5 Location pin
6 Selector fork shaft
7 Circlip
8 Selector fork shaft plug
9 Selector fork shaft
10 Neutral switch plate
11 Spring
12 Neutral contact
13 Screw
14 Selector drum retainer
15 Screw
16 Bearing
17 Circlip
18 Cam plate
19 Cam plate locating pin
20 Special washer
21 Screw
22 Spring
23 Selector drum stopper arm
24 Bolt

26.2a Check that contact is in place and offer up the neutral switch cover

26.2b Secure the switch cover with its three countersunk screws

26.3 Refit the selector drum, forks and shafts

26.5 Refit the selector drum retainer (A) and stopper arm (B)

27 Engine reassembly: refitting the gearbox components

1 Position the half rings which locate the two output shaft bearings in their grooves in the crankcase. A similar groove is provided in the recess for the input shaft's right-hand bearing, and this locates the bearing by means of the large circlip which is fitted to its outer race.
2 Lower the output shaft assembly into the crankcase lower half, ensuring that the selector fork fingers engage in the groove. The input shaft is now positioned in a similar manner. Check that both shafts seat securely.
3 Before proceeding further it is advisable to check gearbox operation. This can be done by turning the selector drum by means of the cam. To facilitate gear engagement, rotate the input shaft to and fro as each gear is selected. Neutral can be identified by noting that the detent arm drops into the shallowest of the cam depressions. When in neutral it should be possible to hold the output shaft stationary whilst the input shaft is turned. From neutral, select each gear in turn, ensuring that all six are available and that they each engage correctly.

27.2 Place the gear clusters in the crankcase, making sure that they locate correctly

28 Engine reassembly: refitting the crankshaft

1 The crankshaft should always be refitted using **new** oil seals. These are vital to the efficient running of all two-stroke engines. If a worn seal is reused, crankcase compression will be lost and performance will suffer. Before commencing reassembly, pack the gap between the seal lips with grease.
2 Position the locating half-ring in its groove in the right-hand main bearing boss. The half ring serves to locate the right-hand bearing, and thus the crankshaft. The remaining bearings are pegged to prevent rotation of the outer races, the pegs sitting in recesses to the front of each one. Lubricate each main bearing with new engine oil.
3 Lower the crankshaft into position, ensuring that the half ring and the three pegs are located correctly. Slide the seal into position over the crankshaft end. The seal should be positioned so that the outer face is flush with the crankcase boss, leaving a small gap between it and the main bearing. When correctly positioned the small bead around the outer face of the seal will locate in the corresponding groove in the casing recess.
4 Fit the right-hand seal in a similar manner, noting that it has a castellated spacing lip on its inner face. This should butt against the outer race of the main bearing, forming an additional method of location. When both seals are in place make sure that the crankshaft assembly is firmly seated along its length.

28.3 Note position of locating pins (circled) and half-ring (arrowed)

29 Engine reassembly: joining the crankcase halves

1 Make sure that the crankcase halves are clean and completely free of grease. To this end it is sound practice to give the jointing faces a final wipe with a clean rag moistened with methylated spirit or clean petrol. Allow the solvent to evaporate completely, then apply a thin film of jointing compound to the gasket face of one half. One of the RTV (room temperature vulcanising) silicone compounds, often sold as 'Instant Gasket' is recommended. Allow the compound to cure for a few minutes and in the meantime fit the two locating dowels to their recesses in the lower casing half.

2 The crankcase upper half can now be lowered into position, noting that the connecting rods must be fed through the crankcase apertures as the two halves meet. As the joint is closed, check that everything locates correctly, then tap the upper casing half down with the palm of one hand to ensure that it locates firmly.

3 There are a total of 16 crankcase securing bolts, each of which is numbered in the correct sequence for tightening. The numbers are cast into the crankcase next to the appropriate hole. When fitting the bolts it should be noted that bolts No 9, 14 and 15 have cable or wiring clips attached to them as shown in the tightening sequence diagram (Fig. 1.10).

4 Fit the upper crankcase bolts first (Nos 9 to 16) and tighten them just enough to secure the crankcase. Turn the unit over on the workbench and install the lower crankcase bolts (Nos 1 to 8). The bolts should now be tightened in two stages and in the sequence shown below:

a) Bolts 9 to 16 to 0.5 kgf m (3.62 lbf ft)
b) Bolts 1 to 8 to 1.0 kgf m (7.23 lbf ft)
c) Bolts 1 to 8 to 2.5 kgf m (18.08 lbf ft)
d) Bolts 9 to 16 to 1.0 kgf m (7.23 lbf ft)

5 The crankcases are now secured and before moving on, check that the crankshaft and the gearbox shafts rotate smoothly with no tight spots. If necessary, separate the crankcase halves and rectify any alignment problem before proceeding further.

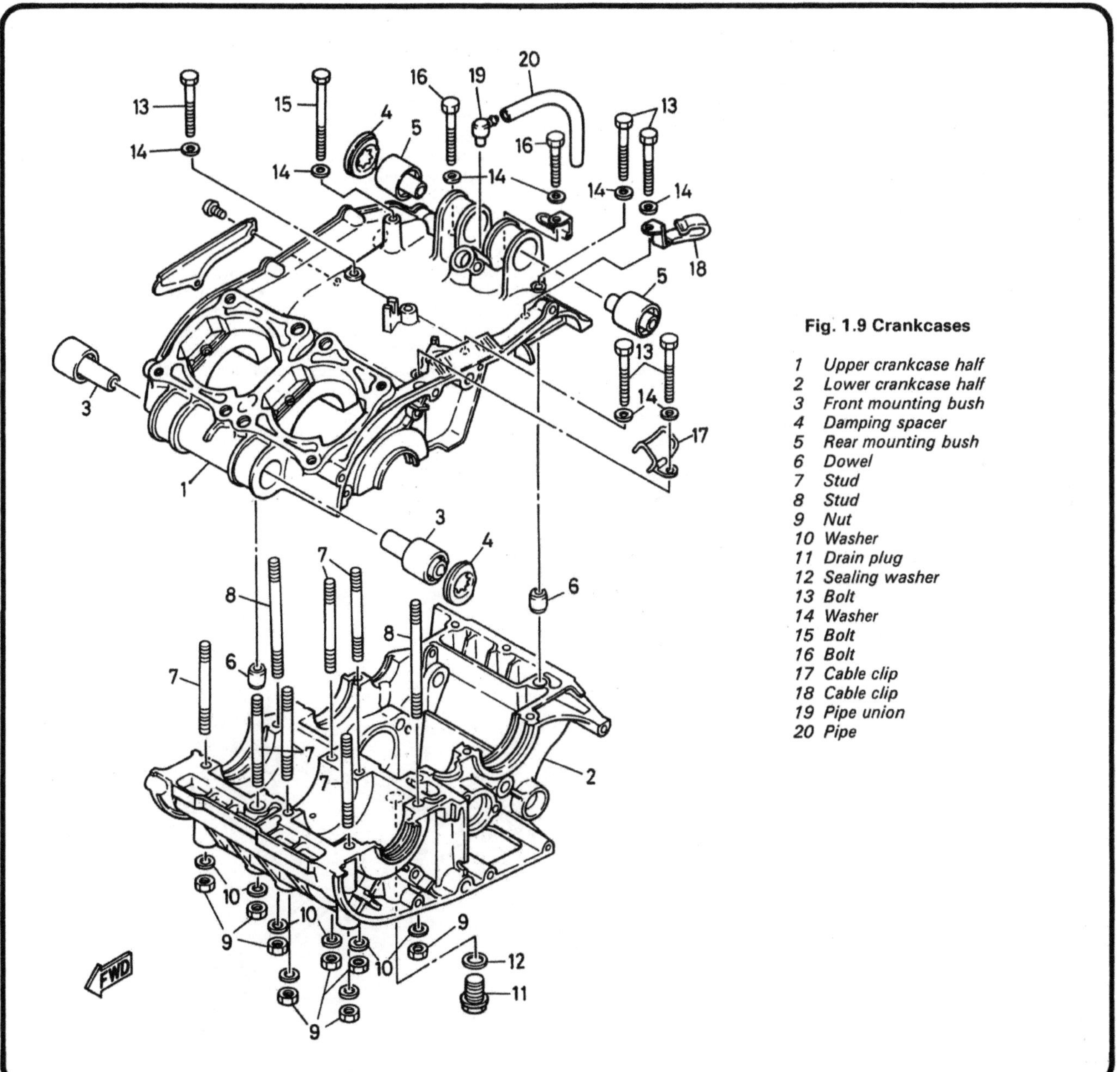

Fig. 1.9 Crankcases

1 Upper crankcase half
2 Lower crankcase half
3 Front mounting bush
4 Damping spacer
5 Rear mounting bush
6 Dowel
7 Stud
8 Stud
9 Nut
10 Washer
11 Drain plug
12 Sealing washer
13 Bolt
14 Washer
15 Bolt
16 Bolt
17 Cable clip
18 Cable clip
19 Pipe union
20 Pipe

29.2 Apply sealant to crankcase joint faces and offer up the casing upper half

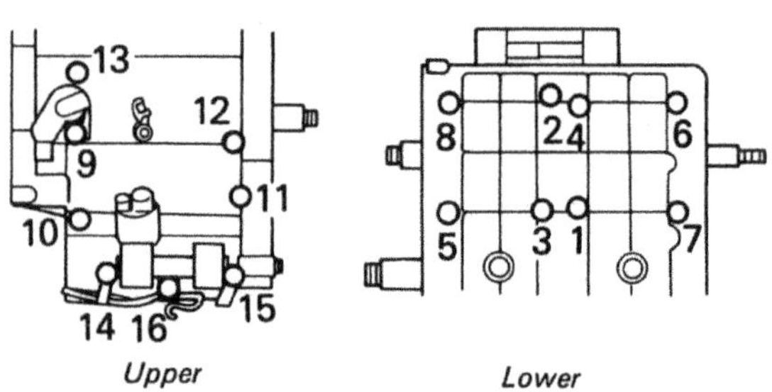

Fig. 1.10 Crankcase bolt tightening sequence (see text)

30 Engine reassembly: refitting and adjusting the gear selector shaft

1 It is advisable to fit a new oil seal to the left-hand end of the gear selector shaft bore, irrespective of its condition. Lever the old seal out with a screwdriver, then tap the new seal into position using a suitably sized socket as a drift. Do not risk damaging the seal lip by hitting the seal directly. Lubricate the seal with a smear of grease before the shaft is installed. The oil seal in the engine left-hand cover should likewise be attended to if it is damaged.
2 Wrap some PVC tape around the splines on the selector shaft to protect the oil seal lip. Slide the shaft into its bore and check that the centralising spring ends engage on the eccentric adjusting screw. When at rest, the claw ends of the selector mechanism should be equidistant from the two adjacent pins (see Fig. 1.11). Check that this setting is correct in each gear and if necessary slacken the locknut on the centralising spring adjuster and adjust it to obtain the correct clearance.

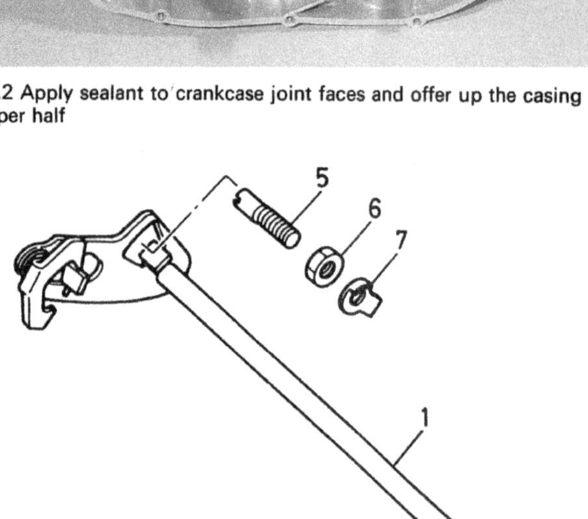

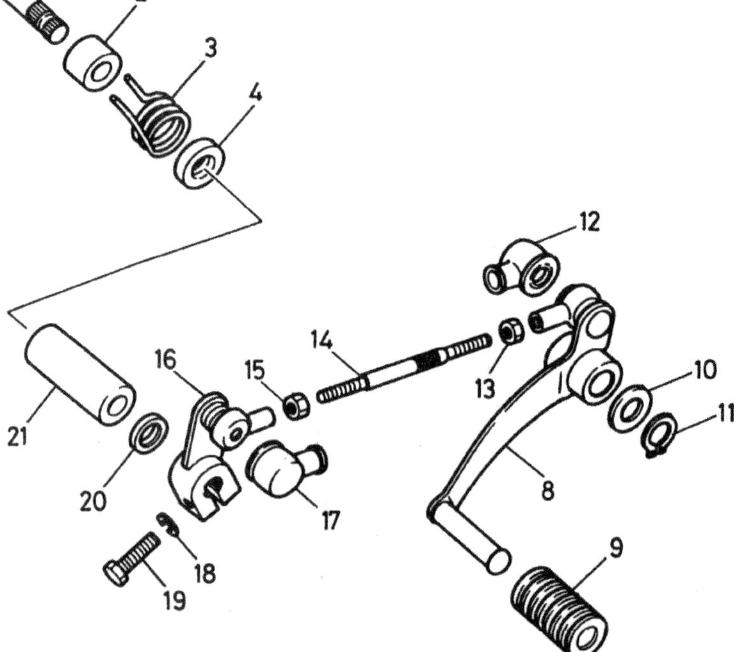

Fig. 1.11 Gearchange mechanism

1 Gearchange shaft
2 Spacer
3 Centralising spring
4 Oil seal
5 Centralising spring adjuster
6 Locknut
7 Tab washer
8 Gearchange lever
9 Lever rubber
10 Washer
11 Circlip
12 Rear boot
13 Locknut
14 Adjusting screw
15 Locknut
16 Front linkage
17 Front boot
18 Spring washer
19 Bolt
20 Washer
21 Spacer

30.2 Refit the selector mechanism and check adjustment

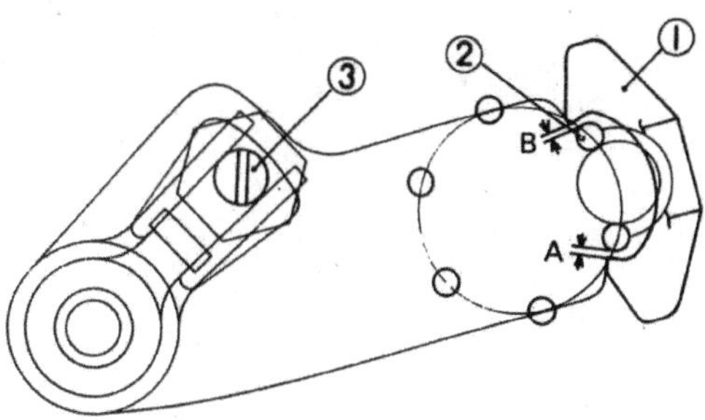

Fig. 1.12 Gearchange selector mechanism adjustment

1 Selector shaft
2 Pins
3 Eccentric adjusting screw
A & B Equal spacing

31 Engine reassembly: refitting the kickstart mechanism, idler pinion and crankcase fittings

1 Check that the kickstart pinion friction clip is in position, then slide the pinion over the shaft. Fit the kickstart return spring over the shaft and engage its inner tang in the shaft cross drilling. Once the spring is located slide the plastic spring guide into position to retain it. The assembly can now be fitted into the casing bore. Grasp the free end of the return spring and hook it over the anchor pin which protrudes from the upper casing half.
2 If it is not already in position the kickstart idler pinion should be fitted next. It is supported on the protruding end of the gearbox output shaft and is preceded by a plain washer. A special washer with an internal flat is fitted next and is secured by a circlip.
3 If it was removed during crankcase overhaul, refit the deflector plate above the input shaft bearing. It is retained by two cross-head screws, the threads of which should be coated with Loctite. The bearing retainer is fitted in a similar manner, noting that it bridges the crankcase halves. Where the cooling system stub was removed this should be refitted using a new O-ring and ensuring that its wire circlip is seated correctly.

32 Engine reassembly: refitting the clutch, primary drive and pump drive pinions

1 Rotate the crankshaft until the keyway is uppermost, then fit the large Woodruff key. The crankshaft primary drive gear can now be slid into place, noting that its shouldered face should be completely smooth and well lubricated where it enters the oil seal. Take care not to force the seal lip inwards when fitting the pinion.
2 Slide the smaller pump drive pinion over the crankshaft end, then fit the Bellville washer and securing nut. It is worth noting that although the pump pinion is relatively lightly loaded it is not keyed to the crankshaft and thus relies on the securing nut being tightened properly. If the nut becomes loose in service, the oil and water pumps would stop, followed swiftly by the engine which, unlubricated and uncooled, would soon seize. Lock the crankshaft as described during dismantling and tighten the securing nut to 6.5 kgf m (47 lbf ft).
3 Slide the large thrust washer over the end of the gearbox input shaft, followed by the clutch bush. The clutch drum can be fitted next, noting that it should engage with the primary drive and kickstart idler pinions. Fit the second thrust washer, followed by the clutch centre, tab washer and clutch centre nut. Lock the clutch, using the same method that was employed during dismantling, and tighten the nut to 6.5 kgf m (47 lbf ft).
4 The clutch plain and friction plates and the rubber damper rings should be coated with engine oil prior to installation. It will be noted that each of the plain plates has a part of its outer edge machined off. This effectively makes the plate become slightly out of balance. This causes each plate to be thrown outwards under centrifugal force and thus prevents clutch noise. To prevent the whole clutch from getting out of balance it is necessary to arrange the plates so that the machined areas are spaced evenly around its circumference. This can be achieved by arranging each cutaway area to be approximately 60° from the previous one.
5 Start by sliding a damper ring over the clutch centre, taking care not to twist it during fitting. A friction plate is fitted next, followed by a plain plate, this process being repeated until all the clutch plates are in position.
6 Slide the long pushrod through the hollow input shaft, noting that the end with the reduced diameter should be fitted first. The single steel ball can be pushed into the shaft bore now, followed by the mushroom-headed pushrod. Offer up the clutch pressure plate, aligning one of its three arrow marks with the corresponding mark on the clutch centre. Fit the clutch springs and secure the assembly by tightening the clutch bolts evenly and firmly in a diagonal sequence.
7 The outer cover should not be refitted until the clutch free play has been set using the screw and locknut adjuster set in the pressure plate centre. If the release lever has been disturbed, refit this as described in Section 35 and then refer to Routine maintenance for details of clutch free play adjustment. Having set the required amount of free play at the clutch and in its operating cable, the outer cover may be refitted. If the engine is removed from the frame, these operations must be left until final installation.

31.3a Pass coolant stub through the crankcase bore ...

31.3b ... and fit the wire circlip and a new O-ring

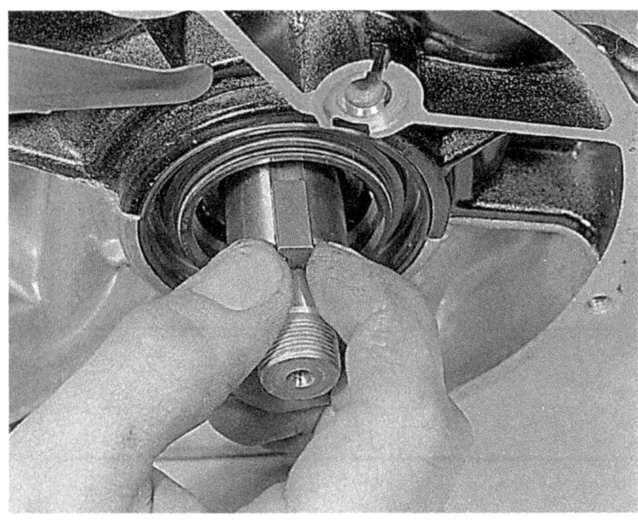

32.1a Fit Woodruff key in its slot in the crankshaft end ...

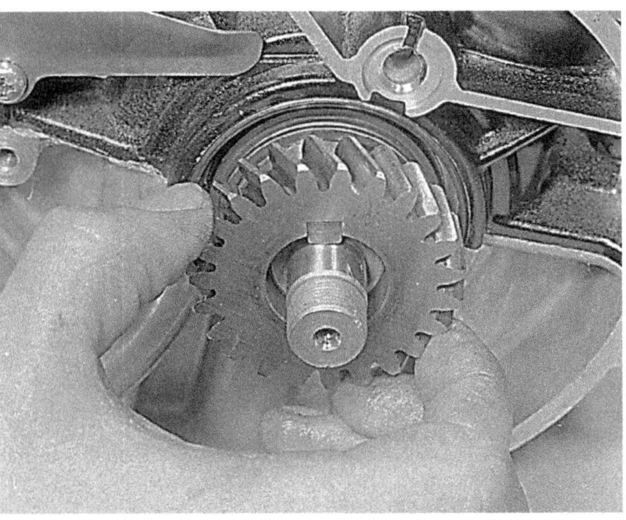

32.1b ... and slide primary drive pinion into place

32.2 Fit the pump drive pinion and secure with Belville washer and nut

32.3a Slide pushrod into the end of the gearbox input shaft ...

32.3b ... followed by the large steel ball

32.3c Place the large thrust washer over the shaft end

32.3d The clutch drum can now be positioned ...

32.3e ... and the inner bush fitted

32.3f Now fit the second thrust washer ...

32.3g ... followed by the clutch centre (note the damper ring, installed)

32.3h Place the tab washer over the shaft end ...

32.3i ... then lock the clutch centre and secure the nut and locking tab

32.4a Install a friction plate over the damper ring, followed by a plain plate ...

32.4b ... a damper ring ...

32.4c ... and a friction plate

32.6a Make sure that one of the arrows on the cluch cover lines up with the arrow mark on the clutch centre

32.6b Fit the springs and bolts, tightening them evenly and progressively

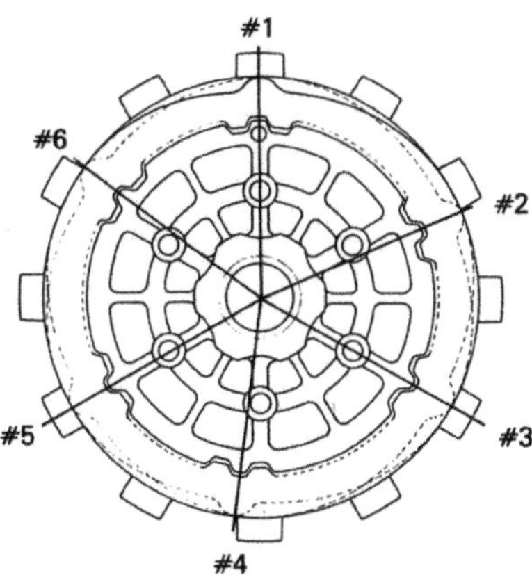

Fig. 1.13 Arrangement of clutch plain plate machined areas

33 Engine reassembly: refitting the alternator, neutral switch and left-hand outer cover

1 Push the alternator stator wiring through its hole in the crankcase and locate the wiring grommet to retain it. The stator can now be offered up and the mounting screws fitted and tightened. Note that the ignition timing is not adjustable, and thus should be correct. If in doubt as to the accuracy of the ignition, refer to Chapter 4 for details.

2 Fit the Woodruff key to the crankshaft keyway, ensuring that it seats correctly. Position the alternator rotor, then fit the plain washer, spring locking washer and securing nut. Lock the crankshaft and tighten the nut to 8.5 kgf m (61.0 lbf ft).

3 Check that the gearbox is in neutral, and where necessary temporarily refit the gear change pedal assembly and select neutral. Offer up the plastic neutral switch cover, ensuring that its contact aligns with the neutral contact on the end of the selector drum. Fit and tighten the three retaining screws. Feed the neutral switch lead behind the stator and through its slot in the casing wall, pushing the rubber guide block into place. Connect the lead to the neutral switch terminal.

4 Slide the metal spacer over the protruding end of the gear selector shaft. If the engine is already in the frame the outer cover can now be refitted, otherwise leave it off until the engine has been installed and the chain and final drive sprocket refitted.

33.1 Install the alternator stator assembly in the casing recess

33.2a Fit the Woodruff key in the crankshaft slot ...

33.2b ... then offer up the rotor

33.2c Fit the plain and spring washers ...

33.2d ... lock the crankshaft and secure the rotor nut

34 Engine reassembly: refitting the YPVS valve, pistons, cylinder barrels and cylinder head

1 Check that the power valve components are clean and free from carbon deposits, paying particular attention to the cleaning groove along the face of the valve, and the valve bore in each barrel. Apply molybdenum-disulphide grease to the power valve bushes, O-rings and oil seals, plus the threads of the long Allen-headed retaining bolt. Offer up the valve halves as shown in the accompanying photographs. Check that the two small dowel pins locate fully. Hold the flats on the end of the valve spindle with a pair of pliers, then fit and tighten the Allen-headed retaining bolt from the other end. Fit the small retainer plate and its single fixing bolt. Repeat the assembly sequence on the remaining valve.

2 Check that the crankcase mouth area is clean and free from grease deposits, then place the cylinder base gaskets over the holding studs. Lubricate the big-end and main bearings with two-stroke oil. Turn the crankshaft to TDC and pack clean rag around the connecting rods so that the crankcase mouths are covered. Lubricate the small-end bearings and slide them into position in the connecting rod eyes.

3 When fitting the pistons it is important to note that they must be fitted in the bore from which they were removed, unless the engine has been rebored in which case new pistons will be fitted. Note that each piston crown carries an arrow mark which should face forward. If the gudgeon pins are tight in the piston bosses it is a good idea to warm the pistons prior to fitting. This will cause the alloy piston to expand more than the steel pin and will make assembly much easier. Hot water at or near boiling point is the best way of heating them with no risk of distortion, but be wary of burns or scalding when using this method – use heavy gloves or some thick rag when handling the hot pistons.

4 Fit each piston in turn, locating the gudgeon pin with **new** circlips. It is false economy to risk reusing old circlips. They may appear to be in good order, and in practice may be quite satisfactory, but in view of their low cost do not run the risk of a weakened circlip breaking or working loose in service. Once the pistons are in place, lubricate the rings with clean two-stroke oil and check that the ring end gaps coincide with the locating pegs.

5 Each barrel has a tapered lead-in at its base to help in guiding the rings into the bore. As the barrel is pushed down over the piston use one hand to feed the rings into the bore. It is important to check that the barrels are exactly square to the crankcase and connecting rods, otherwise there is some risk of ring breakage where the ring ends pass close to the inlet port. This is an important point and warrants the removal of the reed valve units so that a visual check can be made. Once the piston rings have entered the bores correctly the rag padding may be removed from the crankcase mouth and the barrel pushed firmly down onto the base gasket.

6 When both barrels are in position, fit the retaining nuts to the base flange of each. These should be tightened evenly and progressively to a torque setting of 2.5 kgf m (18.0 lbf ft). In practice it proved impossible to get a torque wrench onto some of the nuts, and it was necessary to tighten them by hand using a ring spanner. Align the inner ends of the two power valves and connect them with the semi-circular joint piece. Fit and tighten the two retaining screws. It is worth noting that the screws are in an exposed position and are liable to corrosion. It may be worth replacing the standard screws with Allen screws to aid future removal.

7 Check that the mating surfaces of the cylinder head and barrels are clean and dry, then place the cylinder head gasket in position. Do not use jointing compound on either surface. Offer up the cylinder head and fit the eight retaining bolts finger tight. The head bolts should be tightened in the increasing sequence indicated by the numbers cast into the cylinder head. Initial tightening should be to about 1.0 kgf m (7.2 lbf ft). It will now be necessary to repeat the tightening operation, this time to the final value of 2.8 kgf m (20.0 lbf ft). Note that the cylinder head bolts must be re-tightened after the engine has been run and allowed to cool down. Fit the hose and adaptor between the cylinder head and the crankcase stub. The reed valve assemblies can now be refitted noting that the rubber adaptors must be renewed if they have cracked around the balance pipe stubs. Push the balance pipe into position.

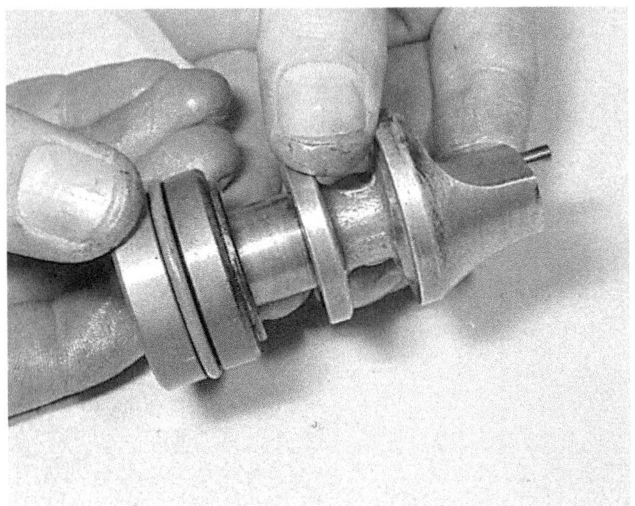

34.1 Clean the valve halves, fit locating pins and new O-rings as required

34.2 Place new cylinder base gaskets over holding studs

34.4a Use new circlips to secure the gudgeon pins ...

34.4b ... leaving end gaps clear of removal slot

34.6 Join the power valves between the two cylinder barrels

34.7a Fit a new cylinder head gasket ...

34.7b ... and lower the head into position

34.7c Tighten the head sleeve bolts in the prescribed sequence

34.7d Reconnect the coolant hose adaptor between the two cylinders

34.7e Refit the balance pipe between the inlet rubbers

35 Fitting the engine/gearbox unit into the frame

1 The engine/gearbox unit is refitted by reversing the removal sequence. As with removal, it is advisable to have an assistant to help manoeuvre the unit into position, but the job is just about feasible unaided if this proves unavoidable. Take care not to damage the paintwork during installation. This can be guarded against by wrapping some card or stiff paper around the more vulnerable areas and taping this in place.
2 Lift the unit into the frame cradle from the right-hand side. It will sit in this position while the mounting brackets are sorted out and positioned. Before the engine is secured, make sure that the final drive chain is looped around the projecting cast boss to the rear of the output shaft. Failure to check this will cause problems later on since it is difficult to get the chain into position with the engine bolted into place.
3 Fit the front and rear mountings in position, fitting the small frame mounting bolts and the large through bolts finger tight. When all are in position, tighten the small bolts to 2.4 kgf m (17.4 lbf ft) and the large through bolts to 6.5 kgf m (47.0 lbf ft). Fit the engine steady bars between the frame brace tube and the underside of the crankcase.
4 Refit the tachometer drive cable to its adaptor at the rear of the crankcase (RD350 LC II model only). The knurled retaining ring should be tightened securely by hand. Fit the final drive chain around the gearbox sprocket and slide the assembly over the splined end of the output shaft. Place the tab washer against the sprocket, then fit the retaining nut noting that its recessed face should be against the tab washer. Lock the rear wheel by applying the brake, then tighten the nut to 6.5 kgf m (47.0 lbf ft). Bend the tab washer over one of the nut's flats and tap it down securely with a hammer and punch.
5 If not already in place, fit the clutch release arm into the crankcase and check that the retainer is in place on the adjacent crankcase bolt. Reconnect the clutch cable, then carry out clutch adjustment as detailed in Routine Maintenance. When adjustment is complete, refit the outer cover using a new gasket and ensuring that the coolant stub engages correctly in its recess.
6 Refit the throttle valves and caps to their respective carburettors. The valves and bodies are handed, but it is possible to get them interchanged and reversed. Guard against this by ensuring that the synchronisation dot on each valve will coincide with the small window in the carburettor body.
7 Manoeuvre the instruments into position between the airbox and inlet adaptors, and secure the retaining clips. Reconnect the oil pump cable and fit the pipe from the oil tank and the two small delivery pipes. The pump cover should be left off until it has been bled and adjusted.
8 Refit the exhaust system using the new sealing rings at each port. Fit the retainer flanges and tighten the two nuts evenly. Fit the silencer mounting bolts to secure each half of the system to the footrest mounting plate. If it was removed, refit the thermostat and housing to the top of the cylinder head, using a new rubber sealing ring and tightening the retaining nuts evenly. Refit the temperature gauge sender unit using a new sealing washer. Fit the spark plugs and caps.
9 Route the alternator output leads across the crankcase and reconnect them next to the battery tray, using the colour coded wiring for guidance. Check that all breather and drain hoses are routed correctly, then refit and connect the battery, observing the correct polarity.
10 Turn the power valve assembly so that the screw heads on the curved joint face upwards. Fit the power valve pulley housing to the left-hand cylinder and refit the pulley. Align the notch in the pulley with the hole in the rear of the housing, then lock the pulley using a 4 mm diameter pin; a drill bit or Allen key can be used for this. Tighten the pulley retaining bolt and then remove the pin.
11 Adjust the pulley cables until all free play is removed, then back off the adjusters by 1/4 turn. Switch on the ignition. The power valve will open and close (this is a self-cleaning function which occurs each time the ignition is turned on). Check that the notch and the hole still align correctly. If necessary, adjust the position of the valve by slackening one cable and tightening the other. Switch the ignition off and then on again to check the setting. When all is well, switch the ignition off, tighten the locknuts and refit the valve covers.

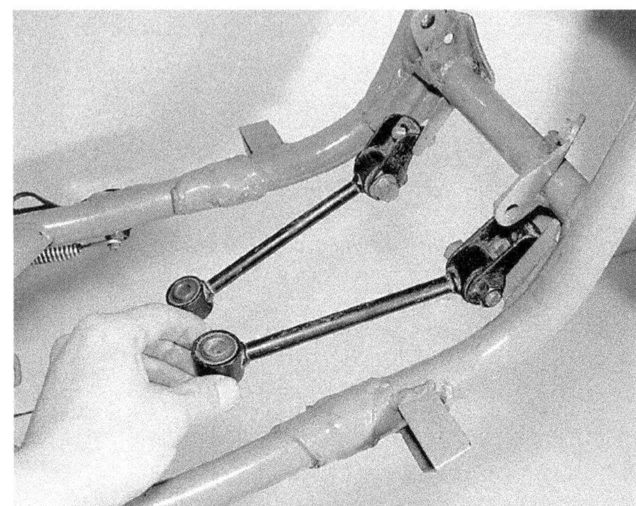
35.1 Fit the engine steady bars loosely in place before engine is installed

35.4a Grease and fit spacer into output shaft seal, if not already in place

35.4b Fit the gearbox sprocket and tab washer ...

35.4c ... tighten securely and bend up the locking tab as shown

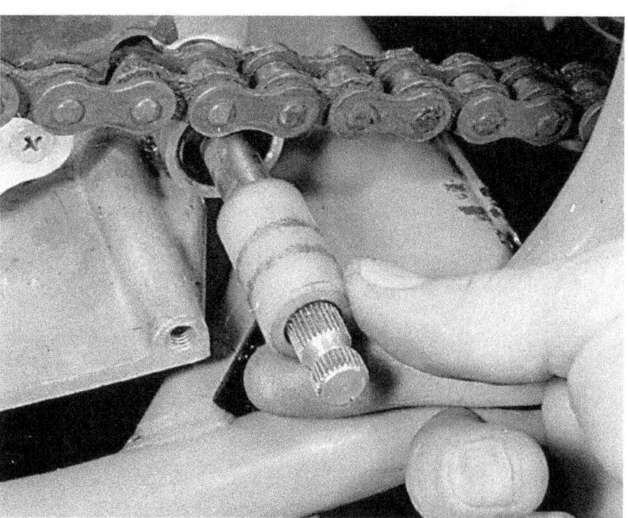

35.4d Do not forget the sleeve over the selector shaft end

35.5a Fit a new gasket to the crankshaft right-hand face ...

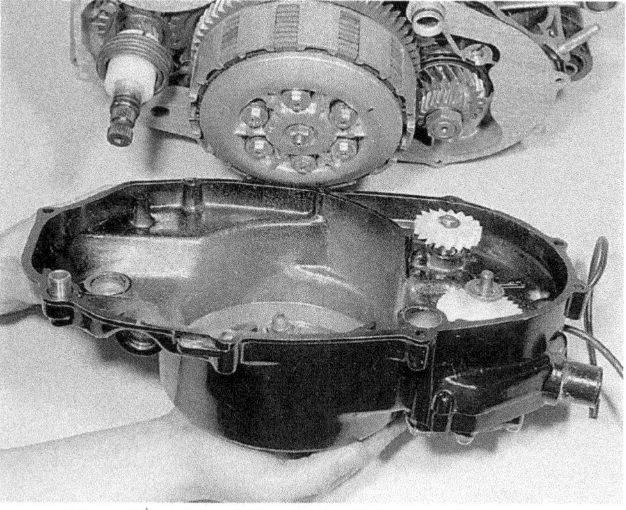

35.5b ... then fit cover, making sure that gears engage correctly

35.6 Refit the carburettors and secure the retaining clips

35.7a Pass oil pump cable through its hole in the crankcase ...

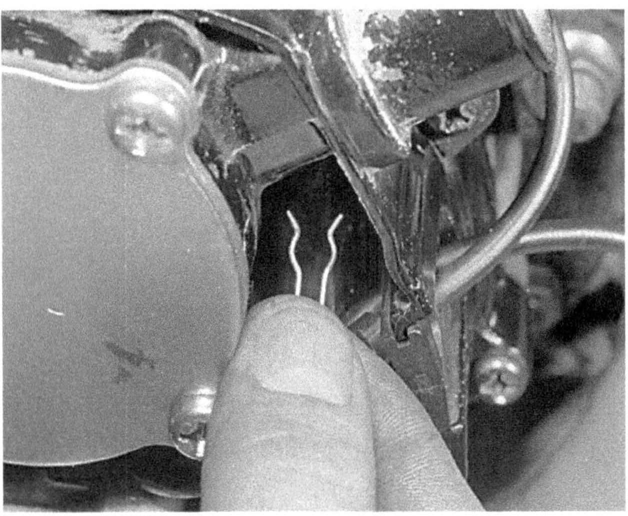

35.7b ... and secure it with its retaining clip as shown

35.7c Hook the cable inner around the pump pulley. Do not fit cover until pump has been bled and adjusted

35.8a Use new sealing rings at the exhaust ports

35.8b Fit and secure the retainer flanges at the front ...

35.8c ... and the silencer mountings at the rear

35.10a Refit the pulley assembly to the end of the power valve

35.10b Pulley can be locked in place and adjustment checked using a twist drill or similar as shown

35.10c Check that cables are located correctly at the servomotor ...

35.10d ... then set the cable adjustment (see text)

35.10e When correctly adjusted, refit the outer cover

36 Engine reassembly: final connections and adjustments

1 Check throttle cable free play and where necessary adjust to give 3 – 7 mm (0.12 – 0.28 in) movement measured at the outer edge of the twistgrip flange, making any adjustment with the in-line adjuster immediately below the throttle twistgrip.

2 Check throttle synchronisation by observing the alignment marks through the inspection windows on the right-hand side of each instrument. Using the adjusters on the carburettor tops, set both throttle valves so that the marks are central in their windows. Open and close the throttle a few times, then recheck.

3 Once synchronisation has been set, check the oil pump cable adjustment as described in Chapter 3, noting that the correct alignment mark is dependent on the model and year of manufacture. The oil pump should now be bled by removing the small bleed screw and allowing the air to be expelled by oil flowing from the tank. When the oil is free of air bubbles, fit and tighten the bleed screw. The oil delivery pipes should be bled once the engine is running as described in Section 37 of this Chapter.

4 Reconnect the hose to the water pump stub on the outer cover and fill the cooling system using a mixture of 50% distilled water and 50% Glycol antifreeze. Do not use ordinary tap water because the impurities contained in it will promote corrosion and furring-up of the system. Fit

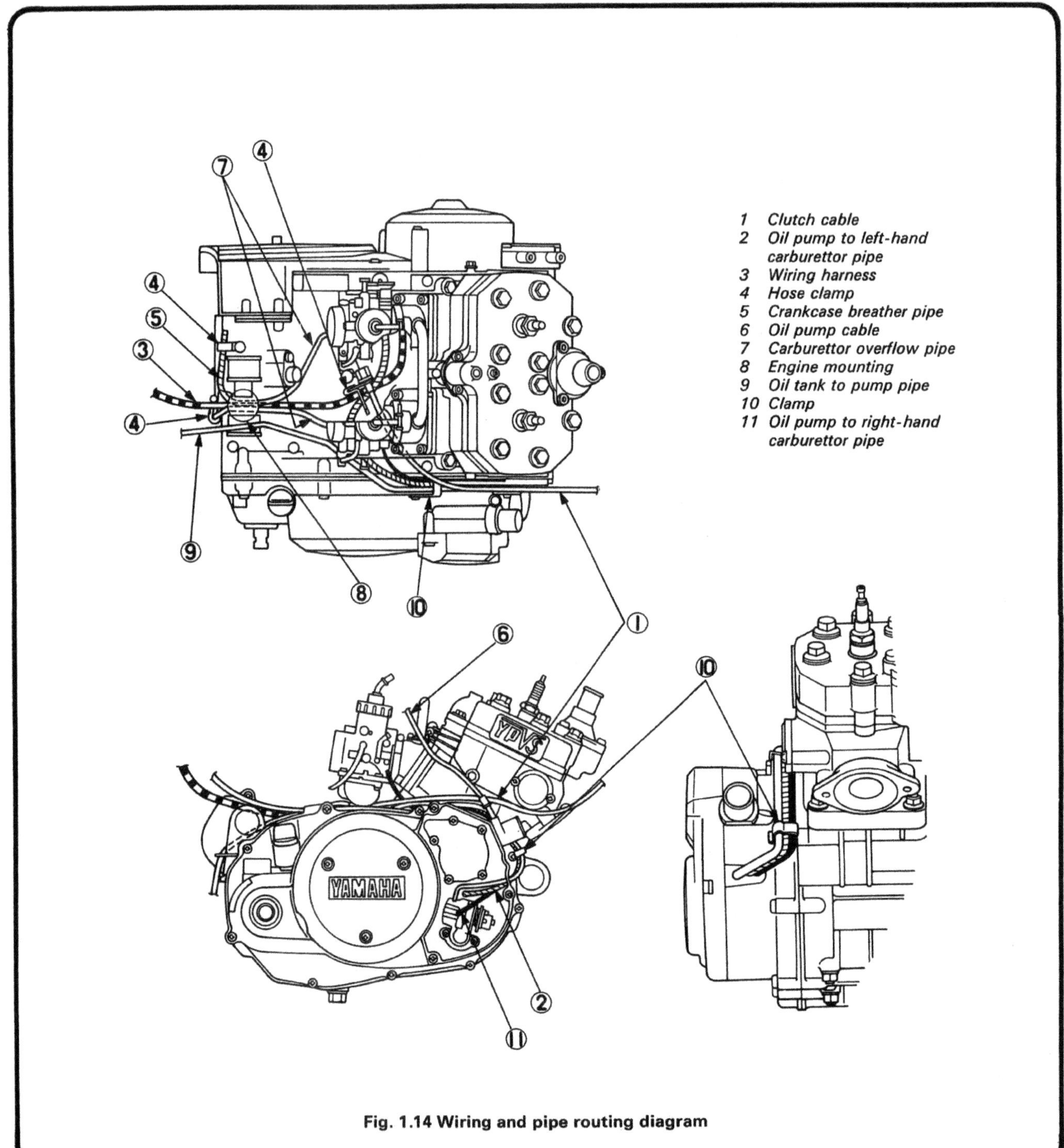

Fig. 1.14 Wiring and pipe routing diagram

the radiator cap and the guard, but do not fit the two right-hand retaining screws. This will allow the guard to be displaced to permit topping up of the cooling system.

5 Remove the transmission oil filler plug and add 1700 cc (2.99 Imp pint) of SAE 10W/30 SE motor oil. The level of the oil should be re-checked after the engine has been run.

6 Fit and secure the left-hand outer cover, and fit the kickstart lever if this is not already in position. Complete reassembly by fitting the fuel tank and pipe and the dual seat. Check around the machine to ensure that all remaining cables and connectors are in place. Where appropriate, readjust the rear chain play and secure the wheel spindle nut. Make a final check of the electrical system by turning the ignition switch on and testing the operation of the various electrical components.

37 Starting and running the rebuilt engine

1 Initial starting may prove a little difficult and it is possible that the oil used during reassembly may cause fouling of the spark plugs. Use the normal cold starting procedure and be prepared for flooding during the first few attempts. If necessary remove and dry the plugs and start again. When the initial start-up is made, run the engine slowly for the first few minutes, especially if the engine has been rebored or a new crankshaft fitted. Check that all controls function correctly and that there are no oil leaks, before taking the machine on the road. The exhausts will emit a high proportion of white smoke during the first few miles, as the excess oil used whilst the engine was reassembled is burnt away. The volume of smoke should gradually diminish until only the customary light blue haze is observed during normal running. It is wise to carry a spare pair of spark plugs during the first run, since the existing plugs may oil up due to the temporary excess of oil.

2 As soon as the engine is running evenly bleed the oil delivery lines by pulling on the pump cable so that the pump stroke is at maximum and the engine is held at a fast idle speed.

3 Remember that a good seal between the pistons and the cylinder barrels is essential for the correct functioning of the engine. A rebored two-stroke engine will require more careful running-in, over a longer period, than its four-stroke counterpart. There is far greater risk of engine seizure during the first hundred miles if the engine is permitted to work hard.

4 Do not tamper with the exhaust sysem or run the engine without baffles fitted to the silencer. Unwarranted changes in the exhaust system will have a very marked effect on engine performance invariably for the worse. The same advice applies to dispensing with the air cleaner or the air cleaner element.

5 Do not on any account add oil to the petrol under the mistaken belief that a little extra oil will improve the engine lubrication. Apart from creating excess smoke, the addition of oil will make the mixture much weaker, with the consequent risk of overheating and engine seizure. The oil pump alone should provide full engine lubrication.

6 Before taking the machine on the road, the cylinder head bolt torque and all oil and water levels should be rechecked. Replace all filler caps and covers. Remember to check the operation of all controls and electrical accessories before taking the machine on the road.

38 Taking the rebuilt machine on the road

1 Any rebuilt machine will need time to settle down, even if parts have been replaced in their original order. For this reason it is highly advisable to treat the machine gently for the first few miles to ensure oil has circulated throughout the lubrication system and that new parts fitted have begun to bed down.

2 Even greater care is necessary if the engine has been rebored or if a new crankshaft has been fitted. In the case of a rebore, the engine will have to be run in again, as if the machine were new. This means greater use of the gearbox and a restraining hand on the throttle until at least 500 miles have been covered. There is no point in keeping to any set speed limit; the main requirement is to keep a light loading on the engine and to gradually work up performance until the 500 mile mark is reached. These recommendations can be lessened to an extent when only a new crankshaft is fitted. Experience is the best guide since it is easy to tell when an engine is running freely.

3 Remember that a good seal between the piston and the cylinder barrel is essential for the correct functioning of the engine. A rebored two-stroke engine will require more careful running-in, over a long period, than its four-stroke counterpart. There is a far greater risk of engine seizure during the first hundred miles if the engine is permitted to work hard.

4 If at any time a lubrication failure is suspected, stop the engine immediately, and investigate the cause. If an engine is run without oil, even for a short period, irreparable engine damage is inevitable.

5 Do not on any account add oil to the petrol under the mistaken belief that a little extra oil will improve the engine lubrication. Apart from creating excess smoke, the addition of oil will make the mixture much weaker, with the consequent risk of overheating and engine seizure. The oil pump alone should provide full engine lubrication.

6 Do not tamper with the exhaust system. Unwarranted changes in the exhaust system will have a marked effect on engine performance, invariably for the worse. The same advice applies to dispensing with the air cleaner or the air cleaner element.

7 When the initial run has been completed allow the engine unit to cool and then check all the fittings and fasteners for security. Re-adjust any controls which may have settled down during initial use.

Chapter 2 Cooling system

Contents

Specifications

Cooling system	
Coolant mixture	50% antifreeze, 50% water
Antifreeze type	Any high quality ethylene glycol mixture with aluminium engine type corrosion inhibitors
Capacity:	
Overall	1.5 litre (2.64 Imp pint)
From low to full	185 cc (0.32 Imp pint)
Reservoir tank capacity	215 cc (0.38 Imp pint)
Radiator	
Radiator:	
Core width	290.6 mm (11.44 in)
Core height	180 mm (7.08 in)
Core thickness	16 mm (0.63 in)
Cap opening pressure	0.9 ± 0.15 kg cm² (12.8 ± 2.13 psi)
Water pump	
Type	Centrifugal impeller
Reduction ratio	32/20 (1.60:1)
Thermostat	
Opening temperature:	
RD350 LC II	Not available
Other models	71° ± 2°C (156° ± 35.6°F)
Fully open temperature:	
RD350 LC II	Not available
Other models	85°C (185°F)
Maximum lift:	
RD350 LC II	Not available
Other models	7.0 mm (0.28 in)

1 General description

The Yamaha LC models are provided with a liquid cooling system which utilises a water/antifreeze coolant to carry away excess energy produced in the form of heat. The cylinders are surrounded by a water jacket from which the heated coolant is circulated by thermo-syphonic action in conjunction with a water pump fitted in the engine right-hand cover and driven via a pinion and shaft from a crankshaft mounted pinion. The hot coolant passes upwards through flexible pipes to the top of the radiator which is mounted on the frame downtubes to take advantage of maximum air flow. The coolant then passes downwards, through the radiator core, where it is cooled by the passing air, and then to the water pump and engine where the cycle is repeated.

The flow of coolant is regulated by a thermostat, a temperature sensitive valve unit contained in a housing on top of the cylinder head. When the engine is cold, the thermostat remains closed, effectively stopping the coolant from circulating through the system. This allows the engine to reach its normal operating temperature rapidly, minimising wear. As the water temperature rises, the thermostat begins to open and the coolant starts to circulate to keep the engine at the optimum temperature.

The complete system is sealed and pressurised; the pressure being controlled by a valve contained in the spring loaded radiator cap. By pressurising the coolant the boiling point is raised, preventing premature boiling in adverse conditions. The overflow pipe from the radiator is connected to an expansion tank into which excess coolant is discharged by pressure. The expelled coolant automatically returns to the radiator, to provide the correct level when the engine cools again.

2 Draining the cooling system

1 It will be necessary to drain the cooling system on infrequent occasions, either to change the coolant at two yearly intervals or to permit engine overhaul or removal. The operation is best undertaken with a cold engine to remove the risk of scalding from hot coolant escaping under pressure.
2 Place the machine on its centre stand and gather together a drain tray or bowl of about 2.0 litres (4.0 pint) capacity, and something to guide the coolant from the cylinder barrel drain plugs into the bowl. A small chute made from thick card will suffice for this purpose, but do not be tempted to allow the coolant to drain over the engine casings – the antifreeze content may discolour the painted surfaces.
3 To gain access to the radiator filler cap it will be necessary to remove the fuel tank. Remove the seat, then release the single retaining bolt at the rear of the tank. Check that the fuel tap is set to the "ON" position, then pull off the left-hand side panel to reveal the fuel and vacuum pipes. Slide the clips down the pipe and prise each one off its stub. Lift the rear of the tank and pull it back to free the mounting rubbers at the front. The radiator filler cap will now be accessible near the steering head.

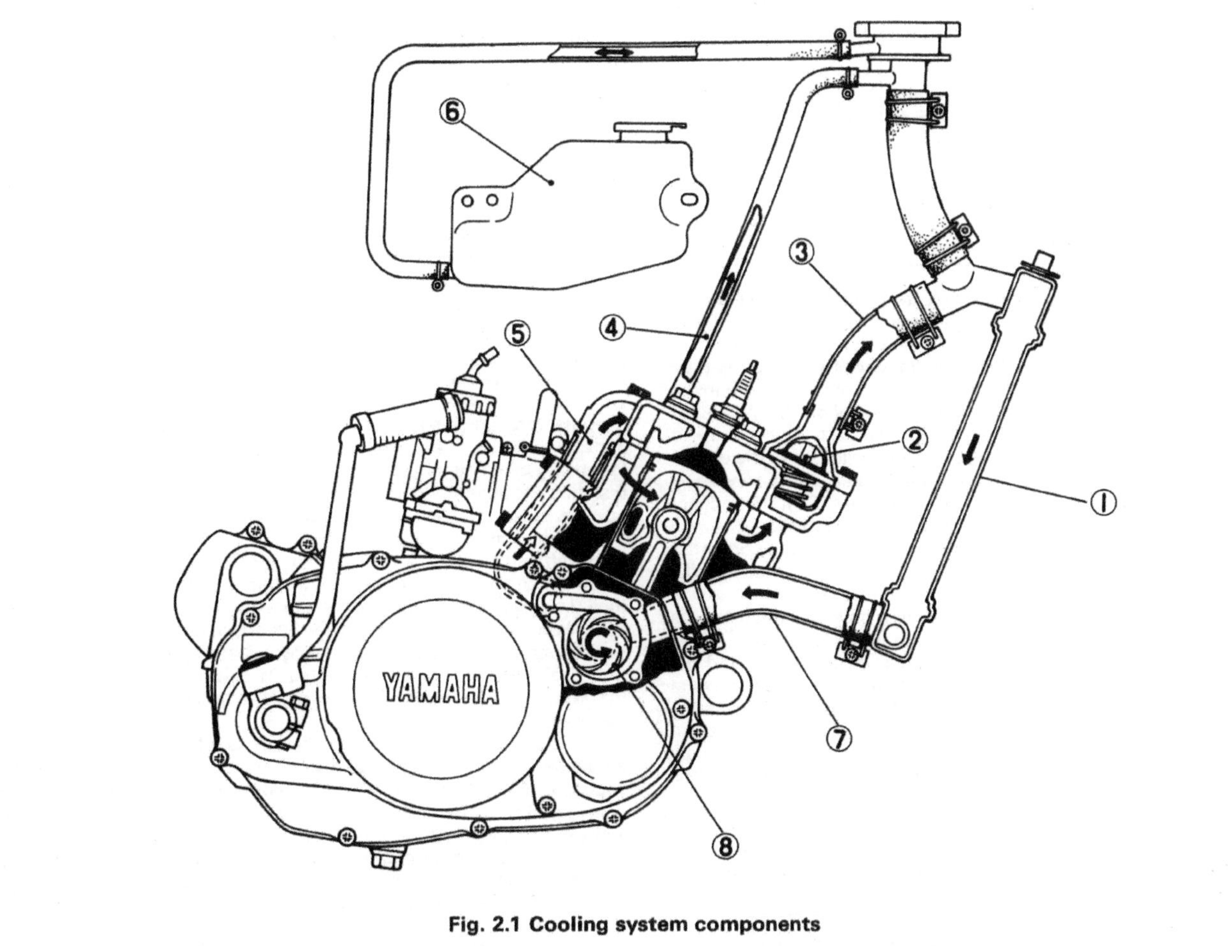

Fig. 2.1 Cooling system components

1 Radiator
2 Thermostat
3 Top hose
4 Bypass pipe
5 Pump to cylinder hose
6 Expansion tank
7 Bottom hose
8 Water pump

4 Slacken and remove each of the cylinder barrel drain plugs in turn, using the chute to guide the coolant clear of the crankcase outer covers and into the bowl. If the system is to be drained fully pull off the pipe from the expansion tank and allow this to drain. Drain any residual coolant by detaching the hose at the front of the right-hand outer cover stub.
5 If the system is being drained before an engine overhaul little else need be done at this stage. If the coolant is reasonably new it can be re-used if it is kept clean and uncontaminated. If, however, the system is to be refilled with new coolant it is advisable to give it a through flushing with tap water, if possible using a hose which can be left running for a while. If the machine has done a fairly high mileage it may be advisable to carry out a more thorough flushing process as described below.

3 Flushing the cooling system

1 After extended service the cooling system will slowly lose efficiency, due to the build up of scale, deposits from the water and other foreign matter which will adhere to the internal surfaces of the radiator and water channels. This will be particularly so if distilled water has not been used at all times. Removal of the deposits can be carried out easily, using a suitable flushing agent in the following manner.
2 After allowing the cooling system to drain, refit the drain plugs and refill the system with clean water and a quantity of flushing agent. Any proprietary flushing agent in either liquid or dry form may be used, providing that it is recommended for use with aluminium engines. NEVER use a compound suitable for iron engines as it will react violently with the aluminium alloy. The manufacturer of the flushing agent will give instructions as to the quantity to be used.
3 Run the engine for ten minutes at operating temperatures and drain the system. Repeat the procedure TWICE and then again using only clean cold water. Finally, refill the system as described in the following Section.

4 Filling the cooling system

1 Before filling the system, check that the sealing washers on the drain plugs are in good condition and renew if necessary. Fit and tighten the drain plugs and check and tighten all the hose clips.
2 Fill the system slowly to reduce the amount of air which will be trapped in the water jacket. When the cooling level is up to the lower edge of the radiator filler neck, run the engine for about 10 minutes at 900 rpm. Increase engine revolutions for the last 30 seconds to accelerate the rate at which any trapped air is expelled. Stop the engine and replenish the coolant level again to the bottom of the filler neck. Refill the expansion tank up to the 'Full' level mark. Refit the radiator cap, ensuring that it is turned clockwise as far as possible.
3 Ideally, distilled water should be used as a basis for the coolant. If this is not readily available, rain water, caught in a non-metallic receptacle, is an adequate substitute as it contains only limited amounts of mineral impurities. In emergencies only, tap water can be used, especially if it is known to be of the soft type. Using non-distilled water will inevitably lead to early 'furring-up' of the system and the need for more frequent flushing. The correct water/antifreeze mixture is 50/50; do not allow the antifreeze level to fall below 40% as the anti-corrosion properties of the coolant will be reduced to an unacceptable level. Antifreeze of the ethylene glycol-based type should always be used. Never use alcohol based antifreeze in the engine.

5 Radiator and filler cap: removal, cleaning, examination and refitting

1 Drain the cooling system as described in Section 2 of this Chapter. On models so equipped, remove the lower fairing section (see Chapter 5 for details).
2 Disconnect the hose from the filler neck and the radiator top and bottom hoses from their respective stubs on the radiator. Release the screws which retain the plastic radiator grille and lift it away.
3 The two radiator mounting bolts can now be removed and the radiator lifted away. The bolts pass through rubber bushes, as does the locating peg on the top of the radiator which is similarly isolated. Note the order of these so that the radiator can be refitted correctly.
4 Remove any obstructions from the exterior of the radiator core, using an air line. The conglomeration of moths, flies and autumnal detritus usually collected in the radiator matrix severely reduces the cooling efficiency of the radiator.
5 The interior of the radiator can most easily be cleaned while the radiator is in-situ on the motorcycle, using the flushing procedure described in Section 3 of this Chapter. Additional flushing can be carried out by placing the hose in the filler neck and allowing the water to flow through for about ten minutes. Under no circumstances should the hose be connected to the filler neck mechanically as any sudden blockage in the radiator outlet would subject the radiator to the full pressure of the mains supply (about 50 psi). The radiator should not be tested to greater than 1.0 kg/cm^2 (15 psi).
6 If care is exercised, bent fins can be straightened by placing the flat of a screwdriver either side of the fin in question and carefully bending it into its original shape. Badly damaged fins cannot be repaired. If bent or damaged fins obstruct the air flow more than 20%, a new radiator will have to be fitted.
7 Generally, if the radiator is found to be leaking, repair is impracticable and a new component must be fitted. Very small leaks may sometimes be stopped by the addition of a special sealing agent in the coolant. If an agent of this type is used, follow the manufacturer's instructions very carefully. Soldering, using soft solder may be efficacious for caulking large leaks but this is a specialised repair best left to experts.
8 Inspect the four radiator mounting rubbers for perishing or compaction. Renew the rubbers if there is any doubt as to their condition. The radiator may suffer from the effect of vibration if the isolating characteristics of the rubber are reduced.
9 If the radiator cap is suspect, have it tested by a Yamaha dealer. This job requires specialist equipment and cannot be done at home. The only alternative is to try a new cap.

6 Hoses and connections: examination and renovation

1 The radiator is connected to the engine unit by two hoses, there being an additional hose between the water pump in the right-hand outer cover and the cylinder head. The hoses should be inspected periodically and renewed if any sign of cracking or perishing is discovered. The most likely area for this is around the wire hose clips which secure each hose to its stub. Particular attention should be given if regular topping up has become necessary. The cooling system can be considered to be a semi-sealed arrangement, the only normal coolant loss being minute amounts through evaporation in the expansion tank. If significant quantities have vanished it must be leaking at some point and the source of the leak should be investigated promptly.
2 Serious leakage will be self-evident, though slight leakage can be more difficult to spot. It is likely that the leak will only be apparent when the engine is running and the system is under pressure, and even then the rate of escape may be such that the hot coolant evaporates as soon as it reaches the atmosphere. Such small leaks may require the use of a special device which will pressurise the system whilst cold and thus enable the leak to be pinpointed. To this end it is best to entrust this work to an authorised Yamaha dealer who will have access to the necessary equipment.
3 In very rare cases the leak may be due to a broken head gasket, in which case the coolant may be drawn into the engine and expelled as vapour in the exhaust gases. If this proves to be the case it will be necessary to remove the cylinder head for investigation. If the rate of leakage has been significant it may prove necessary to remove the cylinder barrels and pistons so that the crankcase can be checked. Any coolant which finds its way that far into the engine can cause rapid corrosion of the main and big-end bearings and must be removed completely.
4 Another possible source of leakage is the stub between the crankcase and the right-hand outer cover. If its O-ring seal becomes damaged or broken it is possible that coolant might find its way into the transmission, and any sign of emulsified transmission oil or water droplets inside the cover should be investigated promptly before corrosion takes place.

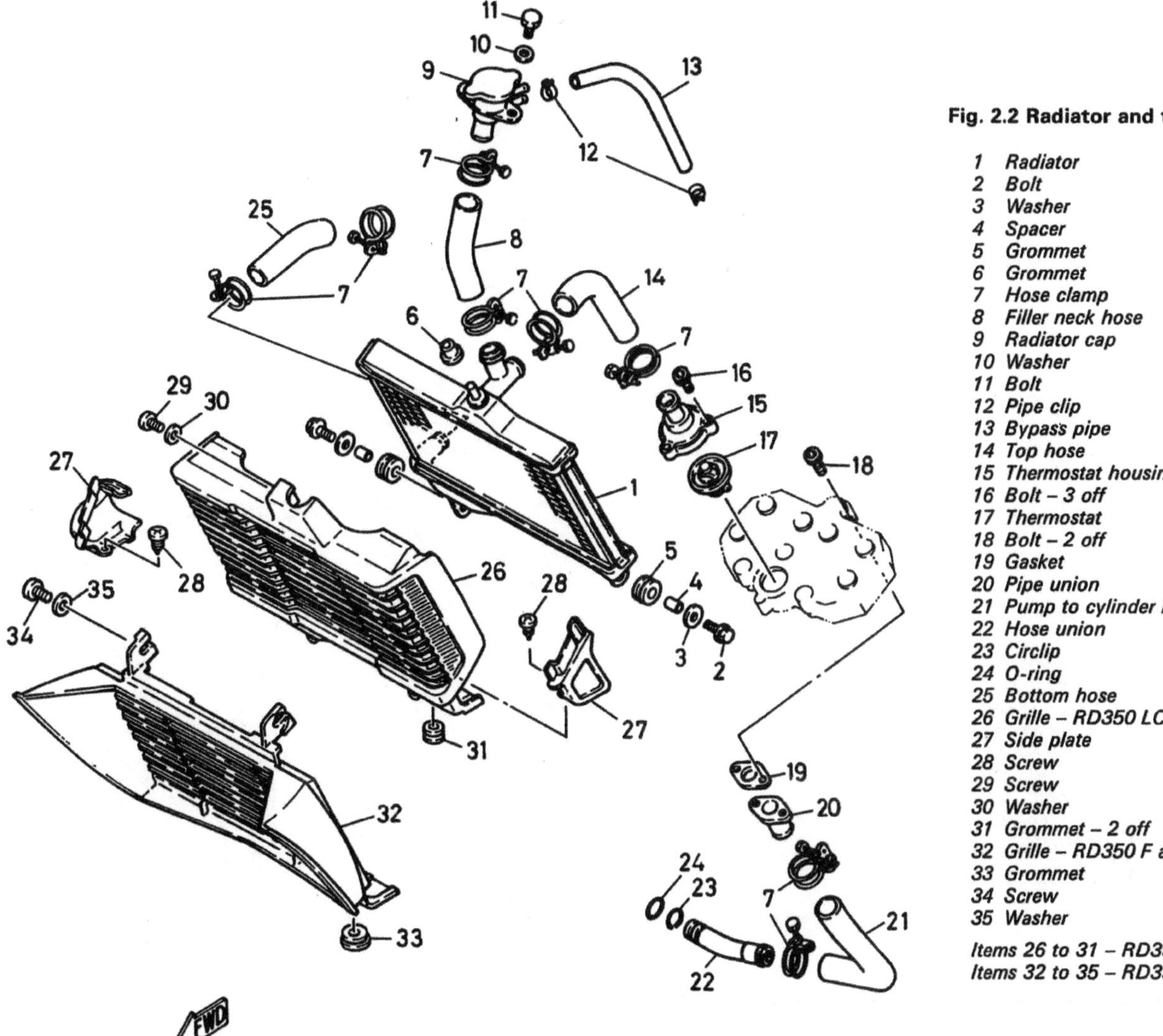

Fig. 2.2 Radiator and thermostat

1 Radiator
2 Bolt
3 Washer
4 Spacer
5 Grommet
6 Grommet
7 Hose clamp
8 Filler neck hose
9 Radiator cap
10 Washer
11 Bolt
12 Pipe clip
13 Bypass pipe
14 Top hose
15 Thermostat housing
16 Bolt – 3 off
17 Thermostat
18 Bolt – 2 off
19 Gasket
20 Pipe union
21 Pump to cylinder hose
22 Hose union
23 Circlip
24 O-ring
25 Bottom hose
26 Grille – RD350 LC II
27 Side plate
28 Screw
29 Screw
30 Washer
31 Grommet – 2 off
32 Grille – RD350 F and N
33 Grommet
34 Screw
35 Washer

Items 26 to 31 – RD350 LC II
Items 32 to 35 – RD350 F and N

7 Water pump: removal and overhaul

1 The water pump will not normally require attention unless its bearing has become noisy if there is obvious leakage of coolant into the transmission oil. To gain access to the pump, drain the coolant and the transmission oil fully, then remove the pump cover followed by the right-hand outer cover itself. Carefully drain the residual coolant from the pump before dismantling commences.
2 The water pump is located immediately above the oil pump, the two sharing a common drive pinion on the crankshaft. To dismantle the water pump it will first be necessary to remove its driven gear by displacing the circlip which secures it to the pump spindle. The locating pin should be removed by pushing it through and out of the spindle.
3 Unscrew the five cover screws and remove the cover and gasket. The impeller and spindle may now be displaced and removed. If the bearing or oil seal is worn or damaged they should be renewed as a set. The two components can be driven out from the oil seal side, having first heated the casing in an oven to 90° – 120°C (194° – 248°F). If using this method it is best to remove all seals, plastic parts and the oil pump first.
4 An alternative method is to pour boiling water (100°C) over the bearing boss area, but it is not advisable to use a blowlamp or other localised heat source in view of the risk of warpage. Once heated, the bearing and seal can be driven out using a suitable round bar or an old socket.
5 The new bearing and oil seal should be greased prior to installation and tapped home using a large socket against the outer race of the bearing. Note that the seal is marked WATER SIDE on one face, and this should face the pump. The bearing serial number should face outward. Tap the bearing and seal home ensuring that they both seat squarely in the casing.
6 Clean the impeller and spindle, being particularly careful to ensure that any corrosion that may have formed around the seal area is removed and the spindle left completely smooth. If the spindle is badly pitted in this area it may be necessary to renew it to avoid rapid seal wear. The spindle should be greased prior to installation and care should be exercised during fitting to avoid damage to the seal face. Complete reassembly by reversing the dismantling sequence, using a new gasket on the pump cover joint. Where practicable, tighten the cover screws to 0.8 kgf m (5.8 lbf ft).

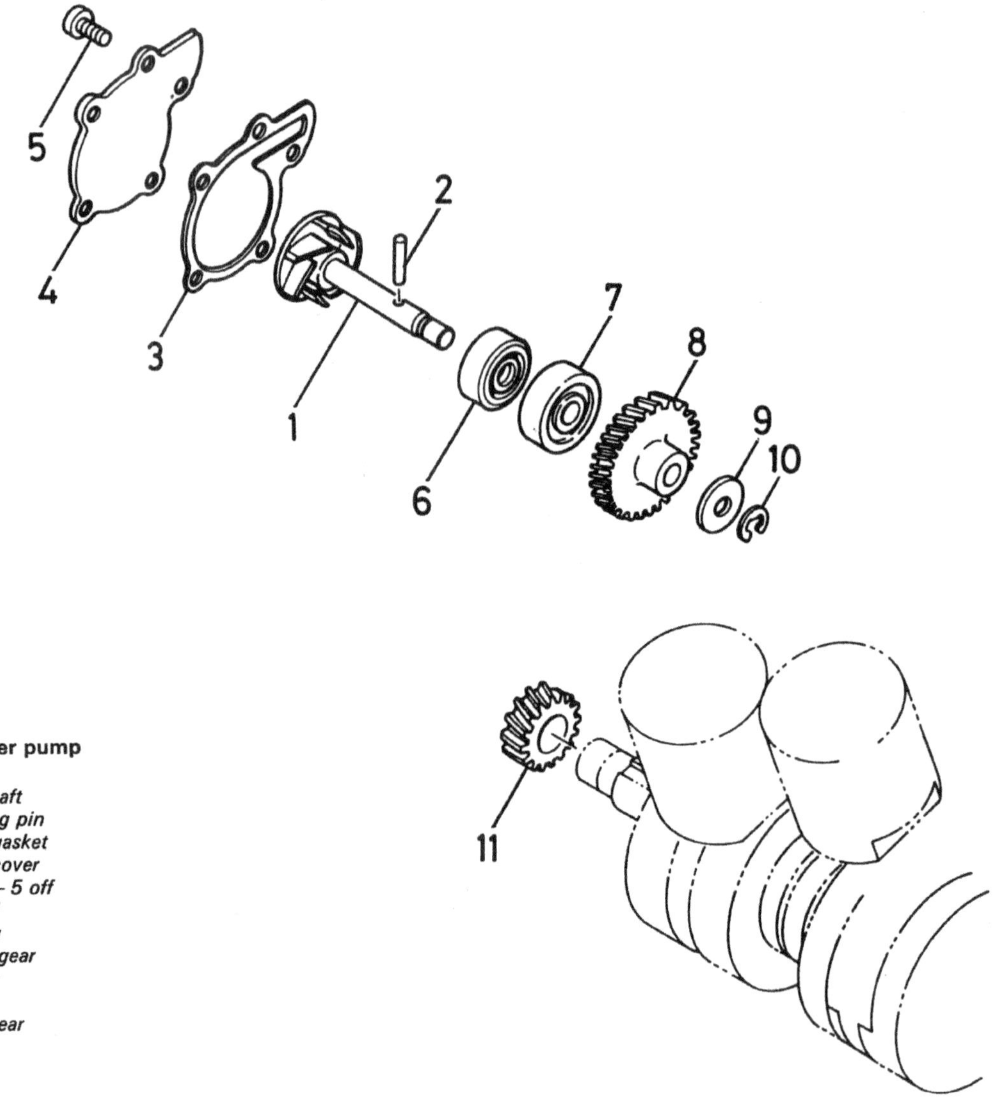

Fig. 2.3 Water pump

1 Driveshaft
2 Locating pin
3 Cover gasket
4 Pump cover
5 Screw – 5 off
6 Oil seal
7 Bearing
8 Driven gear
9 Washer
10 Circlip
11 Drive gear

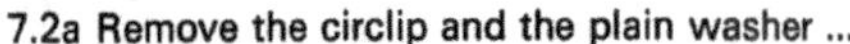

7.2a Remove the circlip and the plain washer ...

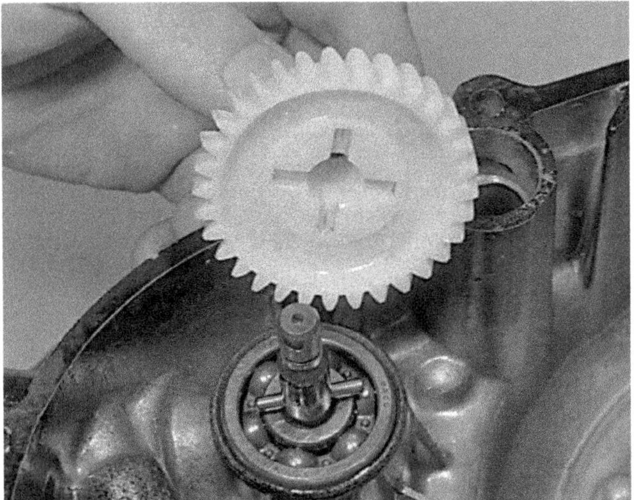

7.2b ... and lift away the white plastic pump pinion

7.2c The driving pin must be displaced and removed to allow the shaft to pass through the bearing

7.2d Remove the pump shaft together with the integral impeller

7.3 Use a new gasket when refitting the pump cover

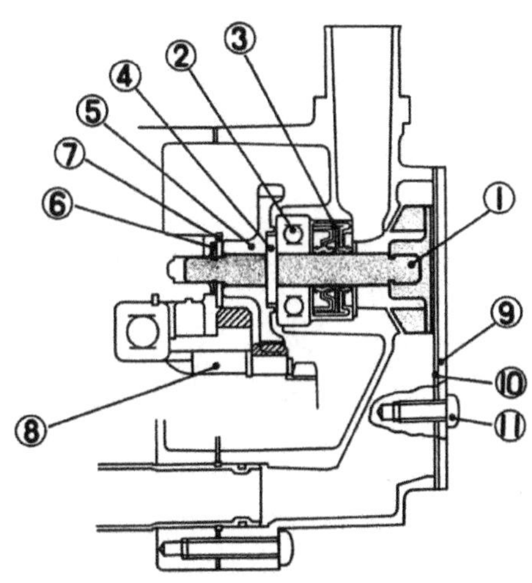

Fig. 2.4 Sectioned view of water pump

1 Driveshaft
2 Bearing
3 Oil seal
4 Locating pin
5 Driven gear
6 Circlip
7 Washer
8 Drive gear
9 Pump cover
10 Gasket
11 Screw

8 Water temperature gauge and sender: testing

Water temperature is monitored by an electrically operated gauge in the instrument panel controlled by a sender unit which screws into the cylinder head water jacket. A description and test procedure of these components will be found in Chapter 7.

9 Thermostat: removal and testing

1 The thermostat is automatic in operation and should give many years service without requiring attention. In the event of a failure, the valve will probably become jammed open, in which case the engine will take much longer than normal to warm up. If, conversely, the valve gets jammed shut, the coolant will be unable to circulate normally and the engine will tend to overheat badly. Neither condition is acceptable, and the fault should be investigated promptly.

2 Before the thermostat can be removed it will be necessary to drain the cooling system as detailed in Section 2. The system need not be drained fully, but the coolant level must be well below the thermostat housing on the cylinder head. Disconnect the hose from the thermostat housing stub, then release the three screws which secure the thermostat housing to the cylinder head.

3 Lift away the housing and remove the thermostat from it, taking care not to damage the rubber seal. If the valve is jammed open this will be obvious during the initial inspection, and further testing is unnecessary; fit a new thermostat and then refit the cover using a new gasket and refill the cooling system.

4 The thermostat can be tested by using an old saucepan or similar and a thermometer capable of reading at least 100°C (212°F). Place the thermostat and the thermometer in the saucepan and fill it with water. Heat up the water and watch the thermostat and the thermometer as the temperature rises. The accompanying chart shows

the point at which the thermostat should begin to open and at which it should be fully open. If the thermostat fails to operate, or if it is significantly outside the range shown in the chart, it should be renewed.
5 Refit the thermostat and its housing by reversing the removal sequence. Use a new rubber seal if the original is damaged and renew the housing gasket as a matter of course. Fill the cooling system, then run the engine to check that the thermostat operates normally and that there are no leaks.

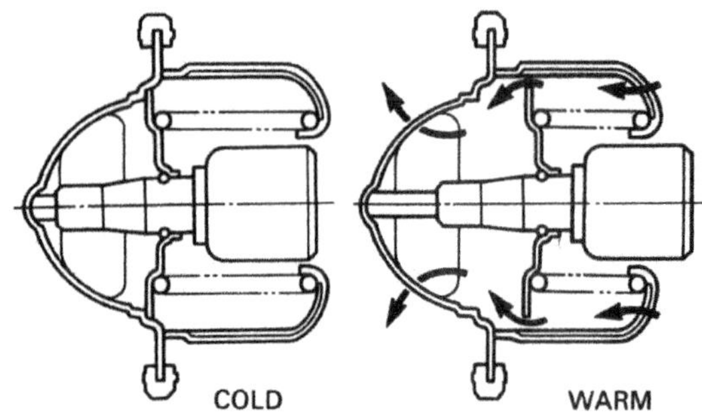

Fig. 2.5 Thermostat operation

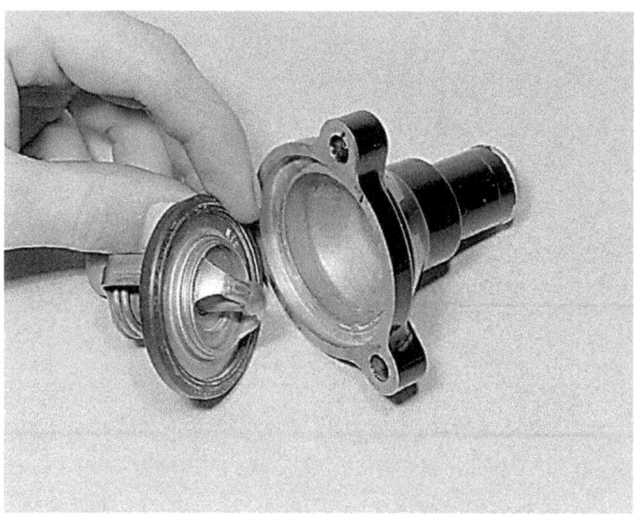

9.3 Check the condition of the rubber seal around the thermostat – renew if damaged

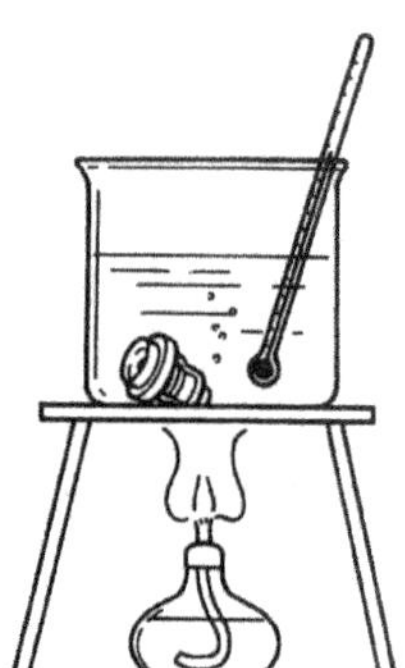

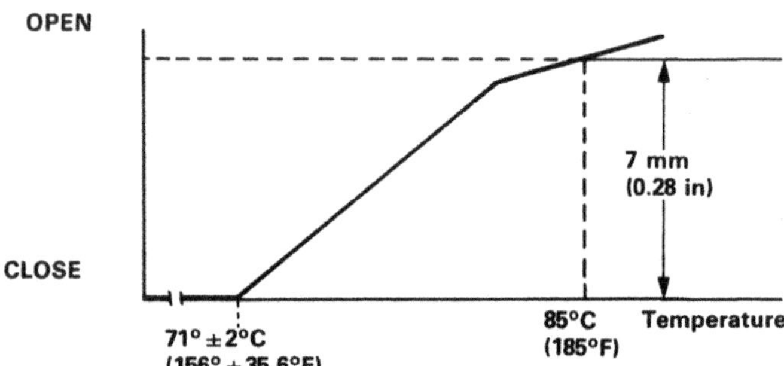

Fig. 2.6 Thermostat test

Chapter 3 Fuel system and lubrication

For information relating to the RD350 F II, N II and R models, refer to Chapter 8

Contents

Specifications

Fuel tank capacity	
Total including reserve	20 litres (4.4 Imp gal)
Reserve	2 litres (0.44 Imp gal)
Fuel grade	Unleaded or leaded (minimum octane rating 95 RON/RM)
Carburettors	
Make	Mikuni
Type	VM26SS
ID mark	31KO
Main jet	240
Air jet	0.7
Jet needle	5K1
Needle clip position	4th groove from top
Throttle valve cutaway	2.0
Pilot jet	22.5
Air screw (turns out)	1 1/4
Starter jet	80
Float height	21 ± 0.5 mm (0.83 ± 0.02 in)
Fuel level	26 ± 1.0 mm (1.02 ± 0.04 in)
Engine idle speed	1200 ± 50 rpm
Engine lubrication	
Type	Pump-fed total loss system (Yamaha Autolube)
Oil tank capacity	1.6 litre (2.8 Imp pint)
Oil grade	Air-cooled 2-stroke engine oil
Pump colour code	Yellow
Minimum stroke	0.10 ± 0.15 mm (0.004 ± 0.006 in)
Maximum stroke	2.05 – 2.27 mm (0.08 – 0.09 in)
Output per 200 strokes:	
Min	0.12 – 0.19 cc (0.004 – 0.007 Imp fl oz)
Max	2.58 – 2.85 cc (0.091 – 0.101 Imp fl oz)
Transmission lubrication	
Capacity:	
At oil change	1.5 litres (2.6 Imp pt)
After rebuild	1.7 litres (3.0 Imp pt)
Oil grade	SAE 10W30 type SE motor oil
Reed valve	
Bend limit	0.5 mm (0.02 in)
Valve lift	10.3 ± 0.2 mm (0.41 ± 0.008 in)

Note: *YPVS power valve is covered in Chapters 1 and 7*

1 General description

The fuel system comprises the fuel tank from which petrol is fed by gravity to the float bowls of the twin Mikuni carburettors via a vacuum-operated fuel tap. The tap has three positions; "ON" being the main feed, "RES" the reserve fuel supply and "PRI" being a priming position to allow the float chambers to be filled after running dry.

The carburettors are of conventional concentric design, the float chambers being integral with the lower part of the carburettor bodies. Cold starting is assisted by a separate starting circuit which supplies the correct fuel-rich mixture when the 'choke' control is operated. The cold start mechanism is fitted to the left-hand carburettor only.

Air entering the carburettors passes through a moulded plastic air cleaner casing, which contains an oil-impregnated foam air filter. This effectively removes any airborne dust, which would otherwise enter the engine and cause premature wear. The air cleaner also helps silence induction noise, a common problem inherent with two-stroke engines.

Engine lubrication is catered for by the Yamaha Autolube system. Oil from a separate tank is fed by an oil pump to small injection nozzles in the inlet tract. The pump is linked to the throttle twistgrip, and this controls the volume of oil fed to the engine.

The exhaust system is a two piece affair, each cylinder having its own expansion chamber with integral exhaust pipe. The system is finished in a heat-resistant matt black coating.

2 Fuel tank: removal and refitting

1 It is unlikely that the petrol tank will need to be removed except on very infrequent occasions, because it does not restrict access to the engine unless a top overhaul is to be carried out whilst the engine is in the frame.

2 Turn the fuel tap lever to the ON or RES position (vacuum operated tap) or the OFF position (gravity-fed tap). Remove both side-panels. Pull the fuel and vacuum pipes off the tap stubs.

3 The petrol tank is secured at the rear by a single bolt, washer and rubber buffer that threads into a strut welded across the two top frame tubes. It is necessary first to remove the dual seat before access is available.

4 When the bolt and washer are withdrawn, the petrol tank can be lifted from the frame. The nose of the tank is a push fit over two small rubber buffers, attached to a peg that projects from each side of the frame, immediately to the rear of the steering head. A small rubber 'mat' cushions the rear of the tank and prevents contact with the two top frame tubes.

5 The petrol tank has a locking filler cap to prevent pilferage of t he tank contents when the machine is left unattended.

3 Fuel tap: removal, dismantling and reassembly

1 The fuel tap is retained by two cross-head screws to the underside of the fuel tank. In the event of a leak or other problem, try to work out the most likely area of the fault before removing the tap; fuel seepage should be obvious during inspection. Before the tap is removed it will be necessary to drain the fuel into a clean metal container for temporary storage. Remove the two retaining screws and lift the tap away, taking care not to damage the filter which projects into the tank.

2 Release the single screw which retains the tap control knob and pull it off the tap spindle. Release the two screws which hold the tap position indicator disc and remove it. The tap rotor assembly is retained by a steel plate, this being held in place by two screws. Remove the screws and withdraw the spindle assembly and its wave washer. If the tap has been leaking, the most likely cause is a worn or damaged O-ring between the tap rotor and the body. This can be purchased separately from the tap assembly, as can the oval O-ring between the tap body and the fuel tank, but no other replacement parts are available.

3 On the opposite side of the tap body is the diaphragm valve assembly. This responds to engine vacuum, turning the fuel supply on only when the engine is running. If the vacuum pipe or the diaphragm become holed, the tap will not operate in any position other than the priming ("PRI") setting. If the fault is not attributable to the vacuum pipe it is normal to renew the tap complete. Note that none of the diaphragm valve parts can be purchased separately. The accompanying photographs show the valve arrangement for information purposes only.

4 The tap is provided with a sediment bowl, in which any fine debris from the tank which has managed to get through the filter gauze, will be trapped, along with any water. The bowl should be periodically removed for cleaning. When refitting the sediment bowl, ensure that the O-ring is in good condition.

5 Before reassembling the petrol tap, check that all the parts are clean, especially the tube which forms the filter and main and reserve intakes.

6 Do not overtighten any of the petrol tap components during reassembly. The castings are in a zinc-based alloy, which will fracture easily if over-stressed. Most leakages occur as the result of defective seals.

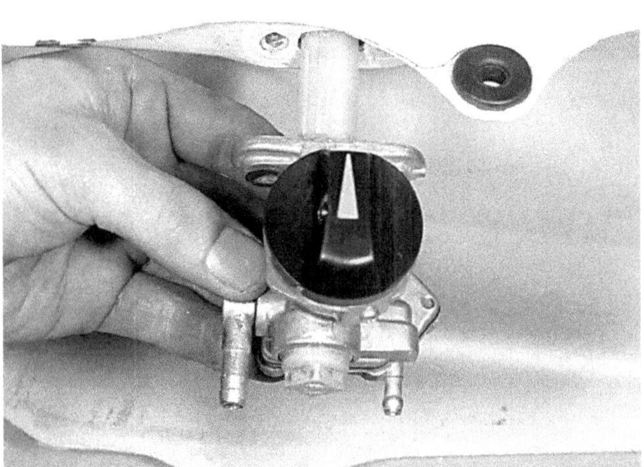

3.1a The fuel tap is retained to the underside of the tank by two bolts

3.1b Note the fuel strainer which projects into the tank from the tap flange

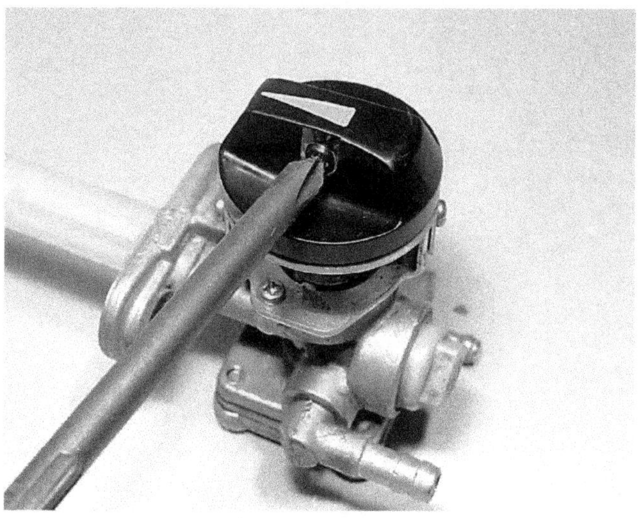

3.2a The tap control knob is secured by a single screw

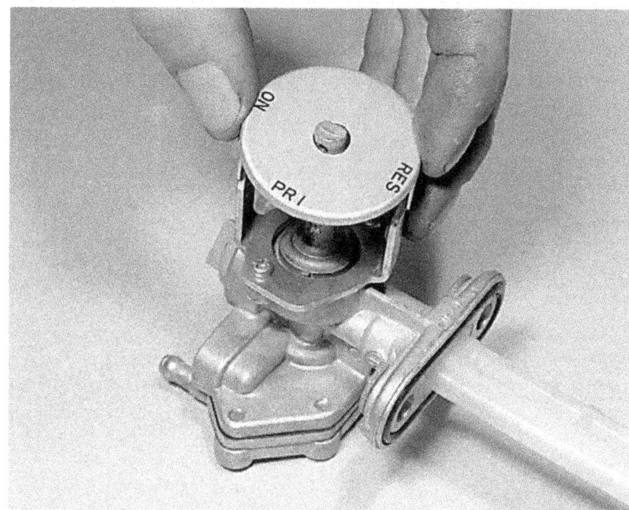

3.2b Remove the two screws and lift away the tap position disc. Note that the 'RES' position is nearest the fuel strainer

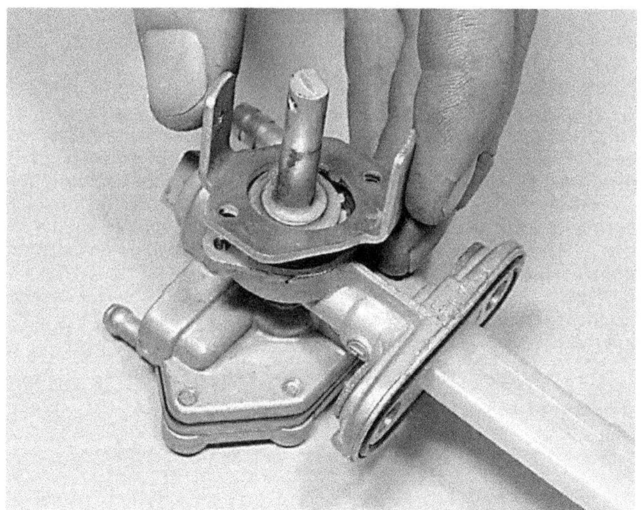

3.2c Tap rotor assembly is retained by two screws

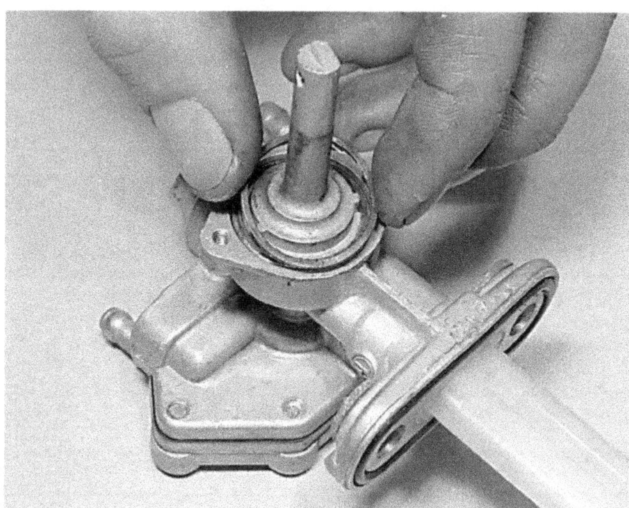

3.2d Lift away the wave washer, followed by the tap rotor

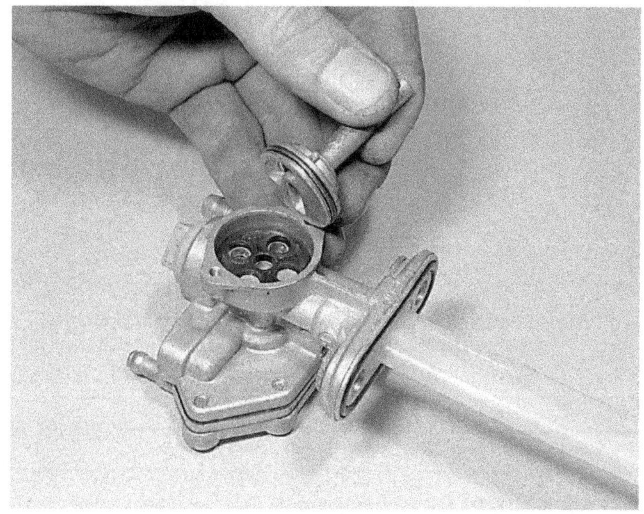

3.2e Renew the O-ring if there have been signs of leakage

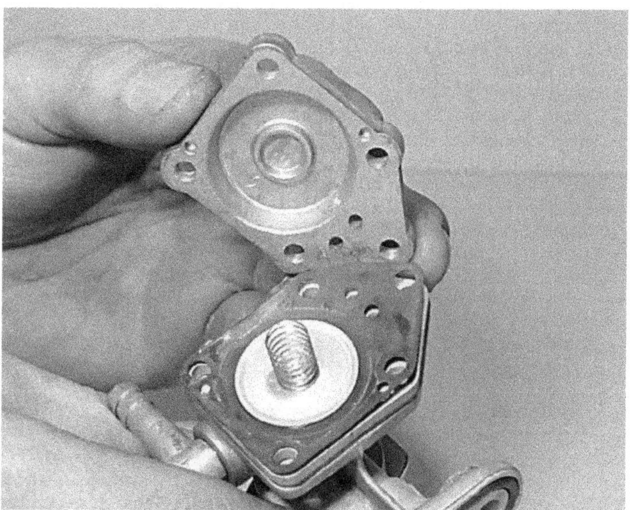

3.3a Cover conceals the diaphragm assembly and spring

3.3b Fuel valve is on inner face of the diaphragm assembly

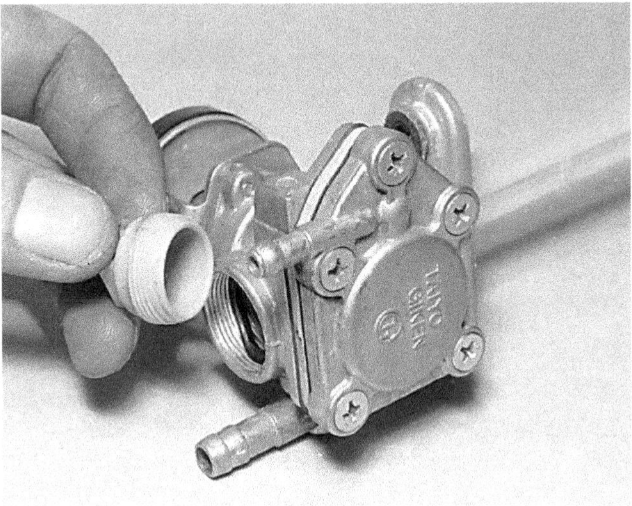

3.4 Sediment bowl is located on the underside of the tap body

4 Fuel and vacuum pipes: examination

1 The fuel and vacuum pipes are made of thin-walled synthetic rubber and are retained at each end by small wire clips. It is good practice to check that the pipes are sound on a regular basis. If split or if they have become hardened due to the effects of engine heat, they should be renewed. Most motorcycle dealers stock tubing of the correct size and material. On no account use natural rubber tubing or plastic tubing in place of the original type; natural rubber is attacked by petrol and will disintegrate, blocking the carburettor jets. Many types of plastic tubing will rapidly become brittle under the influence of fuel, and will soon split or leak.

5 Carburettors: removal and refitting

1 As a general rule, the carburettors should be left alone unless they are in obvious need of overhaul. Before a decision is made to remove and dismantle them, ensure that all other possible sources of trouble have been eliminated. This includes the more obvious candidates such as fouled spark plugs, a dirty air filter element or choked exhaust system baffles. If a fault has been traced back to the carburettors, proceed as follows.
2 Make sure that the fuel tap is turned to the "ON" or "RES" position, then prise off the petrol feed pipes at the carburettor stubs. The oil delivery pipes are removed in a similar manner, noting that the small tubular clips should be displaced first. The pipes can then be eased away from their stubs with the aid of an electrical screwdriver.
3 Slacken the screws of the clips which secure each carburettor to its inlet and airbox adaptors. Each carburettor can now be twisted free of the rubber adaptors and partially removed. This affords access to the threaded carburettor tops, which should be unscrewed to allow the throttle valve assemblies to be withdrawn. It is not normally necessary to remove these from the cables, and they can be left attached and taped clear of the engine. If removal is necessary, however, proceed as follows.
4 Holding the carburettor top, compress the throttle return spring against it and hold it in position against the cap. Once out of the way the cable can be pushed down and slid out of its locating groove. The various parts can now be removed and should be placed with the instrument to which they belong. Do not allow the parts to be interchanged between the two instruments.
5 The carburettors are refitted by reversing the removal sequence. Note that it is important that the instruments are mounted vertically to ensure that the fuel level in the float bowls is correct. A locating tab provides a good guide to alignment but it is worthwhile checking this for accuracy. Once refitted, check the carburettor adjustments and synchronisation as described later in this Chapter. Note too that the oil pump delivery pipes should be bled and the pump adjustments checked after overhaul.
6 **Note:** if the carburettors are to be set up from scratch it is important to check jet and float level settings prior to installation. To this end, refer to the next two Sections before the carburettors are refitted.

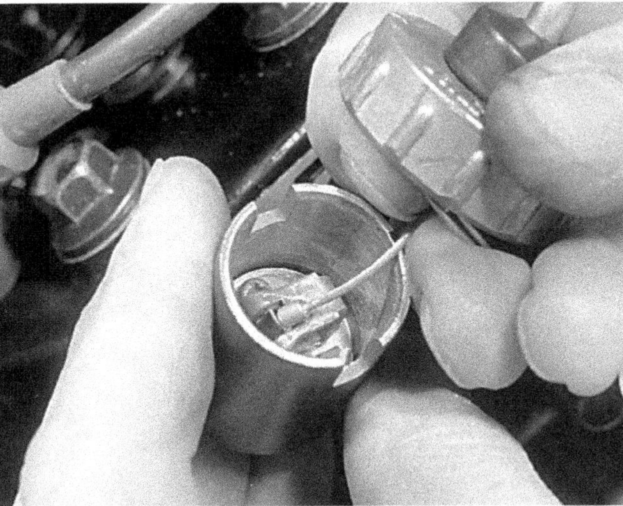

5.4 Compress the return spring, then disengage the inner cable

6 Carburettors: dismantling and reassembly

1 Invert each carburettor and remove the float chamber by withdrawing the four retaining screws. The float chamber bowls will lift away,

exposing the float assembly, hinge and float needle. There is a gasket between the float chamber bowl and the carburettor body which need not be disturbed unless it is leaking.

2 With a pair of thin nose pliers, withdraw the pin that acts as the hinge for the twin floats. This will free the floats and the float needle. Check that none of the floats have punctured and that the float needle and seating are both clean and in good condition. If the needle has a ridge, it should be renewed in conjunction with its seating.

3 The two floats are made of plastic, connected by a brass bridge and pivot piece. If either float is leaking, it will produce the wrong petrol level in the float chamber, leading to flooding and an over-rich mixture. The floats cannot be repaired successfully, and renewal will be required.

4 The main jet is located in the centre of the circular mixing chamber housing. It is threaded into the base of the needle jet and can be unscrewed from the bottom of the carburettor. The needle jet lifts out from the top of the carburettor, after the main jet has been unscrewed. The pilot jet is located in a smaller projection, next to the main jet.

5 The float needle seating is also found in the underside of the carburettor, towards the bell mouth intake. It is secured by a small retainer plate and is sealed by an O-ring. If the float needle and the seating are worn, they should be replaced as a set, never separately. Wear usually takes the form of a ridge or groove, which may cause the needle to seat imperfectly.

6 The carburettor valves, return springs and needle assemblies together with the mixing chamber tops, are attached to the throttle cable. The throttle cable divides into two at a junction box located within the two top frame tubes. There is also a third cable, which is used to link the oil pump with the throttle.

7 After an extended period of service the throttle valves will wear and may produce a clicking sound within each carburettor body. Wear will be evident from inspection, usually at the base of the slide and in the locating groove. Worn slides should be replaced as soon as possible because they will give rise to air leaks which will upset the carburation.

8 The needles are suspended from the valves, where they are retained by a circlip. The needle is normally suspended from the groove specified at the front of this Chapter, but other grooves are provided as a means of adjustment so that the mixture strength can be either increased or decreased by raising or lowering the needle. Care is necessary when replacing the carburettor tops because the needles are easily bent if they do not locate with needle jets.

9 The manually operated choke is unlikely to require attention during the normal service life of the machine. When the plunger is depressed, fuel is drawn throught a special starter jet in the left-hand carburettor by a partial vacuum that is created in the crankcase. Air from the float chamber passes through holes in the starter emulsion tube to aerate the fuel. The fuel then mixes with air drawn in via the starter air inlet to the plunger chamber. The resultant mixture, richened for a cold start, is drawn into the engine through the starter outlet, behind the throttle valve.

10 Before the carburettors are reassembled, using the reversed dismantling procedure, each should be cleaned out thoroughly, preferably by the use of compressed air. Avoid using a rag because there is always risk of fine particles of lint obstructing the internal air passages or the jet orifices.

11 Never use a piece of wire or sharp metal object to clear a blocked jet. It is only too easy to enlarge the jet under these circumstances and increase the rate of petrol consumption. Always use compressed air to clear a blockage; a tyre pump makes an admirable substitute when a compressed air line is not available.

12 Do not use excessive force when reassembling the carburettors because it is quite easy to shear the small jets or some of the smaller screws. Before attaching the air cleaner hoses, check that both throttle slides rise when the throttle is opened.

6.1 Remove the four screws and lift away the float bowl

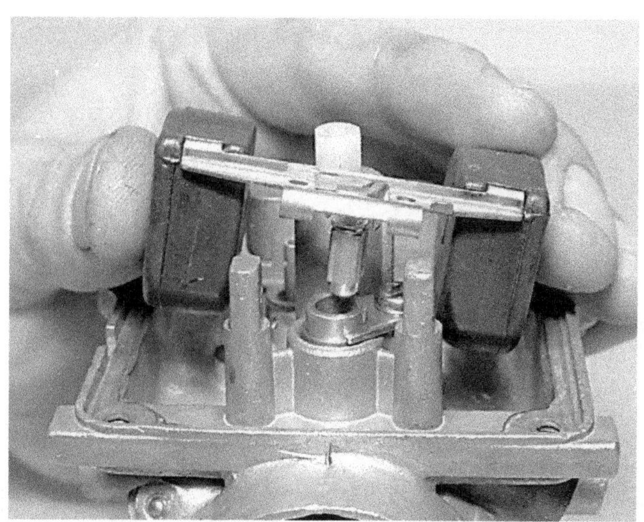

6.2a Remove the pivot pin, then lift away the float assembly

6.2b Release the single screw and retainer ...

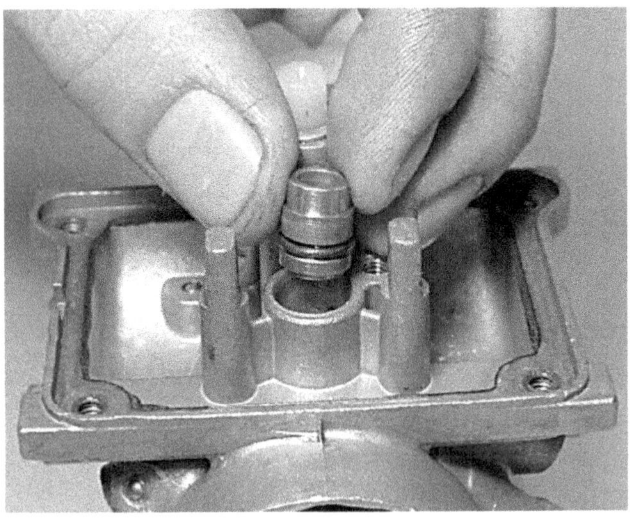

6.2c ... to free the float valve seat. Renew O-ring if worn

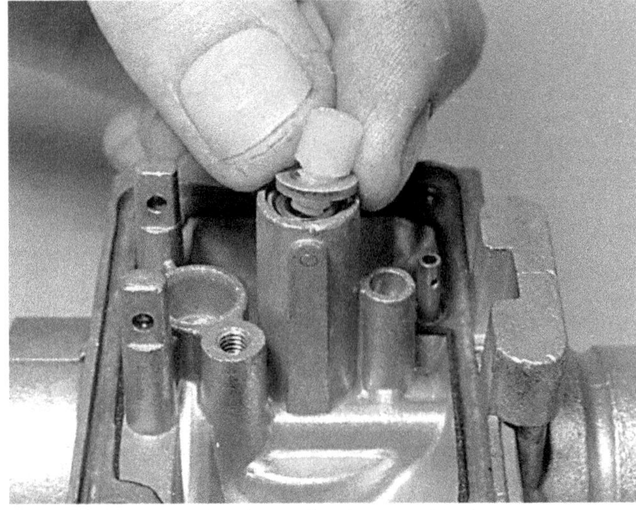

6.4a Main jet screws into base of carburettor. Note plain washer

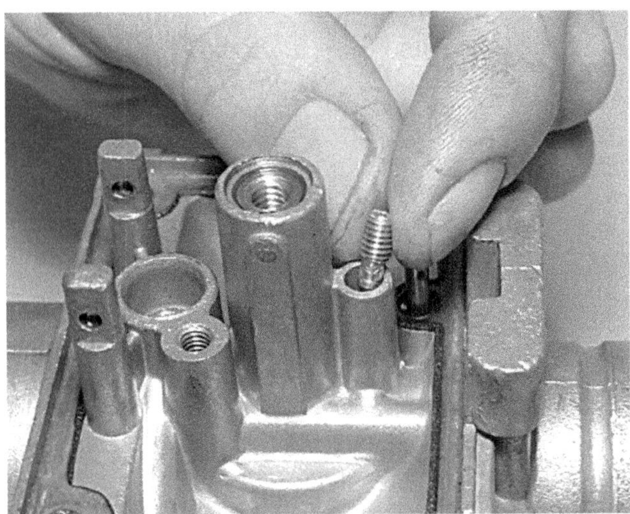

6.4b Pilot jet is fitted into adjacent bore

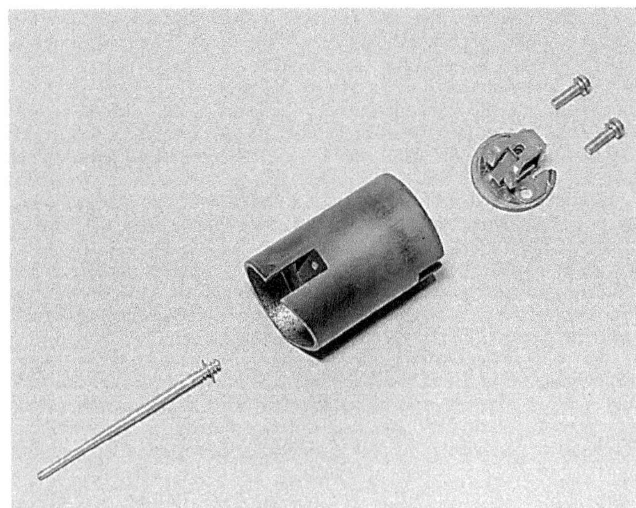

6.8 Remove retainer from inside the throttle valve to free the needle

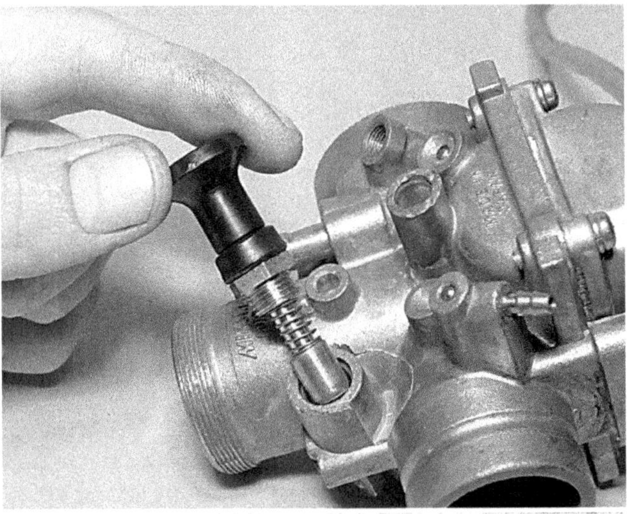

6.9 Choke assembly screws into the carburettor as shown

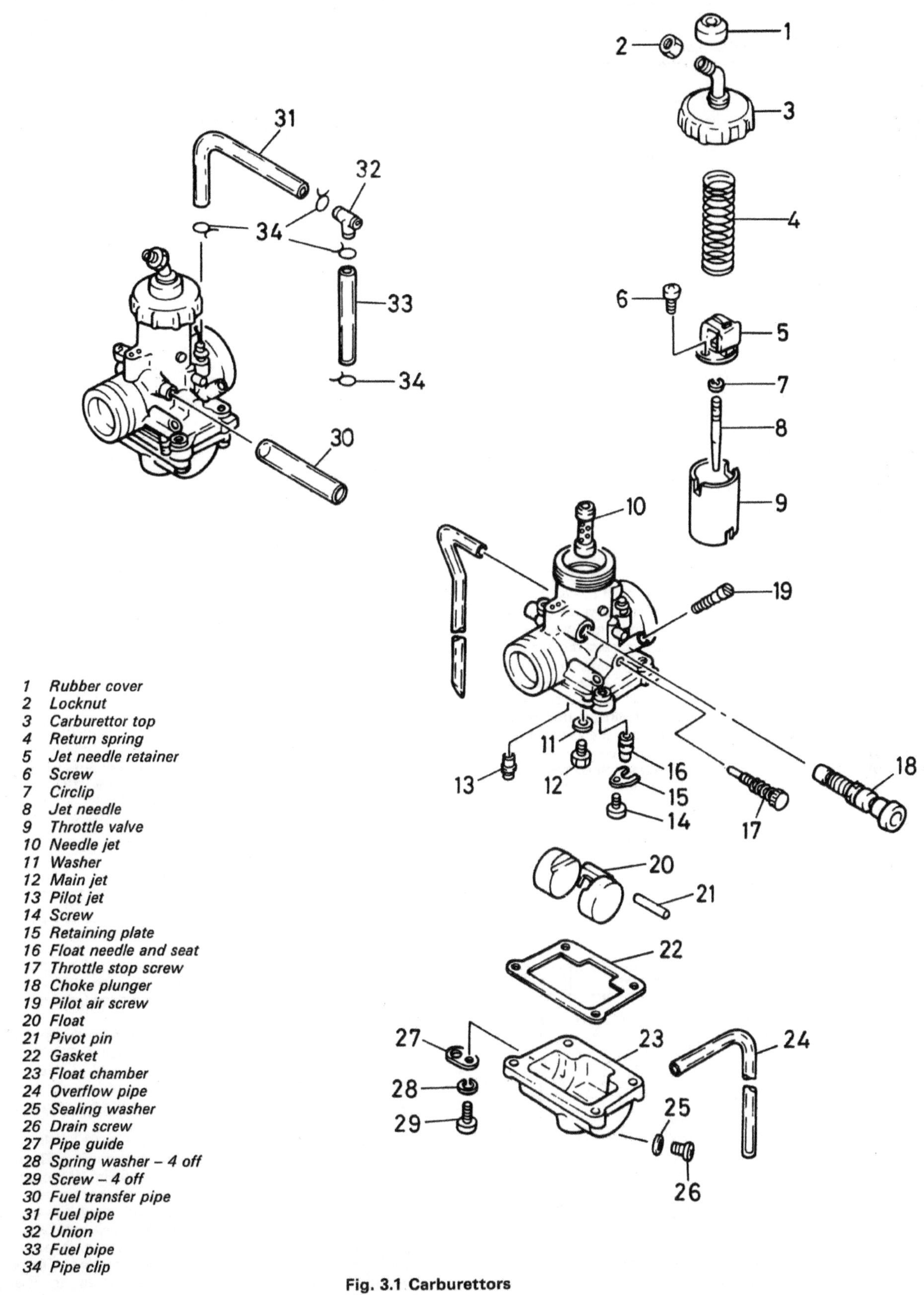

Fig. 3.1 Carburettors

7 Carburettors: adjustment

1 The first step in carburettor adjustment is to ensure that the jet sizes, needle position and float height are correct, which will require the removal and dismantling of the carburettors as described in Section 6.
2 Before any dismantling or adjustment is undertaken eliminate all other possible causes of running problems, checking in particular the spark plugs, ignition timing, air cleaner and the exhaust baffles. Checking and cleaning these items as appropriate will often resolve a mysterious flat spot or misfire.
3 If the carburettors have been removed for the purpose of checking jet sizes, the float level should be measured at the same time. It is unlikely that once this is set up correctly, there will be a significant amount of variation, unless the float needle or seat have worn. These should be checked and renewed as required. With the float bowl removed slowly rotate the carburettor until gravity acting on the floats moves the float until the valve is **just** closed, but not so far that the needle's spring-loaded pin is compressed. Measure the distance between the gasket face and the bottom of the float with an accurate ruler. The correct setting should be 21.0 ± 0.5 mm (0.827 ± 0.020 in). If adjustment is required it can be made by bending by a very small amount, the small tang to which the float needle is attached.
4 When the carburettors are being refitted, set the throttle stop screws as follows. Fit the throttle valve assemblies to their respective instruments and secure the carburettor tops before the bodies are fitted into their adaptors. Identify the throttle stop screws, which will be found projecting at right angles to the main body. These should be slackened off completely to allow the throttle valves to close fully. Check the throttle cable free play, measured at the throttle twistgrip flange and where necessary set this to 3 – 7 mm (0.118 – 0.276 in); cable free play should be set using the adjuster and locknut arrangement located immediately below the twistgrip housing.
5 The throttle stop screws should be screwed slowly inward until they **just** contact the underside of the valves. If this is done carefully it should be possible to feel and see the point at which this occurs. It is worth spending a little time to ensure that the two screws are set accurately. An alternative method is to use a metal rod, the plain end of a drill bit being ideal. Set the throttle stop screws so that the rod is a light sliding fit beneath each throttle valve cut away. With either of the above methods, the object is to ensure that the two screws are set in similar positions. Once set it is important to ensure that in subsequent adjustments each screw is moved by exactly the same amount as the other. The two instruments can now be refitted to the machine.
6 Next, check that the two throttles are synchronised. Unless this is established it will prove impossible to persuade the engine to run evenly, and poor synchronisation will make mixture and throttle stop settings futile. Standing on the right-hand side of the machine open the throttle twist-grip fully and observe the two small windows in the side of each instrument body. A small alignment pip should be visible in each one, and it is important to arrange these so thay they are in accurate synchronisation. Each valve can be adjusted via the independent adjuster in each mixing chamber top.
7 The remaining adjustments are made with the engine running and at normal operating temperature. To this end it may prove necessary to make some provisional idle speed adjustment, remembering to adjust each throttle stop screw by an equal amount to keep the carburettors in balance. Set the pilot air screws to the position shown in the specifications for the appropriate model. Note the pilot air screw on the right-hand carburettor is awkwardly located on its inner face and some degree of dexterity will be called for when adjustment is required. Taking each cylinder in turn, rotate the pilot air screw inwards and outwards from the datum setting until the position is found at which the engine runs fastest. Reduce the idle speed if necessary, and then adjust the second pilot air screw in a similar manner.
8 Check the engine idle speed and adjust both throttle screws equally to bring it to the specified speed. The carburettors should by now be fairly accurately set up and in many instances no further adjustment will be necessary. If, however, there appears to be room for improvement at idle speed, fine adjustment of the throttle stop screws should bring things into balance.

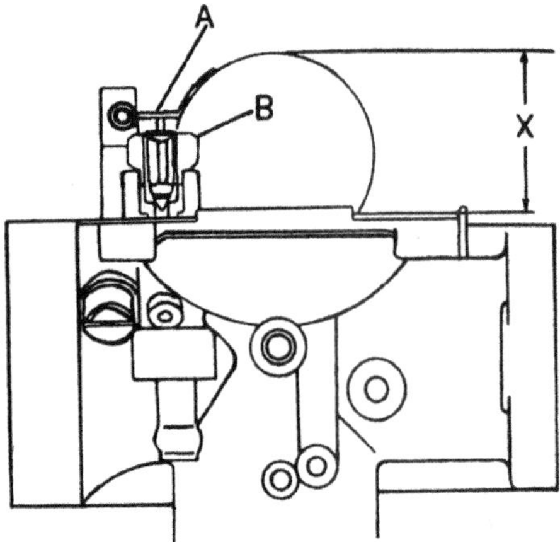

Fig. 3.2 Float height measurement

A Tang
B Float needle housing
X Float height

8 Carburettors: settings

1 Some of the carburettor settings, such as the sizes of the needle jets, main jets and needle position are predetermined by the manufacturer. Under normal circumstances it is unlikely that these settings will require modification, even though there is provision made. If a change appears necessary, if can often be traced to a developing engine fault.
2 As a rough guide, the slow running screw controls the engine speed up to 1/8 throttle. The throttle valve cutaway controls the engine speed from 1/8 to 1/4 throttle and the position of the needle from 1/4 to 3/4 throttle. The main jet is responsible for the engine speed at the final 3/4 to full throttle. It should be added that none of these demarkation lines is clearly defined; there is a certain amount of overlap between the carburettor components involved.
3 Always err on the side of a rich mixture because a weak mixture has a particularly adverse effect on the running of any two-stroke engine. A weak mixture will cause rapid overheating which may eventually promote engine seizure. Reference to Chapter 3 will show how the condition of the sparking plugs can be used as a reliable guide to carburettor mixture strength.

9 Reed valves: examination and renovation

1 The reed valve unit fitted in each inlet port requires little attention during normal use. If there are indications of a fault, such as poor starting or performance, or uneven running, the valves should be removed and checked, otherwise do not disturb them between engine overhauls.
2 The reed valve unit can be unbolted from the inlet port after the carburettors have been removed (see Section 5). Check the rubber sealing face for signs of tears or splits; renew the valve unit if damaged. Examine the reed petals for indications of fatigue, such as cracking near the root end. It is not advisable to remove the petals for inspection purposes.
3 Check the clearance between the reed valve petals and the stopper plate as shown in the accompanying line drawing. The recommended clearance is 10.3 ± 0.2 mm (0.41 ± 0.008 in). If the clearance is only slightly out of specification, carefully bend the stopper plate to correct it. If the error exceeds 0.3 mm (0.012 in) renew the stopper plate.
4 Note the clearance between the reed valve petal and the case. The petal bend limit is 0.5 mm (0.02 in) maximum. In practice the valves should be closed, or almost closed. Valve petals and stopper plates are available as replacement parts from Yamaha dealers. If either is to be renewed, remove the two screws holding the assembly to the case and

lift the stopper and petal away. Always deal with one valve at a time to avoid interchanging parts.

5 Before fitting the new parts, clean and degrease the valve case, reed and stopper plate. Assemble the valve components, noting that the cutaway on the lower corner of both reed and stopper should align. Apply Loctite to the securing screws, tightening them evenly to prevent warpage. If possible, tighten the screws to 0.1 kgf m (0.7 lbf ft).

6 When refitting the assembly to the inlet port, use a new gasket. Tighten the retaining bolts in a diagonal sequence to 1.5 kgf m (11 lbf ft).

10 Exhaust system: cleaning

1 The exhaust system is often the most neglected part of any two-stroke despite the fact that it has a pronounced effect on performance. It is essential that the exhaust system is inspected and cleaned out at regular intervals because the exhaust gases from a two-stroke engine have a particularly oily nature which will encourage the build-up of sludge. This will cause back pressures and restrict the engine's ability to 'breathe'.

2 Cleaning is made easy by fitting the silencers with detachable baffles, held in position by a set screw that passes through each silencer end. If the screw is withdrawn, the baffles can be drawn out of position for cleaning.

3 A wash with a petrol/paraffin mix will remove most of the oil and carbon deposits, but it the build-up is severe it is permissible to heat the baffles with a blow lamp and burn off the carbon and old oil.

4 At less frequent intervals, such as when the engine requires decarbonising, it is advisable also to clean out the exhaust pipes. This will prevent the gradual build-up of an internal coating of carbon and oil, over an extended period.

5 Do not run the machine with the baffles detached or with a quite different type of silencer fitted. The standard production silencers have been designed to give the best possible performance whilst subduing the exhaust note. Although a modified exhaust system may give the illusion of greater speed as a result of the changed exhaust note, the chances are that performance will have suffered accordingly.

6 When replacing the exhaust system, use new sealing rings at the exhaust port joints and check that the baffle retaining screws are tightened fully in the silencer ends.

9.2a Remove the carburettor mounting rubber ...

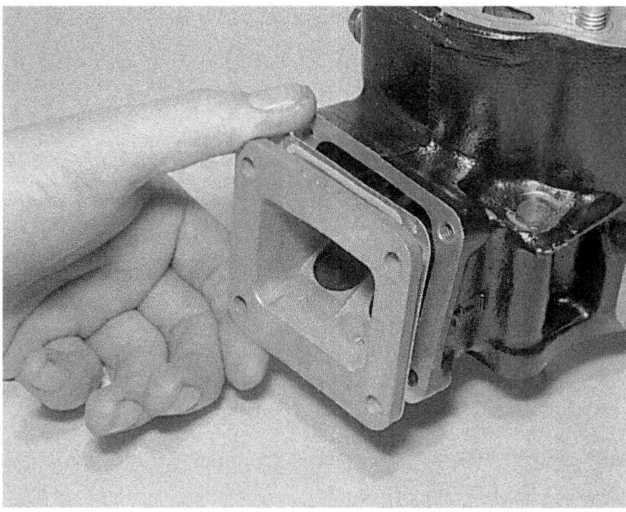

9.2b ... and lift the reed valve clear of the cylinder barrel

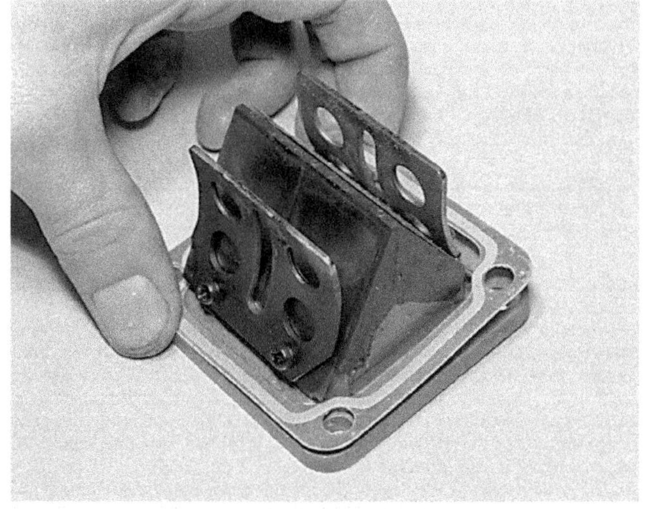

9.3 Check the reed petal and stopper clearances (see text)

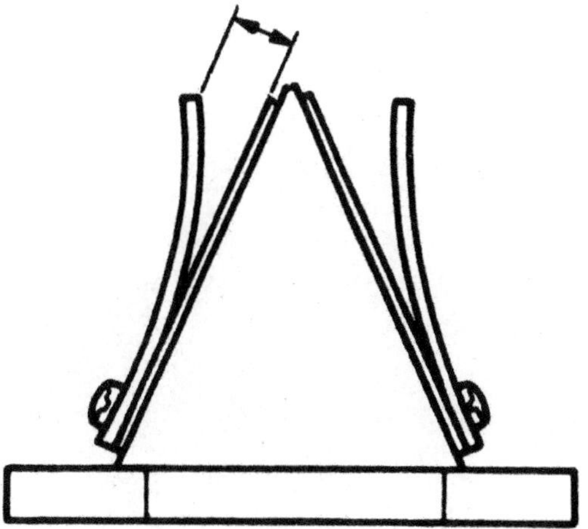

Fig. 3.3 Reed valve stopper plate adjustment

10.2a Exhaust baffles are retained by a screw on underside of silencer

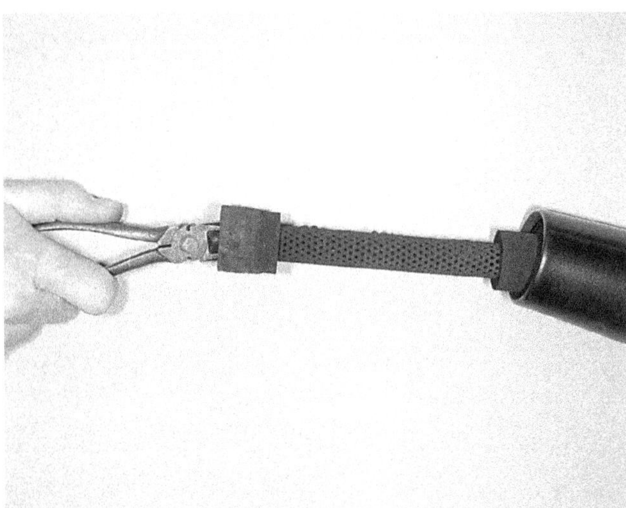

10.2b Grasp baffle with pliers and twist to aid removal

10.6 Use new exhaust port seals to prevent gas leakage

11 Air cleaner: removal and cleaning

1 The air cleaner casing is mounted on the frame beneath the fuel tank. It is connected to a second moulded plastic chamber immediately behind the carburettors. This functions as an intake silencer and conveys the cleaned air to the carburettor intakes via short rubber adaptors.

2 Access to the air cleaner element requires the removal of the fuel tank and seat. The seat can be lifted clear once its retaining latches have been released. Turn the fuel tap to the "ON" or "RES" position and disconnect the fuel feed pipes. Remove the single fixing bolt at the rear of the tank and lift the tank upward and rearwards to free it from its front mounting rubbers.

3 The air cleaner cover is retained by three screws. Remove the screws and lift the cover clear to expose the flat foam element. This can be removed in turn by pulling it out of the casing. Wash the element in clean petrol to remove the old oil and any dust which has been trapped by it. When it is clean, wrap it in some clean rag and gently squeeze out the remaining petrol. The element should now be left for a while to allow any residual petrol to evaporate. Soak the cleaned element in engine oil and then squeeze out any excess to leave the foam damp but not dripping. Refit the element, ensuring that the cover seal has been refitted.

4 Note that a damaged element must be renewed immediately. Apart from the risk of damage from ingested dust, the holed filter will allow a much weaker mixture and may lead to overheating or seizure. It follows that the machine must never be used without the filter in position.

5 The rest of the air cleaner system requires little attention, other than checking that the connecting rubbers are undamaged. When checking these do not omit the adaptors which connect the carburettors to the inlet ports. These are prone to perishing and cracking around the balance pipe stubs and should be renewed if leakage is suspected.

12 The engine lubrication system

1 In line with current two-stroke practice the Yamaha YPVS models utilise a pump-fed engine lubrication system and do not require the mixture of a measured quantity of oil to the petrol content of the fuel tank in order to utilise the so-called 'petroil' method. Oil of the correct viscosity is contained in a separate oil tank mounted on the left-hand side of the machine and is fed to a mechanical oil pump on the right-hand side of the engine which is driven from the crankshaft by a reduction gear. The pump delivers oil at a predetermined rate, via two flexible plastic tubes, unions on the inlet side of the carburettor venturis. In consequence, the oil is carried into the engine by the incoming charge of petrol vapour, when the inlet port opens.

2 The oil pump is also interconnected to the throttle twistgrip so that when the throttle is opened, the oil pump setting is increased a similar amount. This technique ensures that the lubrication requirements of the engine are always directly related to the degree of throttle opening. This facility is arranged by means of a control cable attached to a lever on the end of the pump; the cable is joined to the throttle cable junction box at the point where the cable splits into two for the operation of each carburettor.

13 Removing and refitting the oil pump

1 It is rarely necessary to remove the oil pump unless specific attention to it is required. It should be noted that the pump should be considered a sealed unit – parts are not available and thus it is not practicable to repair it. The pump itself can be removed quite easily leaving the drive shaft and pinion in place in the right-hand outer casing. If these latter components require attention it will be necessary to drain the cooling system and transmission oil so that the right-hand outer casing can be removed. Refer to Chapter 1 for further details. The accompanying photographic sequence describes the procedure for removing the pump drive components.

2 To gain access to the oil pump, remove the screws which secure the pump cover to the right-hand outer casing. With these removed the pump will be clearly visible at the bottom of the pump recess. Do not

disturb the water pump end cover which is immediately above the oil pump.

3 Displace the small spring steel clips which secure oil delivery pipes to the pump outlets, then ease the pipes off the outlet stubs using a small screwdriver. The large feed pipe from the oil tank is removed in a similar fashion, but before removing it have some sort of plug handy to push into the end of the pipe. This will prevent the oil from the tank being lost. Pull on the pump cable inner to rotate the pump pulley. Holding the pulley in its fully open position release the cable and disengage it from the pulley recess.

4 The pump is secured to the cover by two screws which pass throught its mounting flange. Once these have been removed the pump can be removed, noting that it may prove necessary to turn the pump slightly to free it from its drive shaft.

5 Further dismantling is not practicable, and it will be necessary to renew the pump if it is obviously damaged. Maintenance must be confined to keeping the pump clear of air, and correctly adjusted, as described in the following sections.

6 Refit the oil pump to the crankcase cover, using a new gasket at the oil pump/crankcase cover joint. Replace and tighten the two crosshead mounting screws. The remainder of the reassembly is accomplished by reversing the dismantling procedure, but do not replace the pump cover because the oil pump must be bled to ensure the oil lines are completely free from air bubbles. See the following Section.

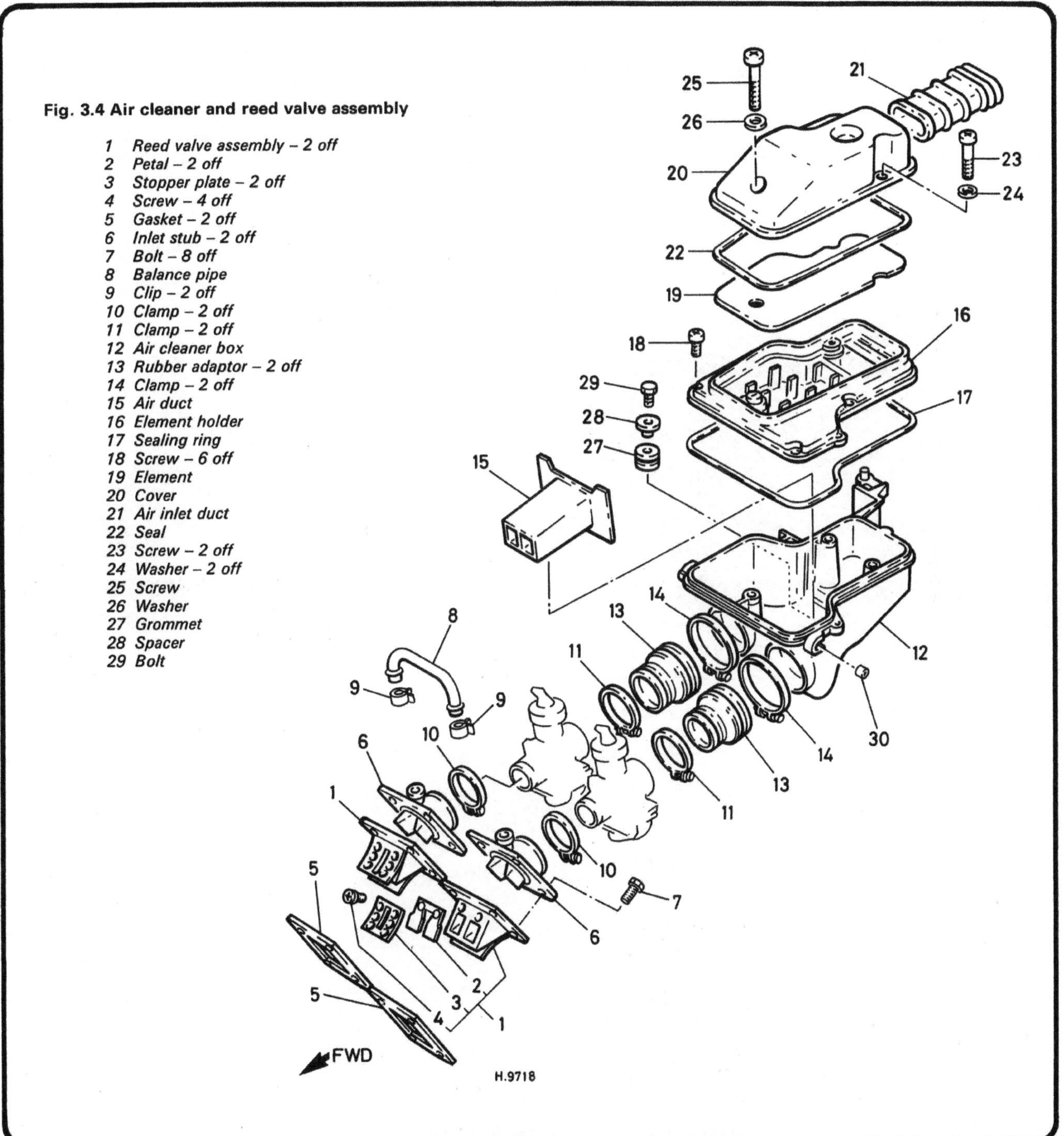

Fig. 3.4 Air cleaner and reed valve assembly

1 Reed valve assembly – 2 off
2 Petal – 2 off
3 Stopper plate – 2 off
4 Screw – 4 off
5 Gasket – 2 off
6 Inlet stub – 2 off
7 Bolt – 8 off
8 Balance pipe
9 Clip – 2 off
10 Clamp – 2 off
11 Clamp – 2 off
12 Air cleaner box
13 Rubber adaptor – 2 off
14 Clamp – 2 off
15 Air duct
16 Element holder
17 Sealing ring
18 Screw – 6 off
19 Element
20 Cover
21 Air inlet duct
22 Seal
23 Screw – 2 off
24 Washer – 2 off
25 Screw
26 Washer
27 Grommet
28 Spacer
29 Bolt

13.4a Oil pump is retained to cover by two screws

13.4b Pump pinion can be removed after circlip has been prised off

13.4c Pinion is located on shaft by this driving pin

13.4d Pump assembly can be removed together with shaft if required

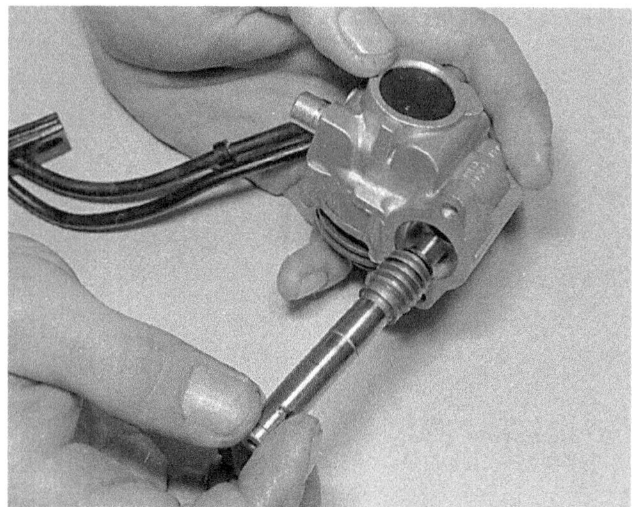

13.4e Pump shaft can be withdrawn for examination

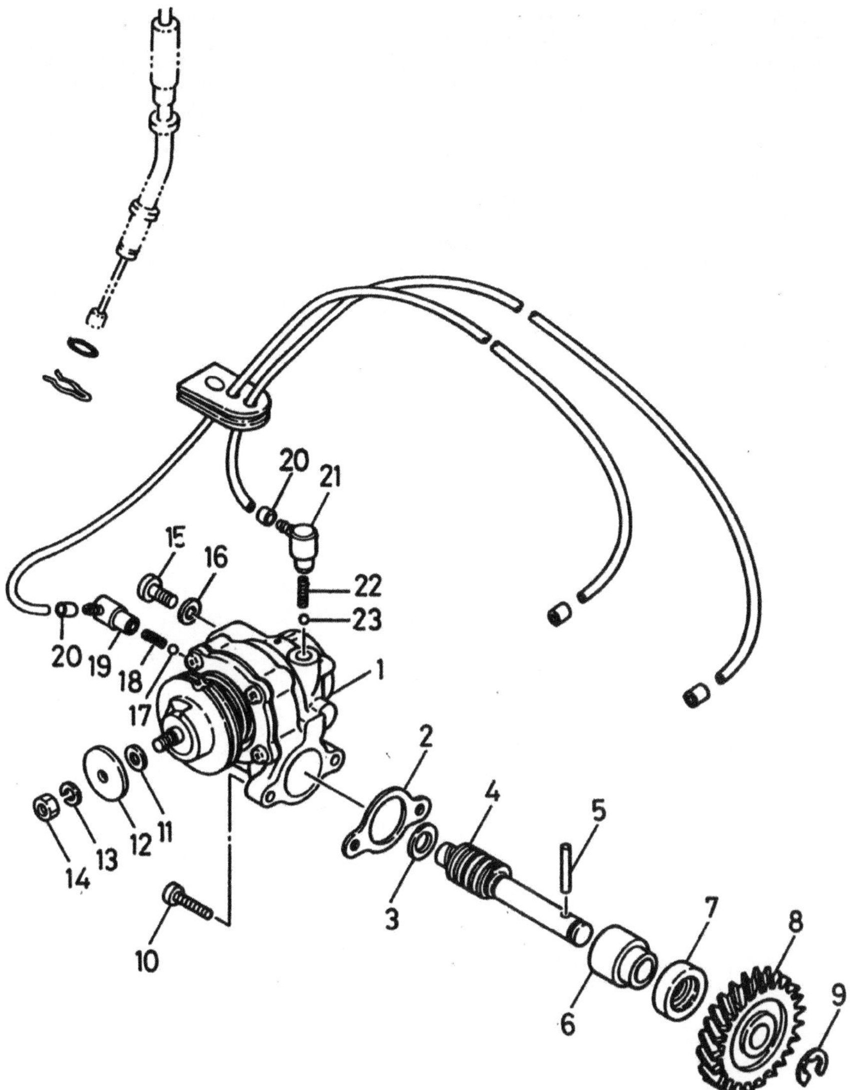

Fig. 3.5 Oil pump

1 *Oil pump*
2 *Gasket*
3 *Washer*
4 *Driveshaft*
5 *Drive pin*
6 *Collar*
7 *Oil seal*
8 *Drive pinion*
9 *Circlip*
10 *Screw – 2 off*
11 *Shim*
12 *Plate*
13 *Spring washer*
14 *Nut*
15 *Screw*
16 *Sealing washer*
17 *Steel ball*
18 *Spring*
19 *Pipe union*
20 *Spring clip*
21 *Pipe union*
22 *Spring*
23 *Steel ball*

14 Bleeding the oil pump

1 It is necessary to bleed the oil pump every time the main feed pipe from the oil tank is removed and replaced. This is because air will be trapped in the oil line, no matter what care is taken when the pipe is removed.
2 Check that the oil pipe is connected correctly, with the retaining clip in position. Then remove the cross-head screw in the outer face of the pump body with the fibre washer beneath the head. This is the oil bleed screw.
3 Check that the oil tank is topped up to the correct level, then place a container below the oil bleed hole to collect the oil that is expelled as the pump is bled. Allow the oil to trickle out of the bleed hole, checking for air bubbles. The bubbles should eventually disappear as the air is displaced by fresh oil. When clear of air, refit the bleed screw. DO NOT replace the front portion of the crankcase cover until the pump setting has been checked, as described in the next Section.
4 Note also that it will be necessary to ensure that the oil delivery pipes are primed if these have been disturbed. Unless this is checked the engine will be starved of oil until the pipes fill. The procedure required to avoid this is to start the engine and allow it to idle for a few minutes whilst holding the pump pulley in its fully open position by pulling the pump cable. The excess oil will make the exhausts smoke heavily for a while, indicating that the pump is delivering oil to the engine.

15 Checking the oil pump adjustment

1 As has been mentioned, the oil pump is regulated by the throttle twistgrip to ensure that the correct amount of oil is supplied to the engine at various engine speeds. It is important to ensure that the pump pulley is correctly synchronised with the carburettor throttle valves, otherwise the oil delivery rate will be incorrect. It follows that pump adjustment must be checked whenever the carburettors have been adjusted, and also that carburettor adjustment must precede pump adjustment. Having checked and adjusted the carburettors, remove the pump cover and proceed as follows.

RD350 LC II

2 Fully open the throttle twistgrip and hold this position; a pair of self-locking pliers clamped lightly around the twistgrip rubber can be useful if carrying out the check unaided. Check that the rectangular

mark on the edge of the pump pulley aligns with the projecting plunger pin. If this is not the case, slacken the cable adjuster locknut and turn the adjuster to obtain the required setting. Open and close the throttle a few times, then recheck the setting. Repeat the adjustment procedure as necessary, then tighten the pump cable adjuster locknut.

RD350 F and N

3 Unlike the LC II model, the pump setting is checked with the throttle **closed**. Look at the pump pulley and identify the three marks near the 9 o'clock position. On either side of a round dot is an engraved line, the shorter and lower of the two being the pump alignment mark. Check that the mark aligns with the pin and if necessary adjust the length of the cable to compensate for any error. Open and close the throttle a few times and check the setting again before securing the adjuster locknut.

Minimum stroke adjustment – all models

4 The minimum stroke setting should not alter readily, but should be checked if there is some question regarding the rate of oil delivery to the engine. Start the engine and allow it to idle. Observe the front of the pump where it will be noticed that the pump adjustment plate moves in and out. Wait until the plate is fully out and stop the engine. Using feeler gauges, measure the gap between the plate and the raised boss of the pulley, making a note of the reading. Start the engine and repeat the check several times, taking the largest gap as the minimum stroke position.

5 The gap must be 0.10 – 0.15 mm (0.004 – 0.006 in). If outside this range, remove the adjuster plate locknut and lift away the plate and shim(s). Add or subtract shims to bring the setting within the prescribed tolerance. Shims can be obtained from Yamaha dealers in 0.3 mm, 0.5 mm and 1.0 mm (0.012 in, 0.020 in and 0.040 in) sizes.

Chapter 4 Ignition system

For information relating to the RD350 F II, N II and R models, refer to Chapter 8

Contents

Specifications

Ignition system	
Make	Nippon Denso
Type	Capacitor discharge ignition (CDI)
Pickup coil resistance (RD350 LC II model) white/red to white/green	115 ohms ± 20%
Pickup coil resistance (other models) white/red to white/green	117 ohms ± 20%
Source coil resistance (RD350 LC II model):	
Brown to green	225 ohms ± 20%
Brown to red	5.3 ohms ± 20%
Source coil resistance (other models):	
Brown to green	113 ohms ± 20%
Brown to red	4.1 ohms ± 20%
Ignition timing:	
@ 1200 rpm	17° BTDC
@ 3500 rpm	27° BTDC
Ignition coil	
Make	Nippon Denso
Type	12900-027
Minimum spark gap	6 mm (0.24 in)
Primary winding resistance	0.33 ohms ± 20% @ 20°C (68°F)
Secondary winding resistance	3.5 K ohms ± 20% @ 20°C (68°F)
CDI unit	
Make	Nippon Denso
Type:	
RD350 LC II	AQAB06
Other models	52Y
Spark plugs	
Make	NGK
Type	BR8ES
Electrode gap	0.7 – 0.8 mm (0.028 – 0.031 in)

1 General description

The Yamaha YPVS models covered by this manual are equipped with a capacitor discharge ignition (CDI) system. The arrangement is entirely electronic, and having no moving parts and can be considered maintenance-free. Unlike the earlier LC models the ignition system is not adjustable, obviating the need for timing checks whenever the alternator stator has been disturbed.

The ignition exciter, or source, coil assembly is housed within the flywheel generator, and provides the power supply for the external CDI unit, which is mounted beneath the right-hand side panel. The system is triggered by a magnet, incorporated in the outer face of the generator rotor, acting upon a pickup coil, known as a pulser, which is outrigged from the stator. The spark plugs are supplied from a single ignition coil, firing in both cylinders simultaneously. This system is known as the 'spare spark' system, as only one cylinder can fire, leaving one spark wasted, or 'spare'.

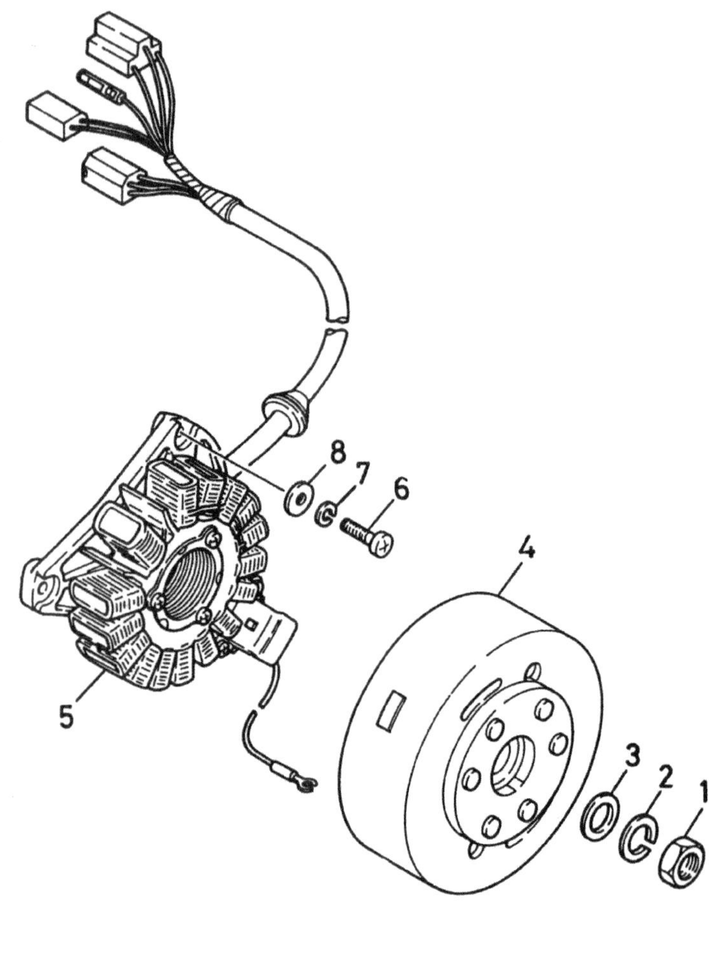

Fig. 4.1 Alternator

1 Nut
2 Spring washer
3 Washer
4 Rotor
5 Stator
6 Screw – 3 off
7 Spring washer – 3 off
8 Washer – 3 off

2 CDI system: operating principles

1 Energy for the ignition system is drawn from the exciter coil. This is mounted on the generator stator, and is integral with the normal alternator windings. It is a two-stage arrangement, having low speed and high speed windings. The low speed windings produce a high output voltage at low engine speeds, this voltage dropping off as the engine builds up speed. The high speed windings, on the other hand, produce little energy at low engine speeds, but the output voltage rises along with engine speed. The two outputs are combined, offsetting each other to give a fairly constant output voltage, this being the sum of the output of each set of windings.
2 The exciter coil assembly feeds the CDI unit, a sealed electronic assembly which forms the heart of the system. This unit contains, amongst other things, a capacitor and a thyristor, or silicon controlled rectifier (SCR). The capacitor is charged with the high voltage output from the exciter coil assembly. The thyristor, or SCR, is in effect an electronic switch. When signally electrically by the pulser, it allows the capacitor to discharge through the primary windings of the ignition coil. This in turn induces a high tension pulse in the secondary windings, which is fed to the spark plugs.
3 The pulser, or pickup, comprises a small coil mounted outside the alternator rotor on a projection from the stator. A permanent magnet embedded in the flywheel rotor is arranged to pass beneath the pulser coil. As the magnet passes the pulser coil, a weak current is generated and it is this that is used to trigger the thyristor in the CDI unit.

3 CDI system: testing and fault diagnosis

1 As stated earlier in this Chapter, the electronic ignition needs no regular maintenance once it has been set up and timed accurately. Occasional attention should, however, be directed at the various connections in the system, and these must be kept clean and secure. A failure in the ignition system is comparatively rare, and usually results in a complete loss of ignition. Usually, this will be traced to the CDI unit, and little can be done at the roadside to effect a repair. In the event of the CDI unit failing, it must be renewed as repair is not practicable.
2 If the CDI unit is thought to be at fault, it is recommended that it be

removed and taken to a Yamaha dealer for testing. The dealer will have the use of diagnostic equipment, and will be able to test the unit accurately and quickly. Testing at home is less practical and cannot be guaranteed to be accurate. At best, it will enable the owner to establish which part of the system is at fault, although replacement of the defective part remains the only effective cure.

3 Care must be exercised when dealing with the CDI unit. Wrong connections could cause instant and irreparable damage to the unit, so this must be an important consideration when removing and replacing the unit. If the unit is to be disconnected and removed for testing, use a short length of insulated wire to short-circuit each pair of terminals to avoid any electric shock from residual energy in the capacitor.

4 For those owners possessing a multimeter and who are fully conversant with its use, a test sequence is given below. It is not recommended that the inexperienced attempt to test the system at home, as more damage could be sustained by the system, and an unpleasant shock experienced by the unwary operator. It should be appreciated that the CDI system is capable, in certain circumstances, of producing sufficient current output to be dangerous to the operator. It follows that an attitude to safety should be adopted similar to that applying when testing household mains electricity. Before performing any tests, the following point must be noted. Disconnect all the connectors in the ignition system, thus isolating the various components.

5 It will be noted that some of the resistance tests require a meter capable of reading in ohms rather than kilo ohms. Many of the cheaper multimeters are only capable of the latter, and thus are of limited use for accurate testing of low-resistance components. When testing the ignition system to isolate faults it is important to follow a logical sequence to avoid wasted effort and money. Use the accompanying flow charts as a guide, referring to the subsequent sections for more detailed information.

4 Wiring, connectors and switches: checking

1 The single most likely cause of ignition failure or partial failure is a broken, shorted or corroded connector, switch contact or wire. Tracing faults of this nature can often prove time consuming but does not require the use of sophisticated test equipment. Although in rare instances a particular fault will not be evident from a physical examination, it is worthwhile checking the obvious points before resorting to expensive professional assistance.

2 Refer to the wiring diagram at the end of the book, trace and check each of the alternator output leads, connector blocks and all connections to the CDI unit, ignition switch and kill switch. Do not ignore the obvious; it can be very frustrating to spend a long time checking the ignition system, only to discover that the kill switch was set to the 'Off' position. The kill switch and some of the wiring connectors are relatively exposed and may have become contaminated by water. To eliminate this possibility, spray each one with a water dispersing aerosol such as WD40.

3 Check the wiring for breakages or chafing paying particular attention to the wiring in areas like the steering head where steering movement causes flexing. The plastic insulation may appear intact even if the internal conductor has broken. If a wire is suspected of having broken internally or shorted against the frame it should be checked using a multimeter as a continuity tester.

5 Pulser (pickup) coil: testing

1 Trace the alternator output wiring back to the connector blocks beneath the seat. Separate the pulser lead connector (White/red lead) and the three-pin connector carrying the leads to the CDI unit (Red, Brown and Black leads).

2 Using an ohmmeter or multimeter set at the ohms x 10 scale, measure the resistance between the White/red pulser lead and the Black earth lead. The pulser can be considered serviceable if the following resistance figures are obtained.

Pulser coil resistance

White/red to white/green lead:

RD350 LC II	115 ohms ± 20% @ 20°C (68°F)
RD350 F and N	117 ohms ± 20% @ 20°C (68°F)

a) No spark is produced, or weak spark

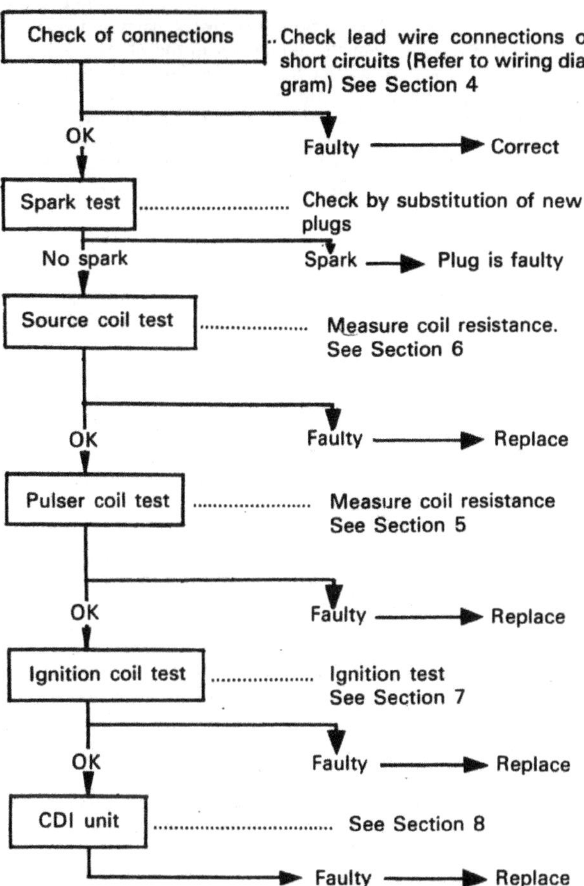

b) The engine starts but will not pick up speed

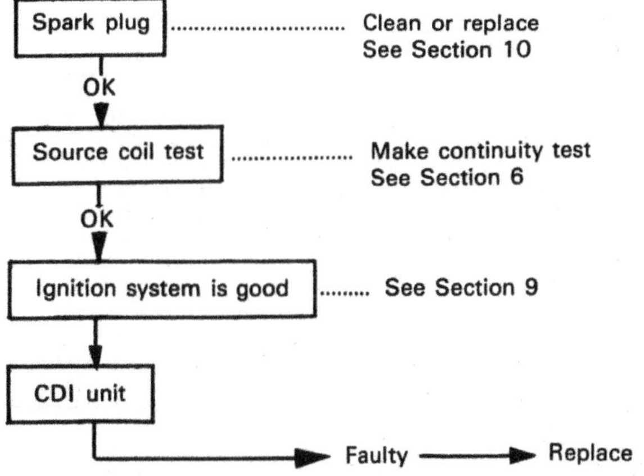

Fig. 4.2 Ignition system fault diagnosis flow chart

6 Source coils: testing

1 The source coils are connected to the CDI unit via three leads (Red, Brown and Green). Trace the wiring back to the three pin connector and separate it. The following resistance test is made on the alternator side of the connector. Set the multimeter to the ohms x 1 scale and measure the resistance between the Red and Brown leads (high speed windings). Next, set the meter to the ohms x 10 scale and measure the resistance between the Brown lead and the Green lead (low speed windings). If the readings obtained differ significantly from those shown below it will be necessary to renew the source coils. These are an integral part of the stator assembly, which must be renewed complete.

Source coil resistances

RD350 LC II:

Red to Brown leads	5.3 ohms ± 20% @ 20°C (68°F)
Brown to Green leads	225 ohms ± 20% @ 20°C (68°F)

RD350 F and N:

Red to Brown leads	4.1 ohms ± 20% @ 20°C (68°F)
Brown to Green leads	113 ohms ± 20% @ 20°C (68°F)

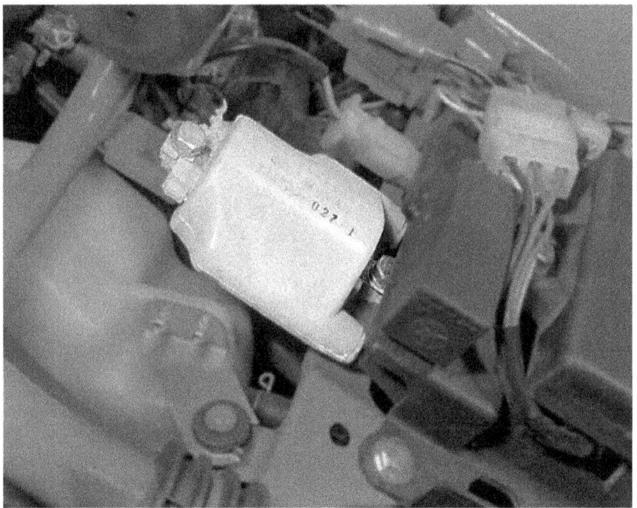

7.1 Ignition coil is bolted to frame below the fuel tank

7 Ignition coil: testing

1 If the ignition coil is suspected of having failed it can be tested by measuring the resistance of its primary and secondary windings. The test can be performed with the coil in place on the frame, having first disconnected the high tension leads at the spark plugs and the low tension lead at the connector block.
2 Connect one of the meter probes to earth and the other to the low tension lead at the connector block. This will give a resistance reading for the primary windings and should be within the limits shown below.

Primary winding resistance
0.33 ohms ± 10% @ 20°C (68°F)

3 Next, connect a meter probe lead to each of the high tension leads to measure the secondary winding resistance and compare the reading obtained with that shown below.

Secondary winding resistance
3.5 kilo ohms ± 20% @ 20°C (68°F)

4 If either of the values obtained differs markedly from the specified resistances it is likely that the coil is defective. It is recommended that the suspect coil is taken to a Yamaha dealer who can then verify the coil's condition and supply a replacement unit where necessary. The coil is a sealed unit and therefore cannot be repaired.

8.1 CDI unit is mounted to the rear of the ignition coil. Note wiring connector at top left of photograph

8 CDI unit: testing

If the tests shown in the preceding Sections have failed to isolate the cause of an ignition fault it is likely that the CDI unit is itself faulty. Whilst it is normally possible to check this by making resistance measurements across the various terminals, Yamaha do not supply the necessary data for these models. It follows that it will be necessary to enlist the help of a Yamaha dealer who will be able to check the operation of the unit by substituting a sound item.

9 Checking the ignition timing

As has been mentioned, the ignition timing is fixed and is thus unlikely to go out of adjustment. The only possible cause of inaccurate ignition timing is an internal fault in the CDI unit or a full or partial failure of the pickup coil windings. Timing marks are provided on the rotor edge, and those possessing a suitable strobe lamp may wish to check the timing in the event of a suspected fault. It will be appreciated that this is of academic interest only; no adjustment being possible.

9.1 Ignition timing can be checked using strobe lamp, but no adjustment is possible

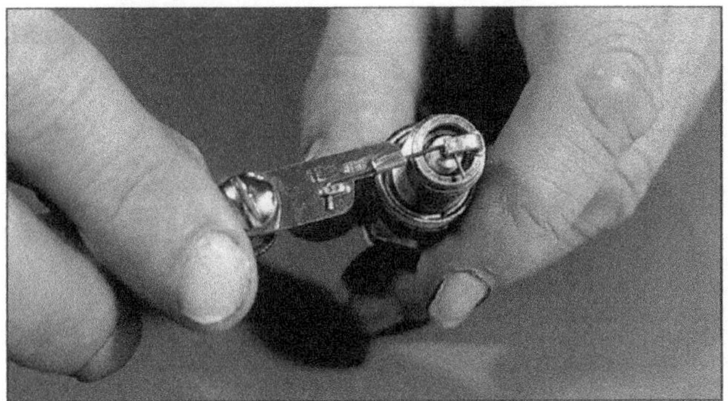

Electrode gap check – use a wire type gauge for best results.

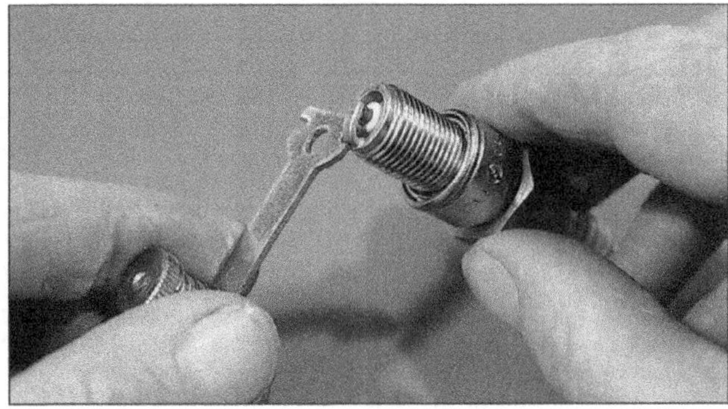

Electrode gap adjustment – bend the side electrode using the correct tool.

Normal condition – A brown, tan or grey firing end indicates that the engine is in good condition and that the plug type is correct.

Ash deposits – Light brown deposits encrusted on the electrodes and insulator, leading to misfire and hesitation. Caused by excessive amounts of oil in the combustion chamber or poor quality fuel/oil.

Carbon fouling – Dry, black sooty deposits leading to misfire and weak spark. Caused by an over-rich fuel/air mixture, faulty choke operation or blocked air filter.

Oil fouling – Wet oily deposits leading to misfire and weak spark. Caused by oil leakage past piston rings or valve guides (4-stroke engine), or excess lubricant (2-stroke engine).

Overheating – A blistered white insulator and glazed electrodes. Caused by ignition system fault, incorrect fuel, or cooling system fault.

Worn plug – Worn electrodes will cause poor starting in damp or cold conditions and will also waste fuel.

10 Spark plugs: checking and resetting the gaps

1 The RD350 YPVS models are fitted as standard with NGK BR8ES spark plugs. The plugs have an internal resistor, denoted by the "R" in the type number, and it is vital that the original grade and type of plug is fitted when replacement is required. There have been instances where the YPVS valve has been upset due to the wrong plug type being used. This is caused by spurious signals being fed to the YPVS microprocessor, which reads the ignition system pulses to gauge the valve opening angle. To avoid problems of this nature, always use the recommended grade of plug.

2 Check the gap of the plug points at every six monthly or 4000 mile service. To reset the gap, bend the outer electrode to bring it closer to, or further away from the central electrode until a 0.7 mm (0.028 in) feeler gauge can be inserted. Never bend the centre electrode or the insulator will crack, causing engine damage if the particles fall into the cylinder whilst the engine is running.

3 With some experience, the condition of the spark plug electrodes and insulator can be used as a reliable guide to engine operating conditions.

4 Always carry spare spark plugs of the recommended grade. In the rare event of plug failure, this will enable the engine to be restarted.

5 Beware of over-tightening the spark plugs, otherwise there is risk of stripping the threads from the aluminium alloy cylinder heads. The plugs should be sufficiently tight to seat firmly on their copper sealing washers, and no more. Use a spanner which is a good fit to prevent the spanner from slipping and breaking the insulator.

6 If the threads in the cylinder head strip as a result of overtightening the spark plugs, it is possible to reclaim the head by the use of a Helicoil thread insert. This is a cheap and convenient method of replacing the threads; most motorcycle dealers operate a service of this nature at an economic price.

7 Make sure the plug insulating caps are a good fit and have their rubber seals. They should also be kept clean to prevent tracking.

Chapter 5 Frame and forks

For information relating to the RD350 F II, N II and R models, refer to Chapter 8

Contents

Specifications

Frame

Type	Tubular, double cradle
Caster angle:	
RD350 LC II	27°
Other models	26°
Trail	96 mm (3.78 in)

Front forks

Type	Oil-damped telescopic, air assisted
Travel	140 mm (5.51 in)
Spring free length:	
RD350 LC II	506.2 mm (19.9 in)
Other models	429.6 mm (16.9 in)
Oil capacity:	
RD350 LC II	253 cc (8.92 Imp fl oz)
Other models	297 cc (10.48 Imp fl oz)
Oil grade:	
RD350 LC II	SAE 10W30 SE motor oil
Other models	SAE 10W fork oil
Oil level:	
RD350 LC II	120 mm (4.72 in)
Other models	106.1 mm (4.18 in)
Air pressure:	
Std	0.4 kg cm² (5.7 psi)
Min	Zero
Max	0.8 kg cm² (11 psi)

Rear suspension

Type	Rising rate (Yamaha Monocross, new type)
Suspension unit travel	40 mm (1.57 in)
Rear wheel travel	100 mm (3.94 in)
Spring free length:	
RD350 LC II	184 mm (7.24 in)
Other models	186 mm (7.32 in)
Gas type	Nitrogen (sealed)
Gas pressure	12 kg cm² (171 psi)
Damping adjuster	Remote, by toothed belt
Swinging arm free play:	
End float	1.0 mm (0.04 in)
Side-to-side	0.1 – 0.3 mm (0.004 – 0.012 in)

Torque wrench settings

Component	kgf m	lbf ft
Steering stem top bolt	8.5	61.0
Top yoke pinch bolts:		
RD350 LC II	2.3	17.0
Other models	2.0	14.0
Lower yoke fork pinch bolts	2.0	14.0
Handlebar section to top yoke:		
RD350 LC II	Not applicable	
Other models	0.9	6.5

Component	kgf m	lbf ft
Handlebar to fork stanchion pinch bolts:		
RD350 LC II	Not applicable	
Other models	2.0	14.0
Fork brace:		
RD350 LC II	Not applicable	
Other models	1.0	7.2
Front wheel spindle	7.5	54.0
Rear wheel spindle:		
RD350 LC II	10.0	72.0
Other models	10.5	75.0
Swinging arm pivot:		
RD350 LC II	7.0	50.0
Other models	9.0	65.0
Rear wheel sprocket	3.3	24.0
Rear suspension unit frame mounting	4.0	28.0
Relay arm to frame mounting	4.0	28.0
Relay arm to links	6.5	47.0
Links to swinging arm	4.0	28.0
Footrest	6.5	47.0
Brake disc mounting bolts	2.0	14.0
Master cylinder hose unions	2.5	18.0
Hydraulic hose to 3-way union	2.5	18.0
Hydraulic hose to caliper	2.5	18.0
Caliper bracket bolts	3.5	25.0
Caliper bleed screw	0.5	4.0

1 General description

The Yamaha RD350 YPVS models employ a conventional welded tubular steel frame. Front suspension is by oil-damped air-assisted coil spring telescopic forks. Rear suspension is by Yamaha Monocross suspension, a rising-rate system controlled by a single central suspension unit. The rear suspension unit is a De Carbon type nitrogen pressurised coil spring unit and features a toothed belt remote preload adjuster.

The RD350 LC II model is equipped with a small handlebar fairing and a separate belly pan beneath the front of the engine and frame. The RD350 F model features a full fairing consisting of a main fairing section, two side sections and a lower section. The RD350 N model is effectively a stripped version of the F model and is unfaired.

2 Fairing: removal and refitting – RD350 LC II and RD350 F models

1 There are a number of occasions where it is either necessary or desirable to remove the fairing from the machine. In the former instance it allows access to areas normally obstructed by the fairing panels, whilst in the latter case, though access may be possible, there may be a risk of damage to the fairing surface. The fairing can be removed quite quickly, so do not be tempted to try to work around it.

RD350 LC II

2 The main fairing section is designed to hinge down to allow access to the headlamp and instruments. Remove the two bolts which retain the fairing moulding to the headlamp brackets and pivot the fairing forward and down. To release the fairing completely, remove the split pin and clevis pin which form the lower pivot and lift the fairing away.
3 The belly pan is secured by four screws, two at the back edge and two just below the radiator. Remove the screws and lift the belly pan away.

RD350 F

4 To release the belly pan, remove the single Allen-headed bolt, spacer and grommet on each side, then remove the three screws on each side which secure the belly pan to the side sections.
5 The side sections are retained to the upper edge of the belly pan by the six screws mentioned above, and by a further four screws and plain washers to the lower edge of the main fairing. Remove the remaining screws and lift away the side sections.
6 The main fairing can be removed complete with the subframe, headlamp, instrument panel and front turn signals. Slacken the knurled ring which retains the speedometer drive cable to the underside of the instrument head. Moving to the right-hand side of the machine, trace and disconnect the wiring to the headlamp, turn signals and instrument panel at the various connectors near the steering head. Release the two mounting bolts which secure the fairing subframe to the steering head and lift the assembly clear of the frame.

All models

7 When refitting the fairing sections, reverse the dismantling sequence, taking care not to strain or crack the plastic mouldings. Fit all fasteners finger tight only, and check that the various sections are correctly aligned. Make sure that all spacers and grommets are fitted correctly. Finally, tighten the fasteners to secure the fairing sections, being careful to avoid overtightening.

2.6a Speedometer cable can be reached from inside fairing (F model)

2.6b Wiring connectors are below cover held by two screws (A). Bolts (B) hold fairing to steering head (F model)

3 Front forks: removal and refitting

1 Place the machine on its centre stand on level ground. In the case of the RD3350 LC II and RD350 F models, remove the main fairing section as described above. Remove the front brake caliper mounting bolts and lift the calipers clear of the front wheels and forks. Tie the calipers to the frame to avoid straining the hoses, and place a wooden wedge between the pads to prevent the pistons from being expelled if the brake lever is accidentally squeezed.
2 Remove the front wheel, referring to Chapter 6 if additional information is required. Remove the front mudguard. On the RD350 LC II model the mudguard is retained by four bolts passing through from the inside of the mudguard and into the side of the fork lower legs. In the case of the RD350 F and RD350 N models, the mudguard incorporates a fork brace and the four fixing bolts pass down from the top of the brace and through the mudguard.
3 On RD350 F and RD350 N models, slacken the pinch bolts securing the handlebar castings to the top of the fork yokes. On all models, release the upper and lower yoke pinch bolts. The fork legs can now be removed by twisting them and pulling downwards. If the fork legs are to be dismantled, it is preferable to slacken the top bolt before the legs are removed completely.
4 Pull the leg down by one or two inches and temporarily tighten the lower yoke pinch bolts to hold them. Remove the fork air caps and release pressure by depressing the valve cores for a few seconds. The top bolts can now be slackened. The RD350 F and RD350 N models have hexagon-headed bolts, whilst that of the RD350 LC II model should be slackened using self-locking pliers on the projecting boss.
5 Installation is accomplished by reversing the removal sequence. Check that the fork oil has been topped up to the correct level and that the top bolts are tightened as the forks are slid into position in the yokes. Note that the top of the stanchion should align with the top face of the upper yoke in the case of the RD350 LC II model, whilst on the RD350 F and RD350 N models it should align with the upper edge of the handlebar castings.
6 In the case of the RD350 LC II model, once in position, tighten the upper yoke pinch bolts to 2.3 kgf m (17 lbf ft) and the lower yoke pinch bolts to 2.0 kgf m (14 lbf ft).
7 On RD350 F and RD350 N models, tighten the lower yoke pinch bolts to 2.0 kgf m (14 lbf ft) then tighten the fork cap bolt to 2.3 kgf m (17 lbf ft). Next, tighten the handlebar pinch bolts and the upper yoke pinch bolts to 2.0 kgf m (14 lbf ft).
8 Complete reassembly by refitting the mudguard, front wheel and brake calipers. Check that the forks operate normally, and remember to set the fork air pressure as described later in this Chapter.

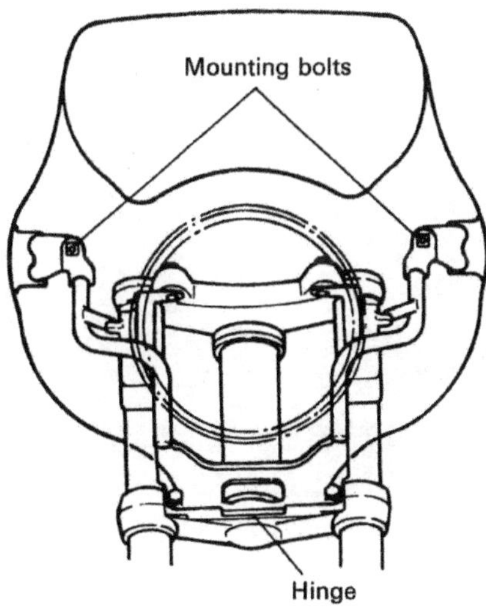

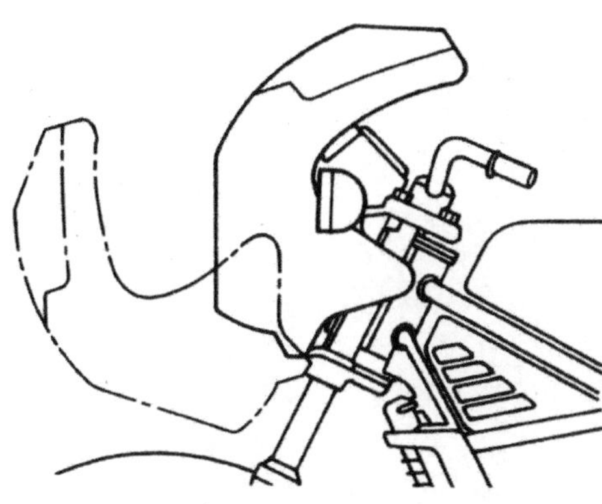

Fig. 5.1 Fairing mountings – RD350 LC II model

3.1 Remove the calipers and free hydraulic hose clips from fork legs

3.2a 'F' and 'N' models have alloy fork brace fitted between fork legs

3.2b Remove outer bolts only to free mudguard and fork brace from fork legs

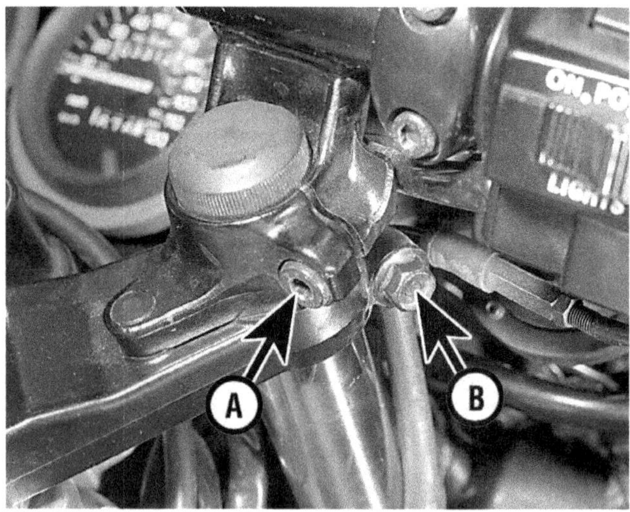

3.3a Slacken handlebar clamp bolts (A) where fitted, and upper yoke clamp bolts (B)

3.3b Slacken lower yoke clamp bolts to free fork legs

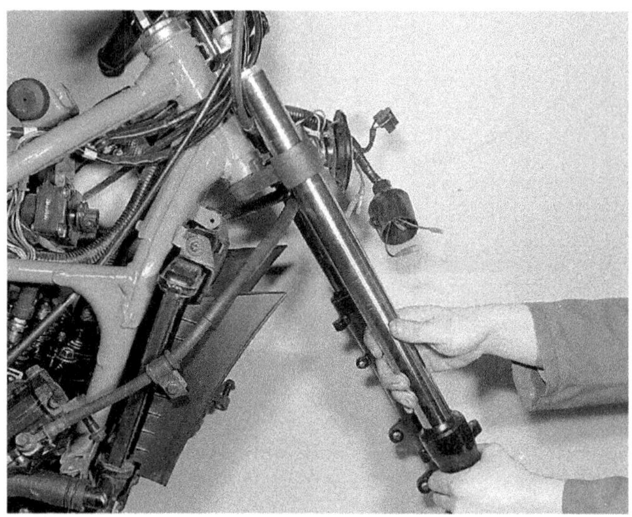
3.4 Slacken fork top bolts before pulling leg down and clear of yokes

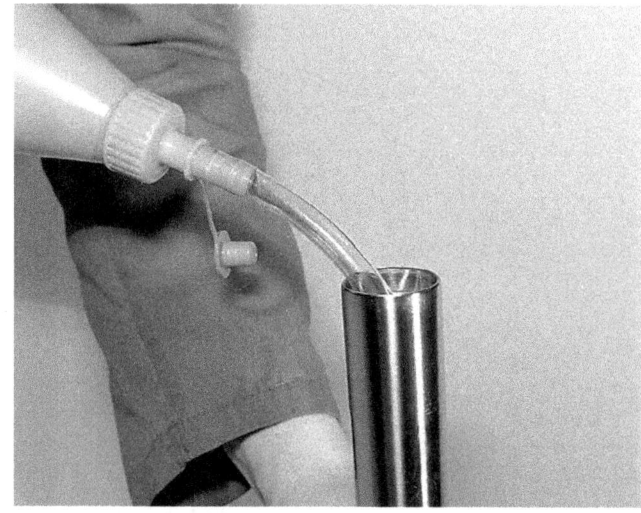
3.5 Top up damping oil to correct level **before** fitting legs into yokes

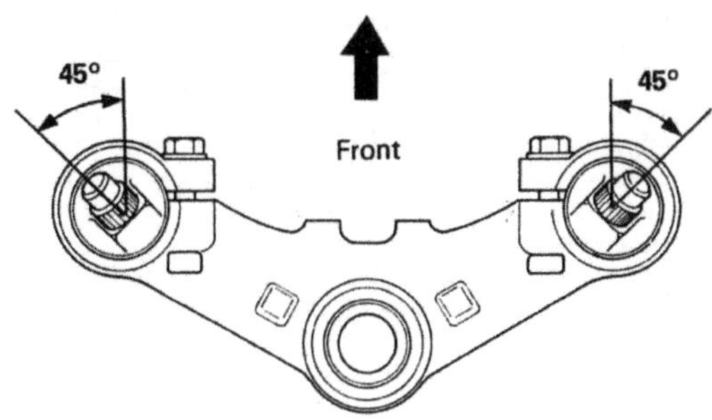

Fig. 5.2 Air valves should be positioned as shown when installing fork legs – RD350 LC II

4 Steering head: dismantling and reassembly

1 Start by removing the fairing (Section 2) where applicable, and the front fork legs (Section 3). Note that in the case of the RD350 F model, the headlamp unit and instrument panel will be removed together with the fairing. On RD350 LC II and RD350 N models, remove the headlamp unit from its shell after releasing the two retaining screws. Lift the unit clear and unplug the headlamp and parking lamp connectors. Unplug the handlebar switch and turn signal wiring at the connectors and push the wiring clear of the headlamp shell. Disconnect the ignition switch and instrument panel wiring.
2 Free the speedometer drive cable from the underside of the instrument panel. On RD350 LC II models release the tachometer cable in the same way. Remove the bolts which retain the instrument panel to the headlamp bracket, lift the panel assembly clear, and place it to one side. Remove the bolts retaining the headlamp bracket to the upper and lower yokes and lift it away.
3 On RD350 F models, much of the preliminary dismantling described above is not necessary. The ignition switch and steering lock assembly can normally be left attached to the upper yoke, but its wiring may need to be disconnected. On all models, free the front brake hydraulic union from the lower yoke, and where appropriate, remove the horn and bracket from the lower yoke. Disconnect the clutch cable to allow more manoeuvring room.
4 Slacken and remove the steering stem top bolt. On RD350 F and RD350 N models, manoeuvre the handlebar sections clear of the upper yoke. The upper yoke can now be lifted away from the steering stem. If necessary, tap the underside of the yoke to jar it free. Note that in the case of the RD350 LC II model, the tubular handlebar together with the control levers and switches will come away with the upper yoke. These can be left in position unless there is some reason that the upper yoke must be removed completely. On all models, try to lodge the controls, cables and hydraulic hoses clear of the steering stem.
5 Before slackening the steering head nut note that the upper and lower races each contain nineteen 1/4 in steel balls. These are uncaged and will tend to drop free as the lower yoke is removed, so spread some rag or an old blanket below the steering head to catch any that drop. Using a C-spanner, slacken and remove the steering stem nut whilst supporting the lower yoke. Lift away the nut, bearing cover and the top cone, then remove the steel balls from the upper race.
6 Carefully lower the yoke and steering stem, trying not to dislodge the balls in the lower race. Remove the balls from the lower race and place them in a container for safe keeping.
7 The steering head assembly should be reassembled in the reverse order of that given for dismantling. When fitting the steel balls to their races, they can be held in place with grease. Check that the correct number is fitted to each bearing race. When fitting the steering stem nut, it must be adjusted so that all perceptible free play is taken up, but no more. It is easy to damage the head races by overtightening. When correctly adjusted, it should be possible to move the steering from lock to lock with the lightest pressure on the handlebar end. Final adjustment can be made after reassembly, by slackening the top bolt and adjusting the steering stem nut as required.

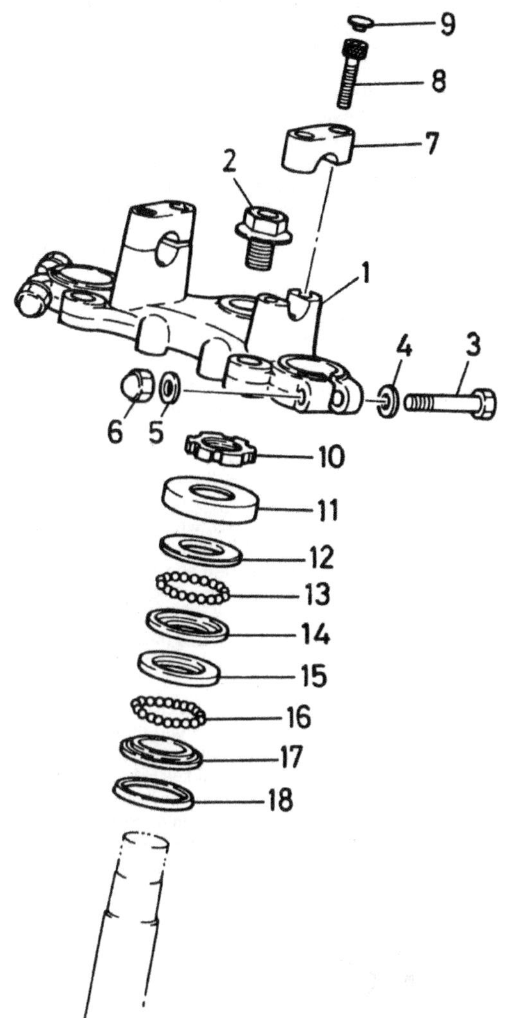

Fig. 5.3 Steering head – typical

1 Upper yoke – RD350 LC II (others similar)
2 Top bolt
3 Bolt
4 Washer
5 Washer
6 Nut
*7 Handlebar clamp – 2 off**
*8 Allen screw – 4 off**
*9 Cap – 4 off**
10 Steering stem nut
11 Bearing cover
12 Top cone
13 Steel balls
14 Top cup
15 Bottom cup
16 Steel balls
17 Bottom cone
18 Dust seal

**RD350 LC II model only*

5 Front forks: dismantling and reassembly

1 With the fork legs removed from the yokes, unscrew the top bolt and lift out the spacer (RD350 F and RD350 N models only) and the fork spring. Invert the leg over a drain tray and "pump" the fork to expel the old damping oil. Slide the dust seal away from the lower leg and remove it. Prise out the wire circlip which retains the oil seal and the plain washer.
2 Once the oil has been drained, slacken the bolt which passes up through the bottom of the lower leg and into the damper rod. It is quite likely that the damper rod will tend to rotate in the lower leg and thus

impede the removal of the bolt. If this problem arises, clamp the assembly in a vice using soft jaws to hold the lower leg by the caliper mounting lugs. Obtain a length of wooden dowel about 1/2 inch in diameter and form a taper on one end.

3 Pass the dowel down the stanchion, having first withdrawn the fork spring. Push the dowel hard against the head of the damper rod to lock it in position whilst an assistant slackens the retaining bolt. If the dowel proves difficult to hold a self-grip wrench or similar can be used to obtain sufficient leverage and pressure. Once the retaining bolt has been removed slide off the dust seal and withdraw the stanchion assembly from the lower leg.

4 To separate the stanchion from the lower leg it will be necessary to displace the top bush and the oil seal. The lower bush should not pass through the top bush, and this can be used to good effect. Push the stanchion gently inwards until it stops against the damper rod seat. Take care not to do this forcibly or the seat may be damaged. Now pull the stanchion sharply outwards until the lower bush strikes the top bush. Repeat this operation until the top bush and seal are tapped out of the lower leg. Invert the stanchion and tip out the damper rod components.

5 The above method usually works well, but we found that on the machine featured in the photographs, the lower bush passed through the top bush and seal, leaving them in place. Should this occur, remove the stanchion, then prise out the oil seal, which should be renewed as a matter of course. The bush may be a little difficult to remove because it sits quite deep in the top of the lower leg and tends to jam on a tapered shoulder. Try working it out of position using a screwdriver with the end bent at right angles as a lever, and taking care not to damage the bush surface or that of the lower leg. If this fails it will be necessary to resort to a bearing extractor; a Yamaha dealer or a local garage may be able to help out with this.

6 After checking the fork components for wear or damage as described in the next section, reassembly can commence. Ensure that all components are clean and that they are fitted to the fork from which they were removed. Always fit a new oil seal; if even slightly worn or damaged the fork will not hold air pressure. The bushes can be reused only if in perfect condition. It is recommended that they too are renewed as a precautionary measure, particularly if the stanchion pulled through the top bush during dismantling. The lower bush is split to allow fitting over the end of the stanchion. Do not open the split any more than is essential to ease it into place.

7 Fit the rebound spring over the damper rod and drop it into place in the stanchion. Pass a length of dowel or a broom handle up through the stanchion to hold the damper rod in place, then fit the damper rod seat. Oil the stanchion and bush, and support it vertically with the damper rod uppermost. Slide the lower leg over the stanchion assembly and refit the damper rod bolt. Use Loctite on the bolt threads and tighten to 2.0 kgf m (14.0 lbf ft).

8 Oil the top bush and slide it down over the stanchion. To fit the bush in its recess it will be necessary to devise an alternative to the tubular drift tool used by Yamaha dealers. The best method is to use a length of tubing slightly bigger in diameter than the stanchion. Place the large plain washer against the bush and then tap it home using the tube as a form of slide hammer. Take care not to scratch the stanchion during this operation; it is best to make sure that the stanchion is pushed fully inwards so that any accidental scoring is confined to the area above the seal.

9 Once the top bush is seated, the new oil seal can be fitted. Fit the plain washer with the chamfered edge facing upwards. Lubricate the seal lips, then slide it down over the stanchion. Take care not to damage the seal lips during fitting. The seal should be tapped gently and squarely into the top of the lower leg, using a tubular drift. Check that the seal remains square to the lower leg and that it seats fully, then fit the retaining clip and slide the dust seal into place.

10 Before fitting the spring, spacer (except RD350 LC II models) and top bolt, add the prescribed amount and grade of damping oil. Check that the oil is at the correct level below the top of the stanchion, with the fork fully compressed and held vertically. When refitting the springs, note that the tighter coils must be uppermost. Once the leg has been refitted in the yokes, check and adjust the air pressure.

5.1a Remove fork top bolt ...

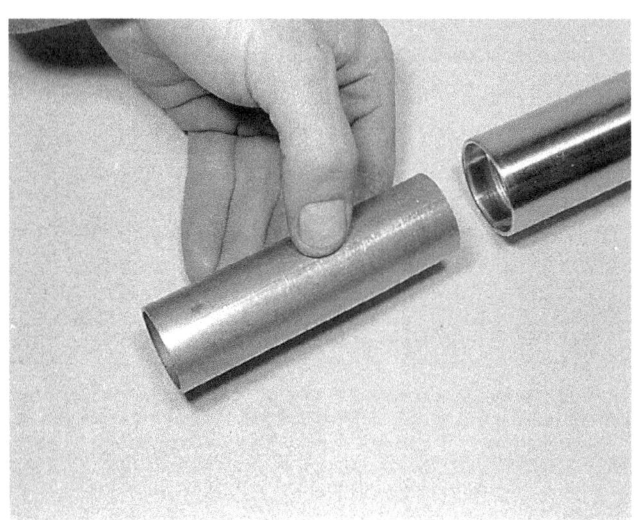

5.1b ... and also spacer, where fitted

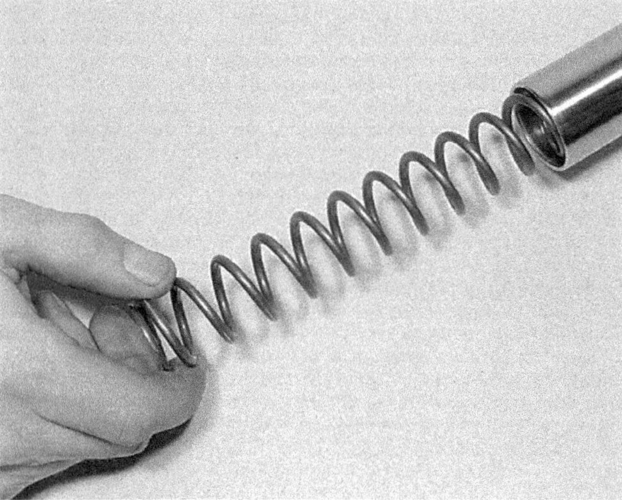

5.1c Remove the fork spring, then drain damping oil

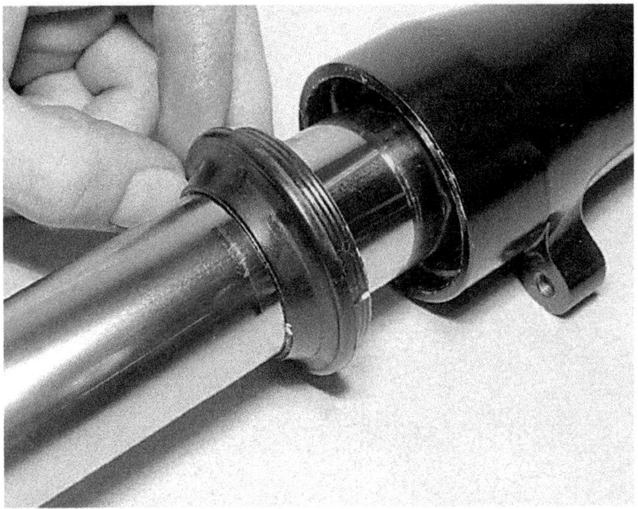

5.1d Prise out and remove the dust seal

5.1e Lever out the wire retaining clip as shown

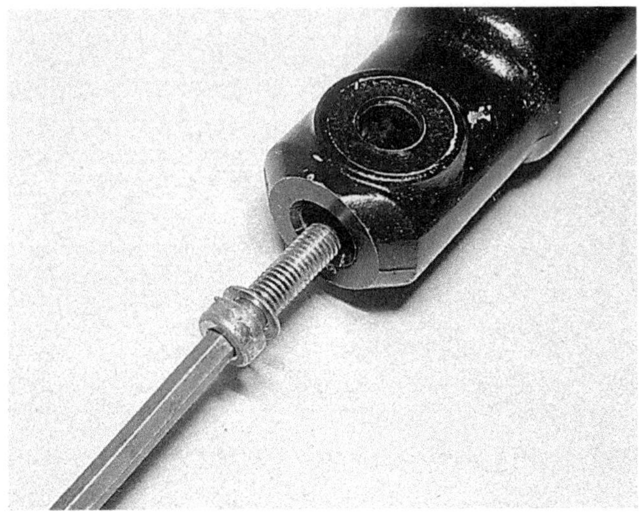

5.3 Slacken and remove the damper rod bolt and separate stanchion and lower leg (see text)

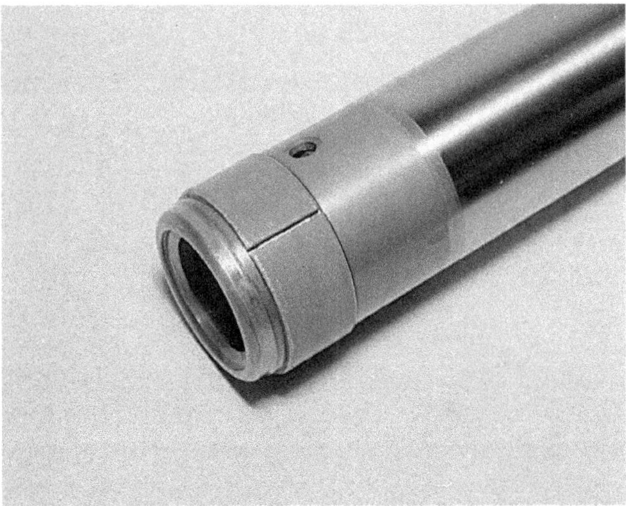

5.6 Bottom bush is split to facilitate removal – do not stretch new bush more than absolutely necessary during fitting

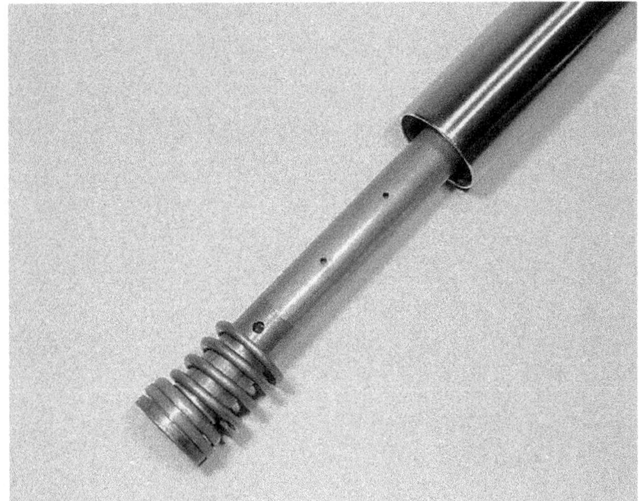

5.7a Fit rebound spring and slide damper rod assembly into the stanchion

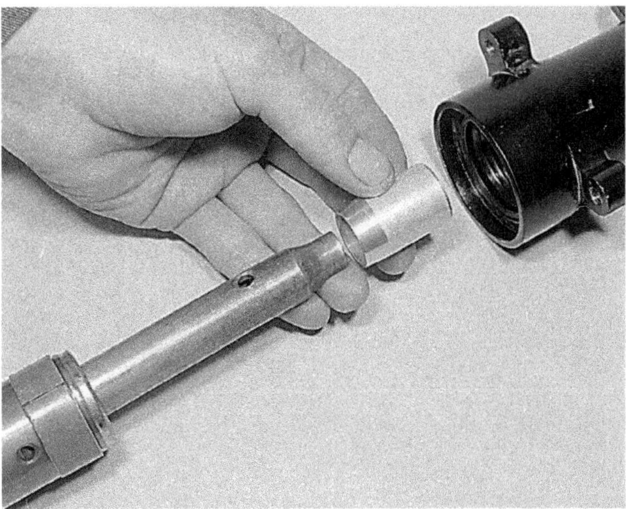

5.7b Fit damper rod seat, then lubricate stanchion ...

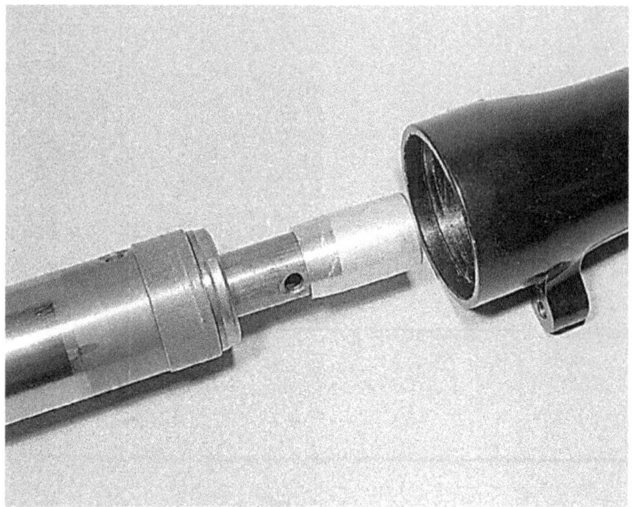

5.7c ... and slide into lower leg

5.8a Lubricate top bush and slide into place

5.8b Fit plain washer and tap bush into top of lower leg

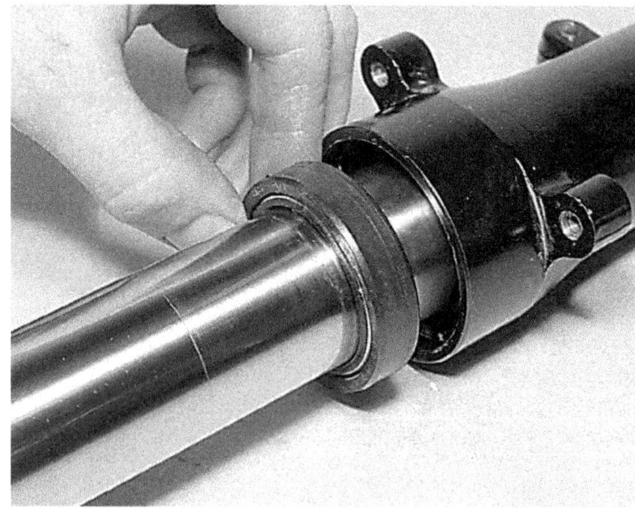

5.9a Lubricate and fit fork oil seal ...

5.9b ... and retain with wire clip

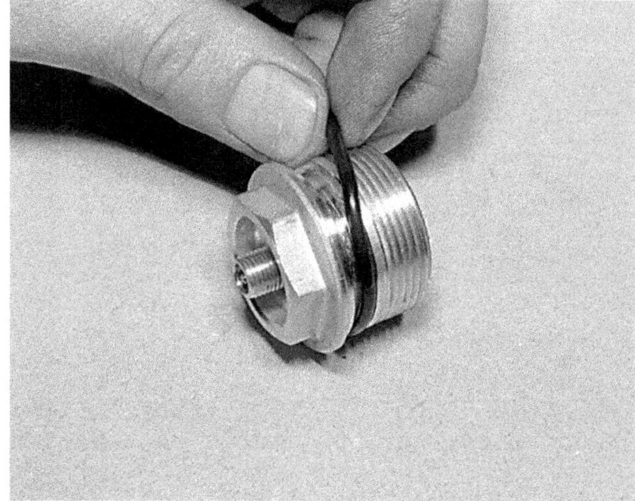

5.10 Check condition of O-ring before adding damping oil and fitting top bolt

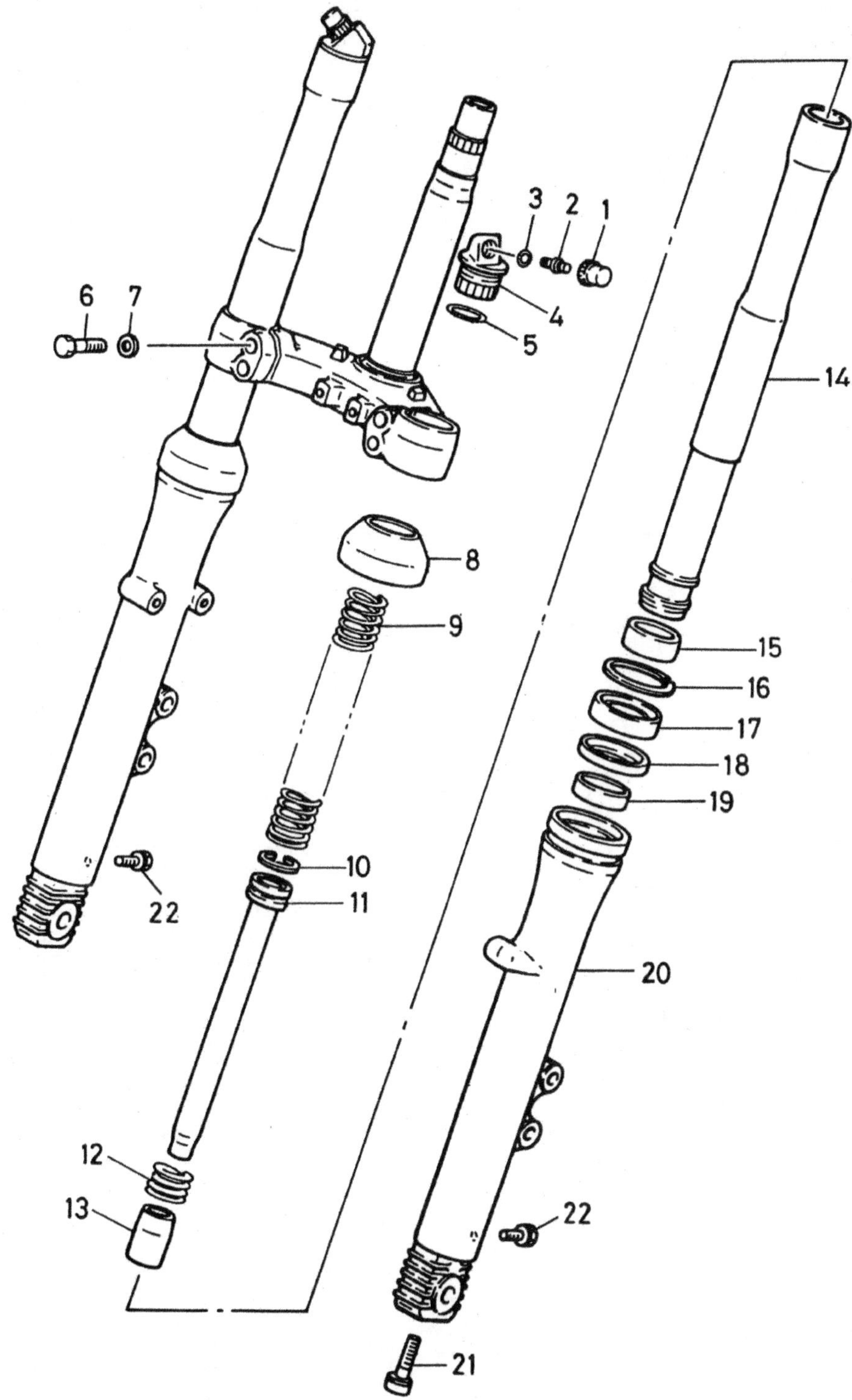

Fig. 5.4 Front forks – RD350 LC II model

1 Cap
2 Air valve
3 O-ring
4 Top bolt
5 O-ring
6 Bolt – 4 off
7 Washer – 4 off
8 Dust seal
9 Spring
10 Piston ring
11 Damper rod
12 Rebound spring
13 Damper rod seat
14 Stanchion
15 Bottom bush
16 Circlip
17 Oil seal
18 Backing ring
19 Top bush
20 Lower leg
21 Allen bolt
22 Drain plug

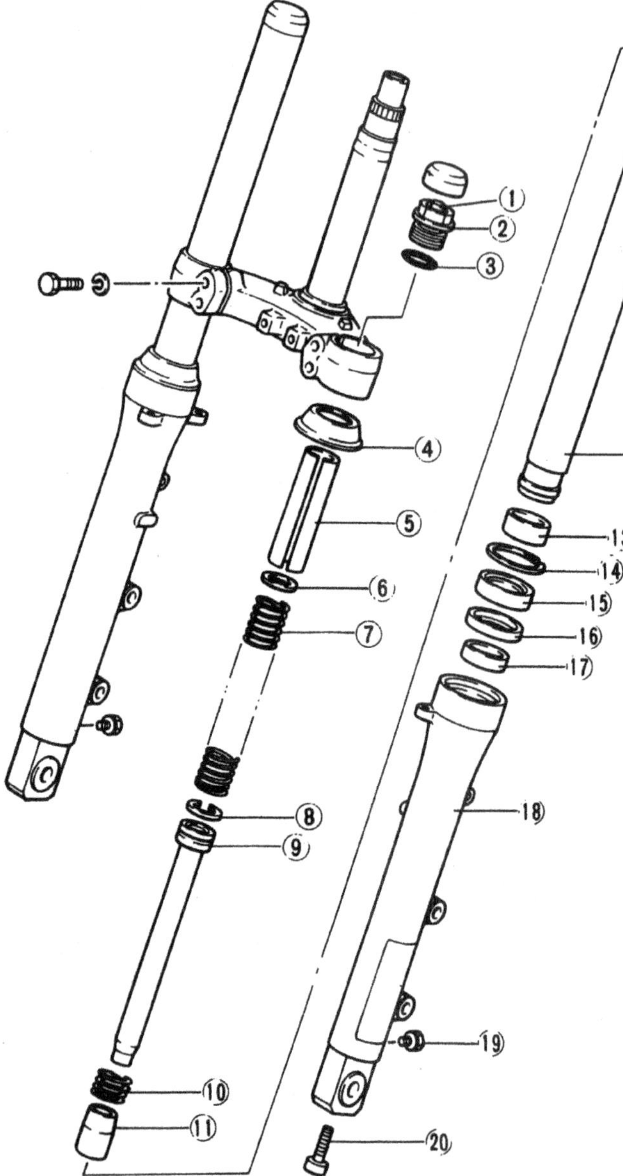

Fig. 5.5 Front forks – RD350 F and N models

1 *Air valve*
2 *Top bolt*
3 *O-ring*
4 *Dust seal*
5 *Spacer*
6 *Washer*
7 *Spring*
8 *Piston ring*
9 *Damper rod*
10 *Rebound spring*
11 *Damper rod seat*
12 *Stanchion*
13 *Bottom bush*
14 *Circlip*
15 *Oil seal*
16 *Washer*
17 *Top bush*
18 *Lower leg*
19 *Drain plug*
20 *Allen bolt*

6 Front forks: examination and renovation

1 The parts most likely to wear are the sliding surfaces of the two bushes. These control the play in the forks and are designed to wear before damage occurs to the stanchion or lower leg. If there are signs of scoring or obvious wear, the bushes must be renewed. Only in extreme cases will the stanchion or lower leg be worn; in these cases the affected item must be renewed.
2 Check the stanchion for signs of scoring. Damage of this type can be caused by dirt trapped below a damaged or worn dust seal and can be avoided by ensuring that it is renewed whenever the oil seal is renewed. If there has been impact damage, check that the stanchions are straight by rolling them on a flat surface. A bent stanchion must be renewed; do not attempt to straighten it.
3 The oil seal should be renewed each time the forks are dismantled. Bear in mind that it must also withstand air pressure. A worn or damaged seal will allow air to leak out of the fork and thus will prevent normal fork operation.
4 The fork springs will become permanently compressed after some years of use. Check the fork spring free length against that shown in the specifications. If below the service limit, renew the springs as a pair.

7 Steering head bearings: examination and renovation

1 Before commencing reassembly of the steering head assembly, examine the steering head races. The ball bearing tracks of the respective cup and cone bearings should be polished and free from indentations, cracks or pitting. If signs of wear are evident, the cups and cones must be renewed. In order for the straight line steering on any motorcycle to be consistently good, the steering head bearings must be absolutely perfect. Even the smallest amount of wear on the cups and cones may cause steering wobble at high speeds and judder during heavy front wheel braking. The cups and cones are an interference fit on their respective seatings and can be tapped from position with a suitable drift.
2 Ball bearings are relatively cheap. If the originals are marked or discoloured they must be renewed. To hold the steel balls in place during reassembly of the fork yokes, pack the bearings with grease. Each race is fitted with nineteen steel balls. Although each race has room for an extra steel ball it must not be fitted. The gap allows the bearings to work correctly, stopping them skidding and accelerating the rate of wear.

8 Steering lock and ignition switch: renewal

The steering lock is combined with the ignition switch, the assembly being retained to the underside of the top yoke by two bolts. In the event of a malfunction, try freeing the mechanism using a maintenance spray such as WD40. If this fails to effect a cure, the lock must be renewed. Repair is not practicable.

9 Frame: examination and renovation

1 The frame is unlikely to require attention unless accident damage has occurred. In some cases, replacement of the frame is the only satisfactory course of action if it is badly out of alignment. Only a few frame repair specialists have the jigs and mandrels necessary for resetting the frame to the required standards of accuracy and even then there is no easy means of assessing to what extent the frame may have been overstressed.
2 After the machine has covered a considerable mileage, it is

advisable to examine the frame closely for signs of cracking or splitting at the welded joints. Rust can also cause weakness at these joints. Minor damage can be repaired by welding or brazing, depending on the extent and nature of the damage.
3 Remember that a frame which is out of alignment will cause handling problems and may even promote 'speed wobbles'. If misalignment is suspected, as the result of an accident, it will be necessary to strip the machine completely so that the frame can be checked and, if necessary, renewed.

10 Rear suspension unit: removal and refitting

1 Place the machine on its centre stand. Remove the dualseat, both side panels, the fairing belly pan (where fitted), the exhaust system and the battery. It may prove advantageous to remove the oil tank to provide easier access, though it should prove possible to work around this. If removing the tank, either drain the oil or disconnect the outlet pipe and plug it. Remember to disconnect the oil level switch leads.
2 Using the tools provided in the machine's toolkit, set the suspension preload adjuster to the softest (1) setting. Should the adjuster prove exceptionally stiff, do not force it and risk damage; wait until the unit has been removed and can be free off properly.
3 Working via the hole provided in the base of the battery case, slacken the screws which secure the adjuster pulley assembly to the frame. Pivot the adjuster pulley towards the suspension unit to release belt tension, then slip the belt off the pulley.
4 Moving to the underside of the machine, slacken and remove the pivot bolt which retains the suspension unit to the relay arm and the two suspension links. Displace and remove the two thrust covers and tap out the tubular inner sleeve to free the suspension unit lower mounting. Slacken and remove the suspension unit upper mounting bolt. The unit is now free and can be removed downwards.
5 Before refitting the unit, check for wear, damage or leakage as described in Section 11. Clean the pivot components to remove all traces of road dirt. All moving parts should be given a thin coating of grease before assembly commences. It is also important to check that the spring preload adjuster works normally. This is prone to seizure due to the effects of corrosion and accumulated road dirt, particularly if it has not been used for some time.
6 It is worth removing the adjuster pulley from the machine so that it can be cleaned and lubricated properly. Remove the circlip which retains the pulley and lift it off its pivot. Clean the pulley teeth and internal bore, and also the pivot. Lubricate the pivot with molybdenum disulphide grease, then refit the pulley and circlip.
7 Clean the suspension unit, paying particular attention to the area around the preload adjuster. If this should seize completely, soak the adjuster in penetrating oil until it can be freed off. Take care not to damage the pulley teeth. Set the adjuster so that the ring cam is on one of its peaks, then lubricate it with graphite or molybdenum disulphide grease. As an alternative, a graphite-based chain lubricant can be used.
8 Install the unit by reversing the removal sequence. Do not forget to refit the adjuster belt, and make sure that it sits squarely and centrally on its pulley. Tension the belt by pulling back the tab at the front of the bracket, and tighten the Allen bolts. Turn the adjuster fully anti-clockwise and check that the adjuster indicates position "1". If necessary, reset the adjuster pulley position, then reset the belt tension. Tighten the upper mounting bolt to 4.0 kgf m (28 lbf ft) and the lower mounting and pivot bolt to 6.5 kgf m (47 lbf ft).

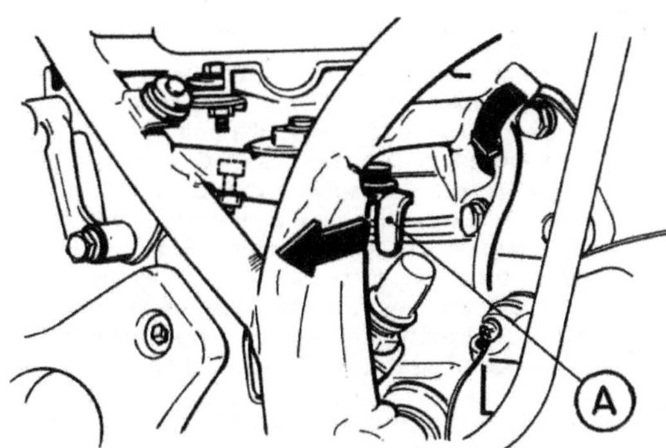

Fig. 5.7 Pull back tab (A) and tighten Allen bolts to tension adjusting belt

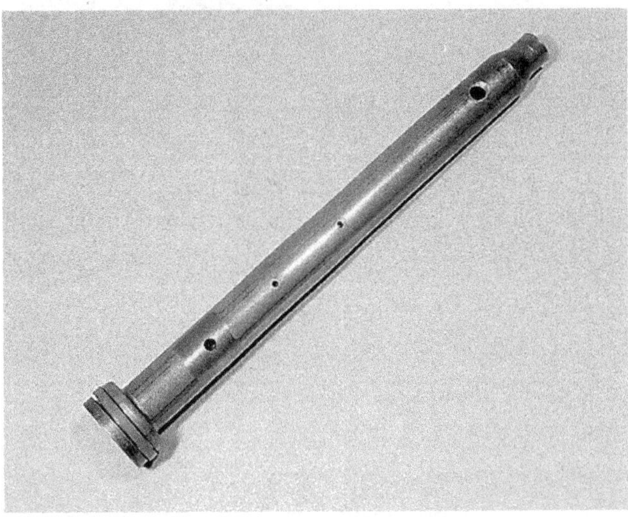

6.1 Check that holes in damper rod are clear of dirt and old oil – damping will be impaired if they are blocked

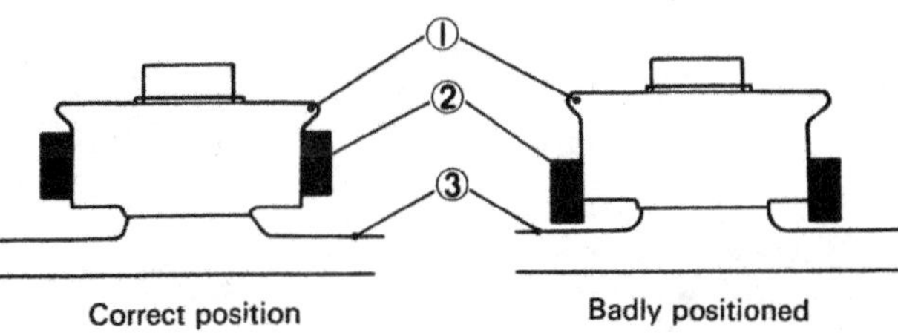

Fig. 5.6 Correct fitting of suspension unit adjuster belt

1 Pulley *3 Bracket*
2 Belt

10.3 Remote adjuster Allen bolts can be reached through hole in battery tray

10.4a Remove suspension unit lower mounting bolt (arrowed)

10.4b Upper mounting bolt can be reached through battery tray

10.6 It is a good idea to remove, strip and lubricate the adjuster

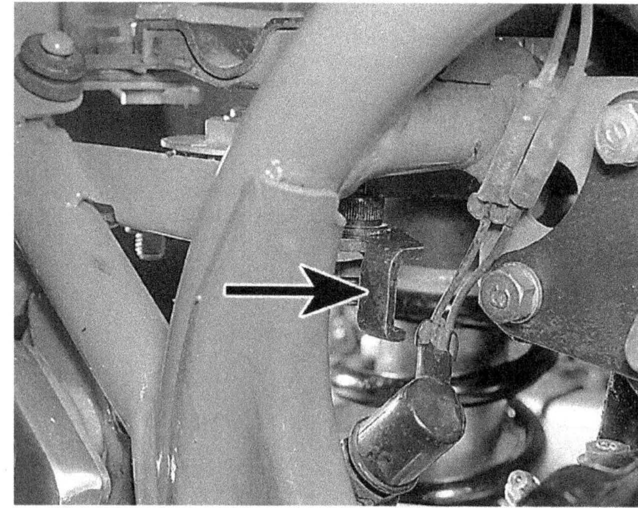
10.8a Tension belt by pressing lever (arrowed) ...

10.8b ... while securing Allen bolt

11 Rear suspension linkage and swinging arm: removal and refitting

1 The swinging arm and its associated linkages can be removed with or without the suspension unit. Given the exposed nature of the various pivots it is preferable to remove the unit as well, so that it too can be checked and lubricated, and this method is described below. Start by removing the seat, side panels, exhaust system and the battery. Remove the rear wheel (see Chapter 6 for details). It will prove helpful to remove the front section of the rear mudguard to gain better access. This is retained by three screws, but it is also necessary to release the fuse box and the CDI unit, and the associated wiring clips first.
2 Release the suspension unit spring preload adjuster and top mounting bolt as described in the previous Section. Before removing the swinging arm, check for free play by pushing and pulling it from side to side. If there is more than 1 mm (0.04 in) movement at the wheel spindle end, the bush or side clearance is excessive and should be investigated.
3 Slacken and remove the relay arm pivot bolt from the frame bracket. Remove the swinging arm pivot shaft nut, then tap the shaft out to release the swinging arm and suspension linkage assembly. Make a

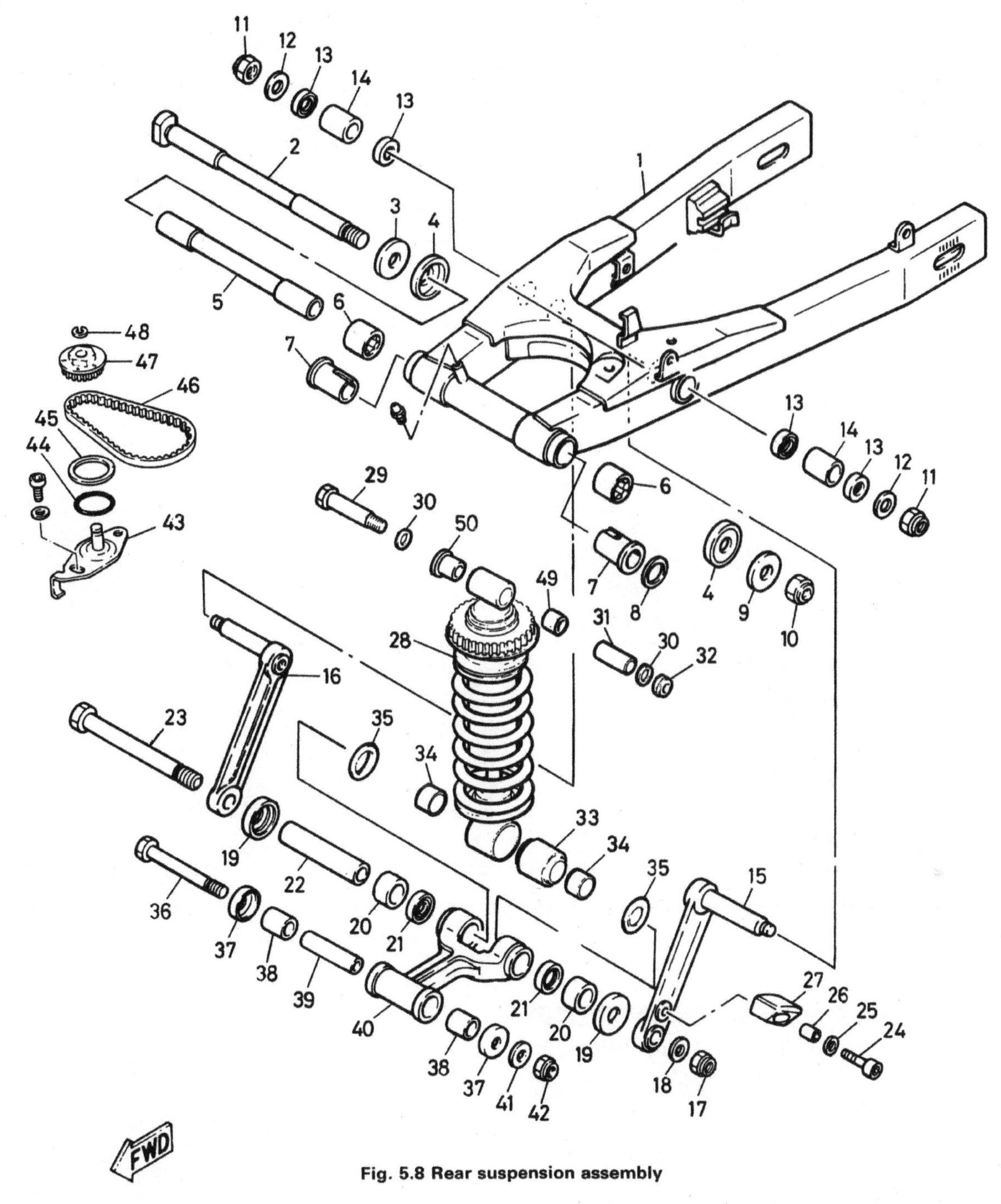

Fig. 5.8 Rear suspension assembly

1 Swinging arm
2 Pivot bolt
3 Washer △
4 Thrust cover
5 Sleeve
6 Bush △
7 Bush □
8 Shim □
9 Washer
10 Nut
11 Nut
12 Washer
13 Oil seal
14 Bush
15 Left-hand suspension link
16 Right-hand suspension link
17 Nut
18 Washer
19 Thrust cover
20 Bush
21 Oil seal
22 Inner sleeve
23 Bolt
24 Allen screw
25 Washer
26 Spacer
27 Chain guide
28 Suspension unit
29 Bolt
30 Washer
31 Sleeve
32 Nut
33 Bush
34 Sleeve
35 Dust seal
36 Bolt
37 Thrust cover
38 Bush
39 Sleeve
40 Relay arm
41 Washer
42 Nut
43 Adjuster pulley bracket
44 O-ring
45 Washer
46 Adjuster belt
47 Adjuster pulley
48 Circlip
49 Bush ○
50 Headed bush ○

△ RD350 N and F only
□ RD350 LC II only
○ 1983 RD350 LC II only

careful note of the position of the various spacing washers and shims. These must be refitted in their original positions. Lower the assembly to the ground and remove it from the machine. Before refitting the suspension components, clean, check and lubricate them as described in Section 13 of this Chapter.

4 Refit the swinging arm, linkage and suspension unit by reversing the removal sequence. Check that all shims, seals and thrust caps are in place and well lubricated. Tighten the relay arm to frame pivot bolt and the link to swinging arm nuts to 4.0 kgf m (28 lbf ft). Tighten the suspension unit lower mounting to relay arm and link bolt to 6.5 kgf m (47 lbf ft) and the swinging arm pivot shaft nut to 7.0 kgf m (50 lbf ft). Lubricate the swinging arm pivot by pumping grease into the grease nipple provided, then check that the suspension moves smoothly and without discernible free play.

12 Rear suspension unit: examination and renovation

1 As mentioned previously, the rear suspension unit is of sealed construction and thus cannot be repaired in the event of failure. Should the damping effect become reduced as a result of wear it is advisable to obtain a new replacement unit well in advance of intended renewal.

2 Should it become necessary to dispose of the cylinder do not just throw it away. It is first necessary to release the gas pressure and the manufacturers recommend that the following procedure is followed.

3 Refer to the accompanying figure and mark a point 10 – 15 mm above the bottom of the cylinder. Place the unit securely in a vice. Wearing proper eye protection against escaping gas and/or metal particles, drill a 2 – 3 mm hole through the previously marked point on the cylinder.

4 Clean the suspension unit, paying particular attention to the area around the preload adjuster. If this should seize completely, soak the adjuster in penetrating oil until it can be freed off. Take care not to damage the pulley teeth. Set the adjuster so that the ring cam is on one of its peaks, then lubricate it with graphite or molybdenum disulphide grease. As an alternative, a graphite-based chain lubricant can be used.

5 It is worth removing the adjuster pulley from the machine so that it can be cleaned and lubricated properly. Remove the circlip which retains the pulley and lift it off its pivot. Clean the pulley teeth and internal bore, and also the pivot. Lubricate the pivot with molybdenum disulphide grease, then refit the pulley and circlip.

11.1a Free wiring clips from rear mudguard ...

11.1b ... and remove YPVS control unit and fuse box from mudguard

11.1c Release two bolts on underside of mudguard ...

11.1d ... and single bolt on left-hand side

11.1e Chainguard is retained by screw and plain washer at front ...

11.1f ... and at rear

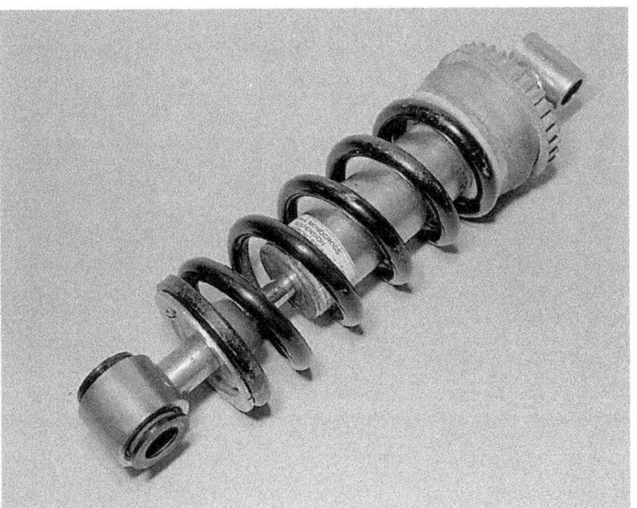

12.4a Clean suspension unit and check for corrosion around spring preload collar

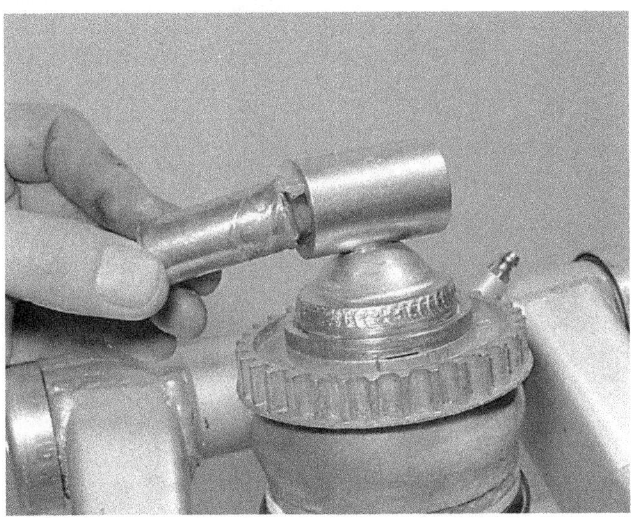

12.4b Clean and grease pivot sleeves before refitting unit

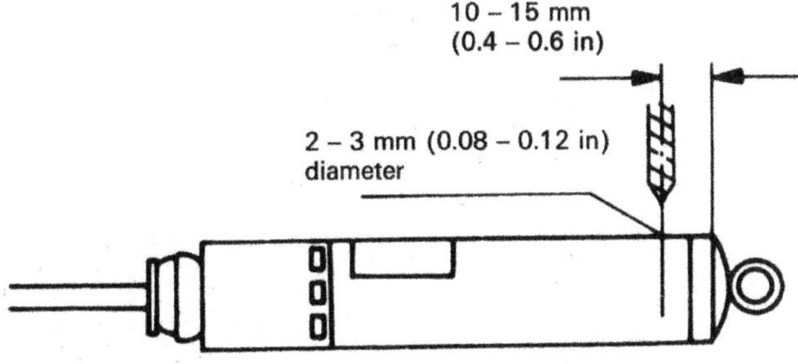

Fig. 5.9 Position of drilling on rear suspension unit

13 Swinging arm and rear suspension linkage: examination and renovation

1 With the swinging arm and its associated linkages removed from the machine, examine the various pivot bolts, sleeves, bushes, seals and thrust caps for signs of wear or damage. If any fault is noted, the affected component should be renewed, together with any related part. For example, if a pivot sleeve shows signs of scoring or other damage, renew it, together with the bushes in which it runs; do not renew the sleeve alone or rapid wear can be expected.

2 Displace and remove the swinging arm pivot sleeve, then degrease and clean it and the swinging arm bore. If the sleeve and the bushes are undamaged and judged to be re-usable, check the swinging arm side clearance as follows. Measure the overall length of the sleeve. This should not have worn significantly under normal circumstances, but if it is not within the range 205.2 – 205.5 mm (8.079 – 8.091 in) it must be renewed, together with the bushes. For details refer to paragraphs 4 to 6 below before proceeding further.

3 Next, measure the distance across the swinging arm bush heads

and subtract this figure from the sleeve length to obtain the side clearance. The specified clearance is 0.1 – 0.3 mm (0.004 – 0.012 in) which can be adjusted using shims fitted between the thrust cap and the bush heads. If additional shims are needed, these can be purchased from Yamaha dealers. When fitting shims, they should be distributed evenly on each side of the pivot. If there is an odd number of shims, put the extra one on the left-hand side.

4 Worn bushes in either the swinging arm or the various linkage pivots can be drifted out of their bores, but note that removal will destroy them; new bushes should be obtained before work commences. The new bushes should be pressed or drawn into their bores, rather than driven into place. In the absence of a press, a suitable drawbolt arrangement can be made up as described below.

5 It will be necessary to obtain a long bolt or a length of threaded rod from a local engineering works or some other supplier. The bolt or rod should be about 1 inch longer than the combined length of the cross tube and one bush. Also required are suitable nuts and two large and robust washers. In the case of threaded rod, fit one nut to one end of the rod and if required, stake it in place for convenience.

6 Fit one of the washers over the bolt or rod so that it rests against the head, then pass the assembly through the cross-tube. Over the projecting end place the bush, which should be greased to ease installation, followed by the remaining washer and nut. Holding the bush to ensure that it is kept square, slowly tighten the nut so that the bush is drawn into the cross-tube. Once it is fully home, remove the drawbolt arrangement and repeat the sequence to fit the remaining bush.

7 When fitting the swinging arm pivot bushes, note that the groove must be fitted within a 90° arc as shown in the accompanying line drawing. Before commencing assembly, check all seals and renew as required, then lubricate the various moving parts as follows. When installation is complete, pump grease into the swinging arm pivot using a grease gun. Wipe off excess grease as it emerges from the ends of the swinging arm. Observe the following greasing requirements:

a) Bushes: Coat the inner surface of all bushes with grease
b) Seals: Fill the lip area of all seals with grease
c) Dust seals: Coat inside and out with grease
d) Thrust caps: Fill the inside with grease
e) Pivot shaft and sleeves: coat outer surface with grease

13.2a Remove end seals and any shims, noting their position

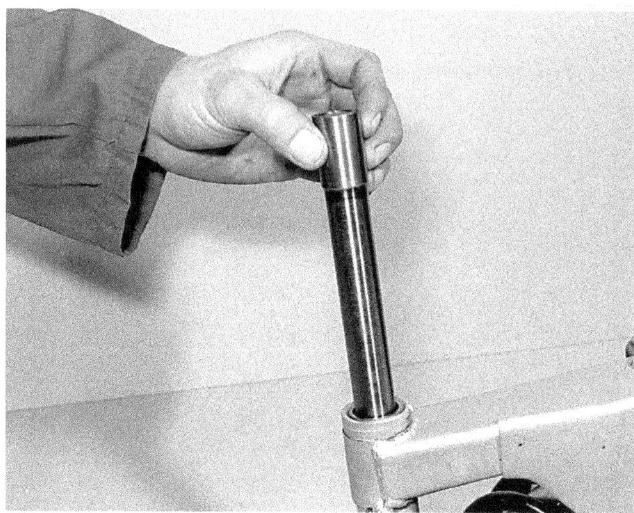

13.2b Slide out the swinging arm pivot sleeve for examination and cleaning

13.3 Do not omit to fit shims of the required thickness (see text)

13.7a Fit new grease seals as required ...

13.7b ... and lubricate with grease ...

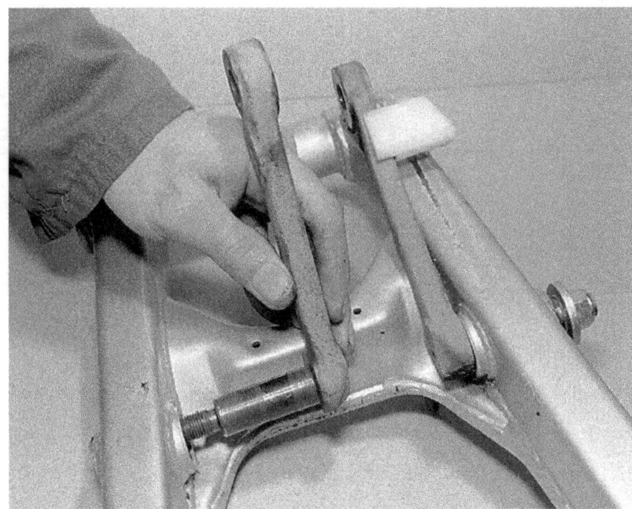
13.7c ... before fitting the link arms

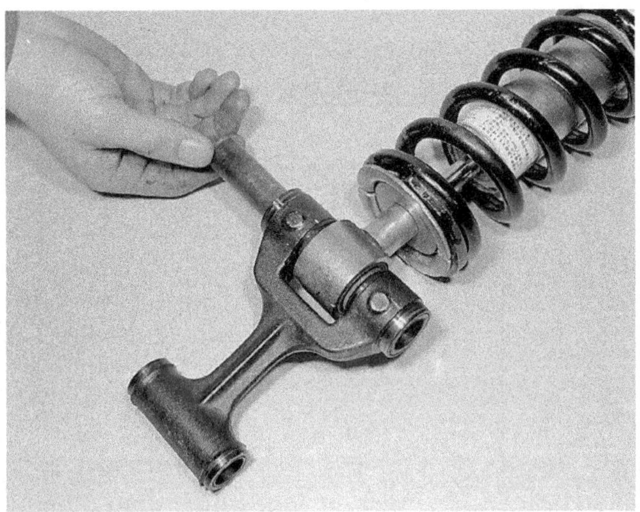
13.7d Grease and fit pivot sleeves ...

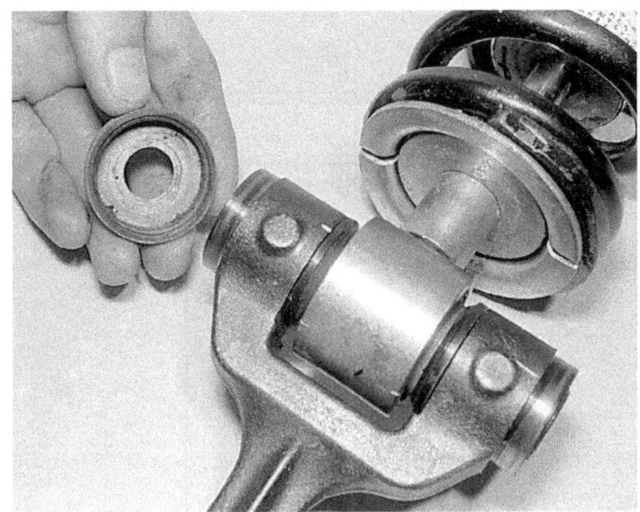
13.7e ... and fit end caps

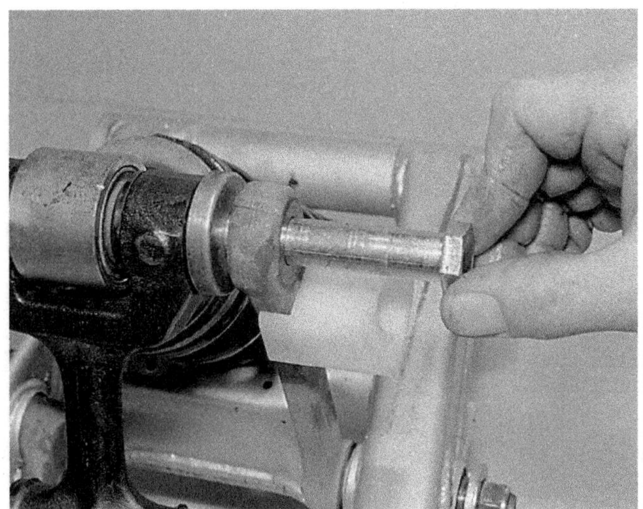
13.7f Fit pivot bolt after greasing ...

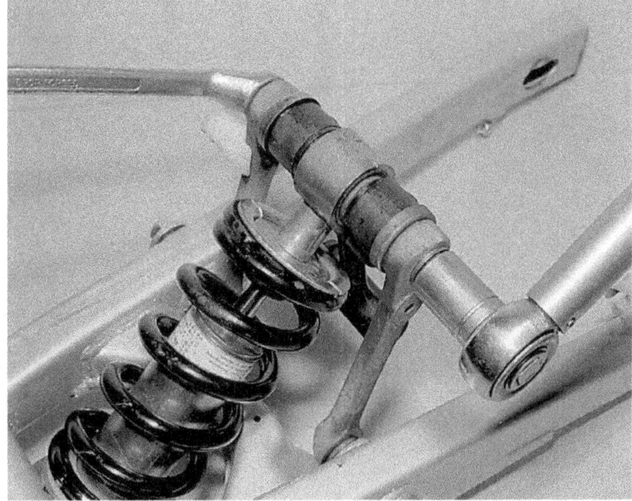
13.7g ... and tighten to prescribed torque setting

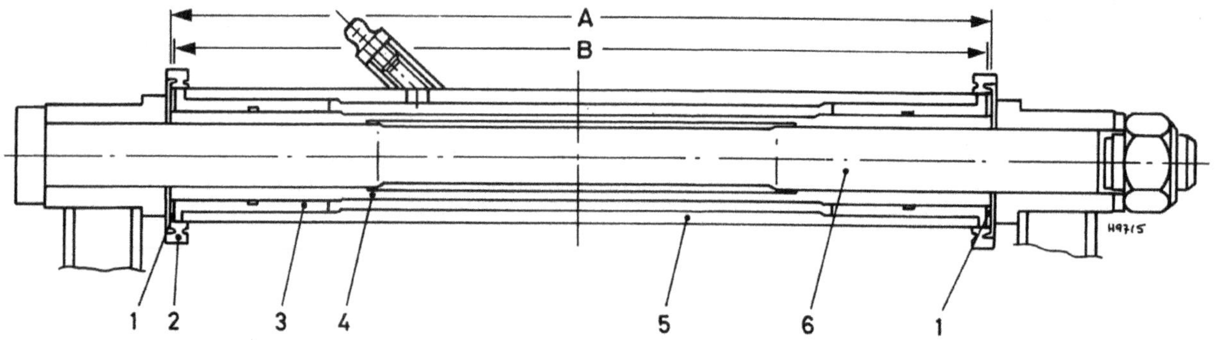

Fig. 5.10 Cross-section of swinging arm pivot components and side clearance measurement datum points – RD350 LC II

1 Shim – as required
2 End cap
3 Bush
4 Pivot sleeve
5 Swinging arm
6 Pivot bolt
A Length of pivot sleeve
B Length across bush heads

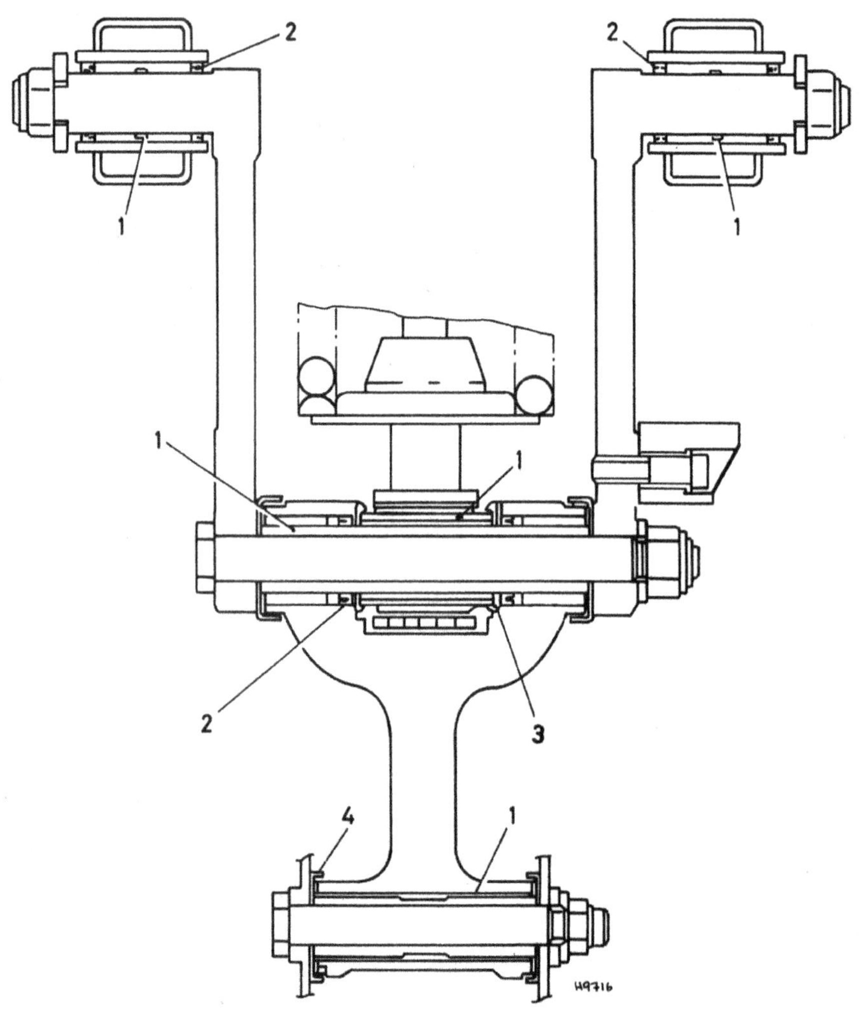

Fig. 5.11 Cross-section of rear suspension linkage pivots

1 Bushes
2 Oil seals
3 Washer
4 Thrust cover

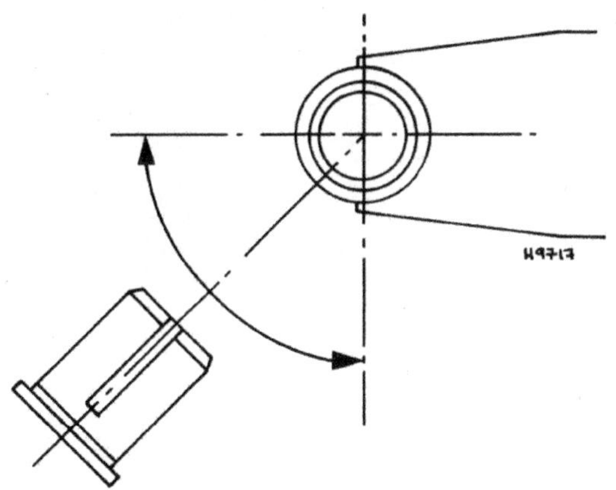

Fig. 5.12 Swinging arm pivot bush groove position – RD350 LC II only

14 Suspension adjustment

1 It is possible to vary the front fork air pressure and rear suspension spring preload to compensate for various loads on the machine. The choice of settings is to some extent discretionary, but as a guide, the following combinations of settings are recommended by the manufacturer.

Loading	Fork air pressure	Rear spring preload
Rider only	0.4 kg/cm², 5.7 psi	2
Rider and luggage	0.6 kg/cm², 8.5 psi	3
Rider and passenger	0.8 kg/cm², 11.0 psi	4
Rider and passenger + luggage	0.8 kg/cm², 11.0 psi	5

Standard rear suspension preload 2
Standard fork air pressure .... 0.4 kg/cm² (5.7 psi)
Maximum fork air pressure .... 0.8 kg/cm² (11.0 psi)

2 When altering the fork air pressure, use only a hand pump and gauge. These are obtainable from motorcycle dealers and are produced specifically for suspension use. Alternatively, use a bicycle tyre pump with a schraeder-type adaptor and a plunger type tyre pressure gauge. When taking pressure readings it should be noted that a small but significant pressure drop will occur each time the check is made. With practice it will be possible to take this into account, allowing one or two psi extra when taking the reading to compensate for the loss of pressure when the gauge is removed. Note that the pressure must be equal between the two legs (within 0.1 kg cm², 1.4 psi) or handling will be impaired. On no account inflate the forks to more than the maximum pressure or the seals may be damaged.

3 When altering the rear suspension spring preload setting, note that the standard setting is position 2, 1 being an extra soft setting, whilst 3, 4 and 5 are harder settings to allow for additional loading. If the adjuster proves too stiff to turn it should be dismantled and lubricated as described elsewhere in this Chapter. Regular operation of the adjuster will help to keep it freed off.

14.2 Check fork air pressures carefully, using pocket gauge – **do not** use airline to pressurise forks

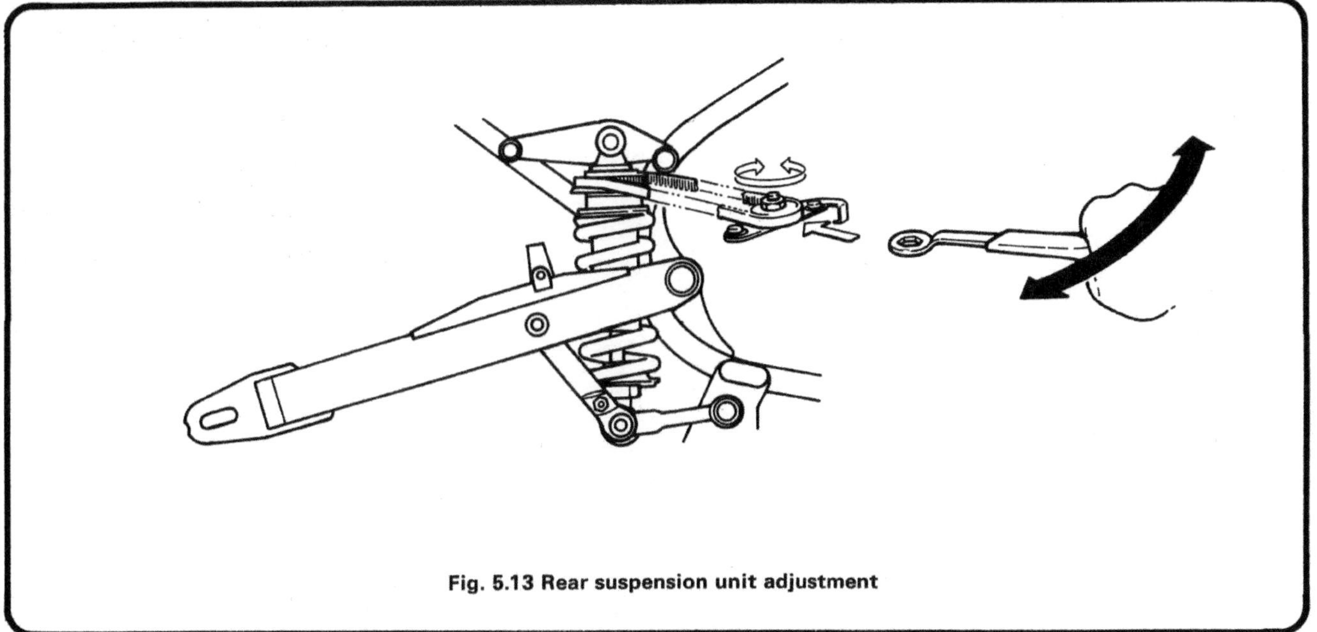

Fig. 5.13 Rear suspension unit adjustment

15 Centre stand: examination and maintenance

1 The centre stand is an important but largely neglected feature of most motorcycles. It is important to check the stand for wear or damage from time to time, as failure of the stand can result in costly repair bills. Check that the stand mounting shaft is secure and in good condition, and that it is kept adequately lubricated.
2 Check that the return spring is in good condition. A broken or weak spring may cause the stand to fall whilst the machine is being ridden, and catch in some obstacle, unseating the rider.

16 Prop stand: examination and maintenance

1 The prop stand is attached to a lug welded to the left-hand lower frame tube. An extension spring anchored to the frame ensures that the stand is retracted when the weight of the machine is taken off the stand.
2 Check that the pivot bolt is secured and that the extension spring is in good condition and not overstretched. An accident is almost certain if the stand extends whilst the machine is on the move.

17.1 Instrument panel mounting bolt (F model)

17 Instrument panel: removal and refitting

1 The instrument panel forms a self-contained unit and is attached to the headlamp brackets via rubber-mounted bolts. On the RD350 F model, the assembly is bolted to the fairing subframe. The panel can be removed after the relevant wiring and instrument drive cables have been disconnected; this will be self-explanatory upon examination.
2 The speedometer on all models is mechanically driven via a flexible cable from the front wheel drive gearbox. In the event of a fault, always check the cable first. If the instrument head is at fault, it will be necessary to fit a new one, unless a specialist repairer can help.
3 The tachometer on the RD350 LC II is mechanically driven from a take-off point near the rear of the crankcase, and can be dealt with in the same way as the speedometer. On all other models the tachometer is electronic, measuring engine speed by monitoring the ignition pulses.

18 Instrument drive cables: examination and maintenance

1 It is advisable to detach the drive cable(s) from time to time in order to check whether the outer coverings are damaged or compressed at any point along their run. Jerky or sluggish movements can be traced to a damaged drive cable.
2 It is not possible to effect a satisfactory repair to a damaged or broken drive cable, and in this event the complete cable must be renewed.

19 Instrument drives: examination and maintenance

1 Drive to the speedometer is taken from a small gearbox mounted on the front wheel and anchored to the left-hand fork leg. The gearbox rarely gives rise to problems provided that it is kept well greased whenever the front wheel is removed. In the event of failure, the gearbox must be replaced as a unit, no individual parts are available.
2 In the case of the RD350 LC II model only, the tachometer is driven from a mechanism incorporated in the engine unit, its takeoff point being near the rear of the crankcase. Part of the drive is contained within the crankcase halves and it is therefore necessary to remove and dismantle the engine unit to gain access to it. Further information on the tachometer drive will be found in Chapter 1.

Chapter 6 Wheels, brakes and tyres

For information relating to the RD350 F II, N II and R models, refer to Chapter 8

Contents

Specifications

	Front	Rear
Wheels		
Type	Cast aluminium alloy	Cast aluminium alloy
Size	MT2.15 x 18	MT2.50 x 18
Maximum runout at rim:		
Radial	1.0 mm (0.04 in)	1.0 mm (0.04 in)
Axial	0.5 mm (0.02 in)	0.5 mm (0.02 in)
Tyres		
Size	90/90 – 18 51H	110/80 – 18 58H
Pressures (cold):		
Up to 90 kg (198 lb) load	26 psi	28 psi
90 – 211 kg (198 – 428 lb) load	32 psi	40 psi
High speed riding	28 psi	32 psi
Brakes		
Type	Twin disc brake	Single disc brake
Disc diameter	267 mm (10.5 in)	267 mm (10.5 in)
Disc thickness	5.0 mm (0.19 in)	5.0 mm (0.19 in)
Service limit	4.5 mm (0.18 in)	4.5 mm (0.18 in)
Pad thickness:		
RD350 LC II	6.8 mm (0.27 in)	6.8 mm (0.27 in)
Service limit	0.8 mm (0.03 in)	0.8 mm (0.03 in)
Other models	5.5 mm (0.22 in)	5.5 mm (0.22 in)
Service limit	0.5 mm (0.02 in)	0.5 mm (0.02 in)

Master cylinder bore ID	15.87 mm (0.62 in)	12.70 mm (0.51 in)
Caliper bore ID	38.18 mm (1.50 in)	38.18 mm (1.50 in)
Brake fluid type	DOT 4, 3 or SAE J1703	DOT 4, 3 or SAE J1703

Torque wrench settings

Component	kgf m	lbf ft
Front wheel spindle	7.5	54.0
Rear wheel spindle:		
RD350 LC II	10.0	72.0
Other models	10.5	75.0
Rear wheel sprocket	3.3	24.0
Brake disc mounting bolts	2.0	14.0
Master cylinder hose unions	2.5	18.0
Hydraulic hose to 3-way union	2.5	18.0
Hydraulic hose to caliper	2.5	18.0
Caliper bracket bolts	3.5	25.0
Caliper bleed screw	0.5	4.0

1 General description

The Yamaha RD350 YPVS models are fitted with cast aluminium alloy wheels carrying tubeless tyres. The front brake is a twin hydraulic disc unit, the rear brake being a single hydraulic disc.

2 Front wheel: examination and renovation

1 Carefully check the complete wheel for cracks and chipping, particularly at the spoke roots and the edge of the rim. As a general rule a damaged wheel must be renewed as cracks will cause stress points which may lead to sudden failure under heavy load. Small nicks may be radiused carefully with a fine file and emery paper (No 600 – No 1000) to relieve the stress. If there is any doubt as to the condition of a wheel, advice should be sought from a reputable dealer or specialist repairer.
2 Each wheel is covered with a coating of lacquer, to prevent corrosion. If damage occurs to the wheel and the lacquer finish is penetrated, the bared aluminium alloy will soon start to corrode. A whitish grey oxide will form over the damaged area, which in itself is a protective coating. This deposit however, should be removed carefully as soon as possible and a new protective coating of lacquer applied.
3 Check the lateral and radial run out at the rim by spinning the wheel and placing a fixed pointer close to the rim edge. If the maximum run out is greater than that specified the manufacturer recommends that the wheel be renewed. This is, however, a counsel of perfection; a run out somewhat greater than this can probably be accommodated without noticeable effect on steering. No means is available for straightening a warped wheel without resorting to the expense of having the wheel skimmed on all faces. If warpage was caused by impact during an accident, the safest measure is to renew the wheel complete. Worn wheel bearings may cause rim run out. These should be renewed.

3 Front disc brake: pad renewal – RD350 LC II

1 The brake pads can be removed for inspection or renewal without disturbing the caliper unit or hydraulic system. In view of the relative ease of the operation it is recommended that the pads are removed for examination, rather than attempting this with them in position.
2 Slacken and remove the pad retaining bolt between the bleed nipple and hose union. The pads should now be free and can be slid out. If they prove reluctant to come free, gently lever the caliper piston inwards by about 1 mm to give a little extra clearance.
3 Examine the pads for signs of wear, damage or contamination. The backing metal of each pad has a raised section near each end. If the friction material is worn down to this indicator, renew the pads as a set. **Do not** rely on the central groove in the friction material as a reliable indication of the degree of wear; the pads tend to wear more at one end than the other. If there are signs that the friction material is beginning to crack, renew the pads. Where the pad surface appears damp, suspect a fluid leak from the caliper. This must be rectified immediately, before new pads are fitted.
4 If the jaw area of the caliper is heavily coated with brake dust it is suggested that the caliper is removed from the fork leg to allow better access for cleaning. Remove the accumulated dust using a small paint brush dipped in methylated spirit, and work outside or in a well ventilated position. Take care not to inhale any of the asbestos-based friction material.
5 Due to the amount of road salt used in the UK, lubrication of certain areas of the pads and caliper is required to ensure that the pads move freely in the caliper. Before installing the pads, apply a thin film of Duckhams Copper 10 or equivalent to the following areas:

a) To the edges of the metal backing on the pads.
b) To the pad retaining pins or bolt shaft.
c) To the areas of the caliper where the pads rub.
d) To the threads of the caliper mounting bolts.

Apply a thin film of Shin-Etsu G-40M or equivalent silicone grease to the following:

e) Exposed areas of the caliper pistons.
f) The areas of the pad backing plates that contact the piston(s).

Caution: *Do not use too much Copper 10, and do not apply it to the pad anti-rattle spring. Make sure that none contacts the brake discs or the pad friction material.*
6 Fit the new pads by reversing the removal sequence, ensuring that they locate correctly against the shims. Note that the manufacturer recommends that the shims and the retaining bolts are renewed each time new pads are fitted.

4 Front disc brake: pad renewal – RD350 F and N

1 The RD350 F and N models employ twin piston brake calipers in place of the single piston type previously fitted. The procedure for pad renewal thus differs slightly from that described for the RD350 LC II. Start by slackening the two Allen-headed pad retaining pins. Next, remove the caliper mounting bolts and lift the caliper clear of the disc. Remove the pad pins and lift away the pads and shim.
2 Examine the pads for signs of wear, damage or contamination. The backing metal of each pad has a raised section near each end. If the friction material is worn down to this indicator at either end, renew the pads as a set. Do **not** rely on the central groove in the friction material as a reliable indication of the degree of wear; the pads tend to wear more at one end than the other.
3 If there are signs that the friction material is beginning to crack, renew the pads. Where the pad surface appears damp, suspect a fluid leak from the caliper. This must be rectified immediately, before new pads are fitted.
4 Remove the accumulated dust around the caliper jaw area using a small paint brush dipped in methylated spirit, working outside or in a well ventilated position. Take care not to inhale any of the asbestos-based friction material.
5 Lubricate the brake pad and caliper areas detailed in Paragraph 5 of Section 3.
6 Fit the new pads by reversing the removal sequence, ensuring that they locate correctly against the central anti-rattle shim. Note

5.1a To release the rear caliper pads, first slacken the pad retaining pins ...

5.1b ... then release the caliper from its mounting and remove the pad pins fully

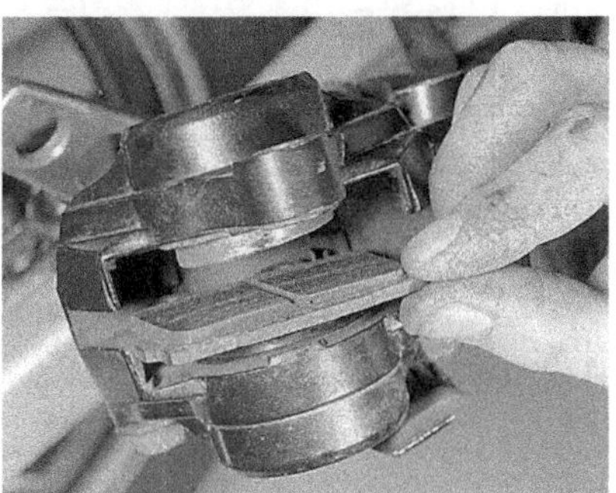

5.1c The pads can then be withdrawn from the underside of the caliper

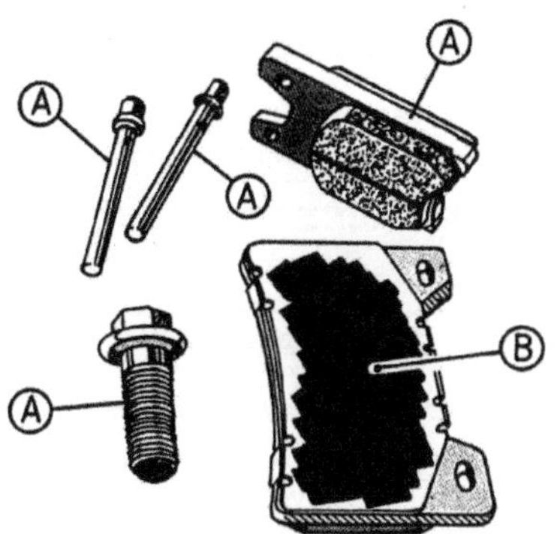

Fig. 6.1 Brake pad lubricant application points
A Apply Duckhams Copper 10 or equivalent
B Apply Shin-Etsu G-40M or equivalent silicone grease

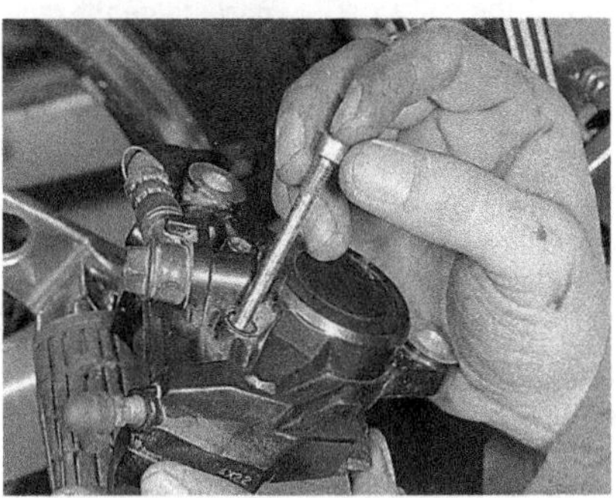

5.1d When refitting pads, hold them against the pad spring while the pins are inserted

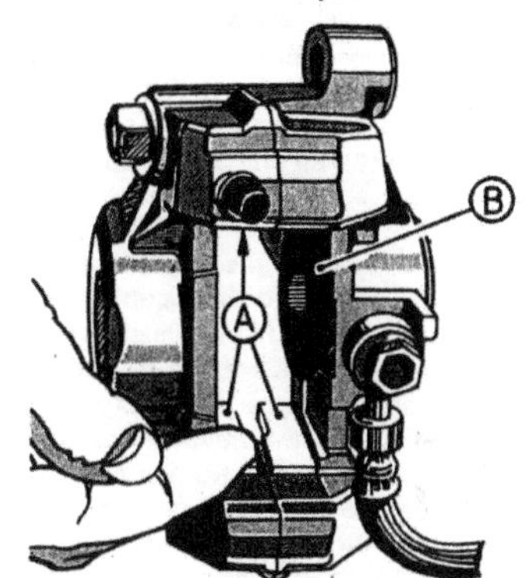

Fig. 6.2 Brake caliper lubricant application points
A Apply Duckhams Copper 10 or equivalent
B Apply Shin-Etsu G-40M or equivalent silicone grease

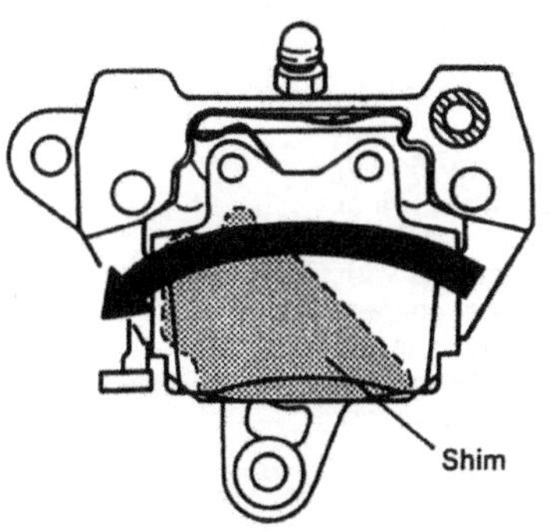

Fig. 6.3 Front brake pad shim installation – RD350 F and N
Arrows show direction of wheel rotation

that each pad has a thin backing shim. Each of these has a chamfered corner; this must face upwards and to the rear of the machine.
7 Fit the pad pins finger tight, then refit the caliper to the fork leg. Note that to accommodate the new pads it will probably be necessary to push the pistons back into the caliper body. Fit and tighten the caliper mounting bolts to 3.5 kgf m (25 lbf ft), then tighten the pad retaining pins to 1.0 kgf m (7.2 lbf ft).

5 Rear disc brake: pad renewal – all models

1 Rear pad renewal is basically similar to front pad renewal described in Section 4, and the remarks in the preceding Section can be applied in most respects. Note, however, that in the case of the RD350 LC II model, the caliper is provided with an inspection window concealed by a clip-on cover. If this is prised off and the two pad retaining pins removed the pads can be lifted away, leaving the caliper in position. On later machines the pads can be removed after releasing the caliper mounting bolts.
2 Lubricate the brake pad and caliper areas detailed in Paragraph 5 of Section 3.

6 Front brake caliper: removal, overhaul and refitting – RD350 LC II

1 The caliper units should be removed and dismantled whenever there is reason to suspect fluid leakage. Remember that your life, and that of other road users, depends on the condition of the braking system more than any other component. When working on the brakes, keep everything absolutely clean, and always work on one caliper at a time to avoid interchanging parts between them.
2 The caliper unit is of the single piston floating type, the caliper body being free to move sideways along a support pin in relation to the caliper mounting bracket. When the handlebar lever is squeezed the piston is displaced pushing the moving pad against the disc. This then causes the caliper body to move slightly in the opposite direction until the fixed pad exerts equal pressure on the opposite face of the disc.
3 If the caliper unit warrants removal for inspection or renovation, it is first necessary to remove and drain the hydraulic hose. Disconnect the union at the caliper. Have a suitable container in which to catch the

fluid. At this stage, it is as well to stop the flow of fluid from the reservoir, by holding the front lever in against the handlebar. This is easily done using a stout elastic band, or alternatively, a section cut from an old inner tube.

4 **Note:** Brake fluid will discolour or remove paint if contact is allowed. Avoid this where possible and remove accidental splashes immediately. Similarly, avoid contact between the fluid and plastic parts such as instrument lenses, as damage will also be done to these. When all the fluid is drained from the hose, clean the connections carefully and secure the hose end and fittings inside a clean polythene bag, to await reassembly. As with all hydraulic systems, it is most important to keep each component scrupulously clean, and to prevent the ingress of any foreign matter. For this reason, it is as well to prepare a clean area in which to work, before further dismantling. As in any form of component dismantling, ensure that the outside of the caliper is thoroughly cleaned down.

5 The caliper unit is attached to the fork leg by two bolts, which, when removed, will allow the unit to be lifted away. If the caliper is being removed with the front wheel in position, it should be lifted clear of the disc. Remove the brake pads as described in the preceding Section, exposing the piston. The piston may be driven out of the caliper body by an air jet – a foot pump if necessary. Remove the piston seal and dust seal, if necessary, from the caliper body. Under no circumstances should any attempt be made to lever or prise the piston out of the caliper. If the compressed air method fails, temporarily reconnect the caliper to the flexible hose, and use the handlebar lever to displace the piston hydraulically. Wrap some rag around the caliper to catch the inevitable shower of brake fluid.

6 Lay the caliper components out on a clean work surface to await examination. Clean the caliper internals with new brake fluid of the specified type; never use normal cleaning solvents because these will cause the seals to swell and degrade. Carefully examine the piston seal and dust seal and renew if damaged; note that Yamaha recommend that these seals be renewed every two years as a matter of course. Examine the piston and caliper bore for any signs of scoring on their working surfaces or evidence of damage from corrosion. If found, such damage will necessitate the renewal of the damaged components. Check the fit of the caliper body on the mounting bracket pins. There should be no sign of excessive free play, or of sticking due to corrosion. If corrosion has occurred on the pins this can be polished away with fine emery paper. Note that corrosion in this area is often indicative of failed dust seals; check and renew the seals if necessary.

7 On reassembly, lubricate the piston seals and piston surface with fresh brake fluid and carefully insert the piston in the caliper. Lubricate the caliper and pad areas detailed in Section 3, Paragraph 5 of this Chapter. Continue the reassembly sequence in a reversal of the dismantling sequence, noting that the system must be filled and bled as described in Section 12.

7 Front brake caliper: removal, overhaul and refitting – RD350 F and N models

1 The general procedure for dealing with the front brake calipers on the later models is similar to that outlined above for the RD350 LC II, but the new caliper design necessitates a few detail changes in the approach to the overhaul. These are outlined below.

2 Start by disconnecting the brake hoses at the caliper unions and drain the hydraulic system as described above. Remove the calipers from the fork leg after removing the two bolts which retain each one. Note: **on no account** slacken or remove the caliper bridge bolts which secure the two halves of the caliper body; these do not need to be separated during overhaul. Remove the pad pins and lift away the pads, their backing shims and the anti-rattle shim.

3 Before proceeding any further, clean thoroughly the outside of the caliper using a rag soaked in methylated spirit. Do not use petrol or similar solvents; this will damage the dust seals and must be kept well away from the caliper. It will be noted that the RD350 F and N models use opposed piston calipers in place of the earlier single piston design, and thus there are two pistons and sets of seals to each unit. It is important that care is taken to fit pistons to the bore from which they were removed.

4 To displace the pistons it is necessary to hold one in place while the other is expelled using compressed air. Yamaha recommend that the right-hand piston is held in place using a pair of slip-joint pliers with rag to protect the piston surface. Compressed air is then applied to the fluid inlet to expel the piston. When work on the first piston is complete, refit it and then expel the remaining piston in a similar manner. In the absence of compressed air, temporarily refit the hydraulic hose, and gradually "pump" the pistons out of their bores. Whichever method is employed take great care to avoid getting fingers trapped by the emerging pistons.

5 Lay the caliper components out on a clean work surface to await examination. Clean the caliper internals with new brake fluid of the specified type; never use normal cleaning solvents because these will cause the seals to swell and degrade. Deal with each piston assembly separately to avoid interchanging them. If both are removed from the caliper at the same time it is advisable to clean them with methylated spirit and then to mark them 'L' or 'R' using a spirit-based marker pen on the recessed outer face of each one. Examine each piston and caliper bore for any signs of scoring on their working surfaces or evidence of damage from corrosion. If found, such damage will necessitate the renewal of the damaged components.

6 On reassembly, fit new piston fluid and dust seals as a matter of course, lubricating the seals and piston surfaces with fresh brake fluid, then carefully insert each piston in the caliper. Lubricate the caliper and pad areas detailed in Section 3, Paragraph 5 of this Chapter. Continue the reassembly sequence in a reversal of the dismantling sequence, noting that the system must be filled and bled as described in Section 12.

8 Rear brake caliper: removal, overhaul and refitting – all models

1 The rear brake caliper fitted to all models is substantially similar to the front caliper used on the RD350 F and N models. Apart from detail differences relating to the mounting arrangement and position, the caliper is identical in operation, and the information given in Section 7 can be applied.

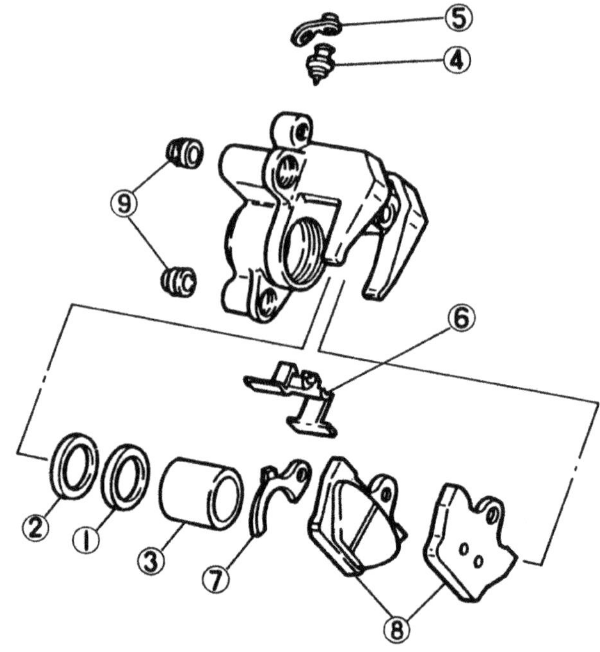

Fig. 6.4 Front brake caliper – RD350 LC II model

1	*Dust seal*	6	*Anti-rattle spring*
2	*Fluid seal*	7	*Shim*
3	*Piston*	8	*Brake pads*
4	*Bleed nipple*	9	*Dust seal*
5	*Cap*		

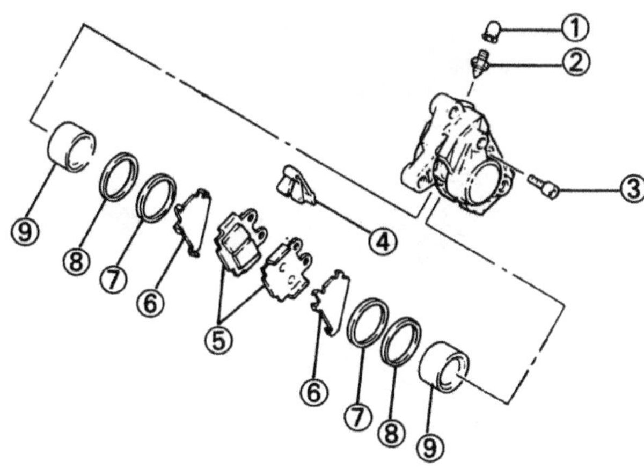

Fig. 6.5 Front brake caliper – RD350 F and N models

1 *Cap*
2 *Bleed nipple*
3 *Allen bolt – 2 off*
4 *Anti-rattle spring*
5 *Brake pad*
6 *Shim*
7 *Dust seal*
8 *Fluid seal*
9 *Piston*

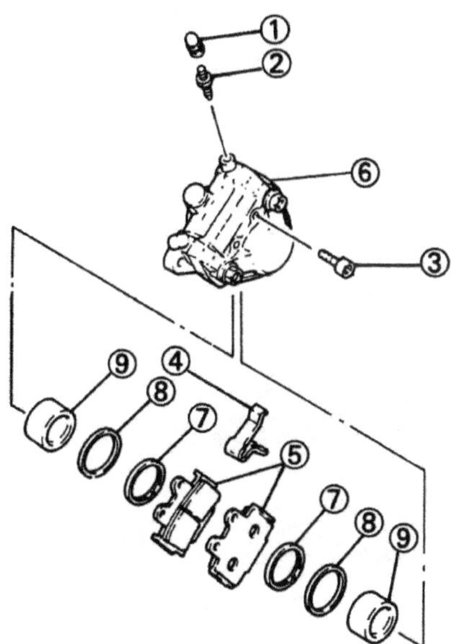

Fig. 6.7 Rear brake caliper – RD350 F and N models

1 *Cap*
2 *Bleed nipple*
3 *Allen bolt – 2 off*
4 *Anti-rattle spring*
5 *Brake pad*
6 *Caliper*
7 *Dust seal*
8 *Fluid seal*
9 *Piston*

9 Front brake master cylinder: removal, overhaul and refitting

1 The master cylinder forms a unit with the hydraulic fluid reservoir and front brake lever, and is mounted by a clamp to the right-hand side of the handlebars.

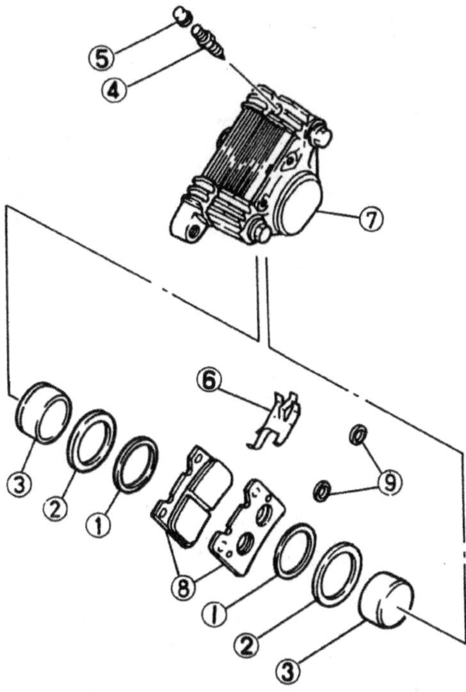

Fig. 6.6 Rear brake caliper – RD350 LC II model

1 *Dust seal*
2 *Fluid seal*
3 *Piston*
4 *Bleed nipple*
5 *Cap*
6 *Anti-rattle spring*
7 *Caliper*
8 *Brake pad*
9 *Seals*

2 The unit must be drained before any dismantling can be undertaken. Place a suitable container below the caliper unit and run a length of plastic tubing from the caliper bleed screw to the container. Unscrew the bleed screw one full turn and proceed to empty the system by squeezing the front brake lever. When all the fluid has been expelled, tighten the bleed screw and remove the tube.

3 Select a suitable clean area in which the various components may be safely laid out, a large piece of white lint-free cloth or white paper being ideal.

4 Remove the locknut and the brake lever pivot bolt to free the lever. As it is lifted away note the small spring which is fitted into the end of the lever blade. Trace back and disconnect the brake switch wiring.

5 Remove the two bolts which hold the master cylinder clamp half to the body and then lift the master cylinder away. Remove the cover and empty the reservoir. If it is still connected, remove the banjo bolt and free the hydraulic hose from the master cylinder body.

6 Pull off the dust seal from the end of the piston bore to expose the piston end and the circlip which retains it. Remove the circlip to free the piston. If the piston tends to stick in the bore it can be pulled clear using pointed-nose pliers. As the piston is removed the main seal and spring will be released.

7 Examine the piston and seals for scoring or wear and renew if imperfect. Excessive scoring may be due to contaminated fluid, and if this is suspected, it is probably worth checking the condition of the caliper seals and piston. Note that the master cylinder seals, piston and spring are supplied as a set – they cannot be obtained individually.

8 Reassemble carefully, using hydraulic fluid as a lubricant on seals and piston, reversing the dismantling sequence. Make sure the rubber boot is fitted correctly, and that the unit is clamped securely to the handlebars. Reconnect the hydraulic hose, tightening the banjo union bolt to the recommended torque setting. Refill the reservoir remembering to top up after the system has been bled by following the procedure given in Section 12 of this Chapter.

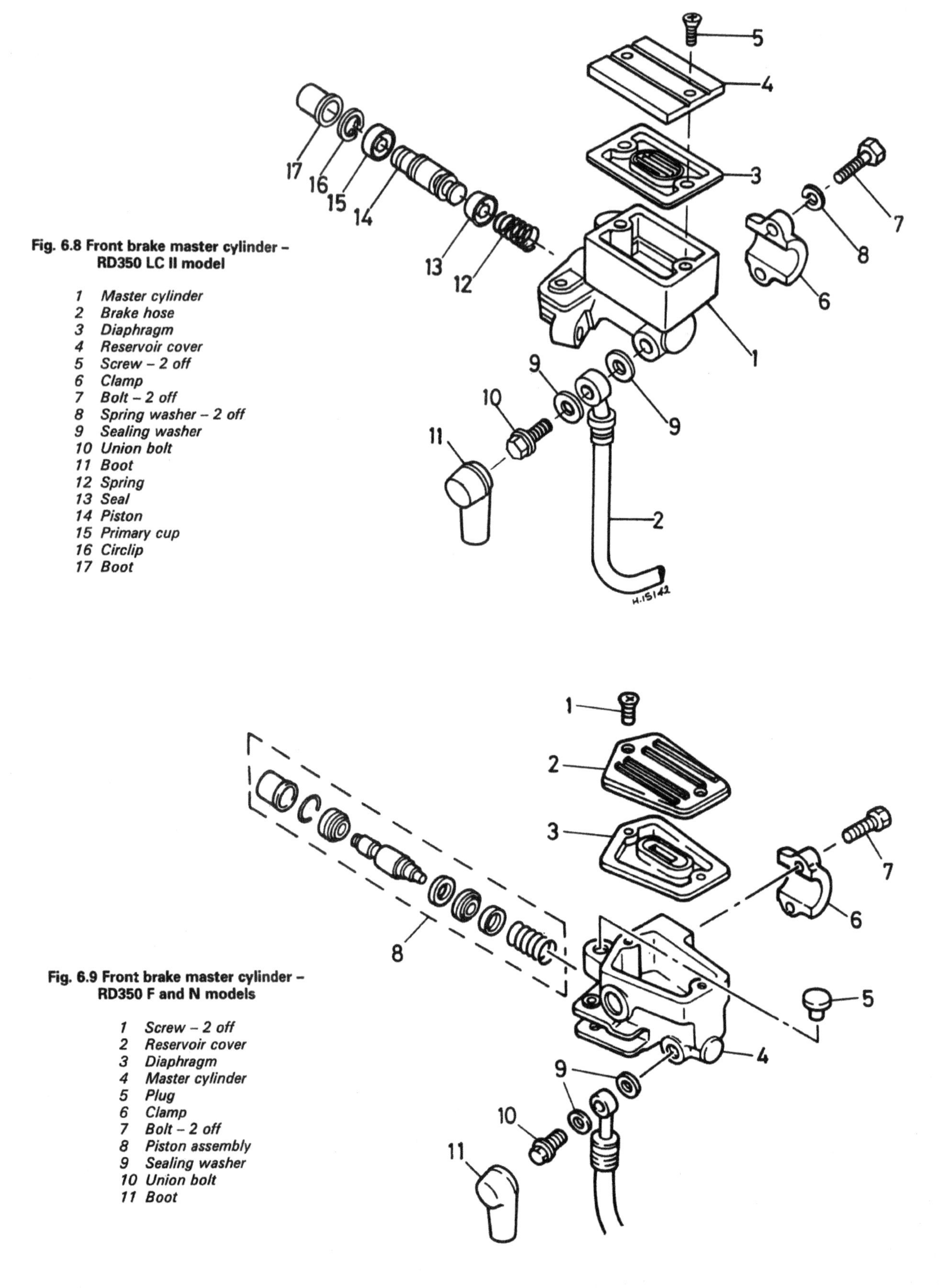

Fig. 6.8 Front brake master cylinder – RD350 LC II model

1 Master cylinder
2 Brake hose
3 Diaphragm
4 Reservoir cover
5 Screw – 2 off
6 Clamp
7 Bolt – 2 off
8 Spring washer – 2 off
9 Sealing washer
10 Union bolt
11 Boot
12 Spring
13 Seal
14 Piston
15 Primary cup
16 Circlip
17 Boot

Fig. 6.9 Front brake master cylinder – RD350 F and N models

1 Screw – 2 off
2 Reservoir cover
3 Diaphragm
4 Master cylinder
5 Plug
6 Clamp
7 Bolt – 2 off
8 Piston assembly
9 Sealing washer
10 Union bolt
11 Boot

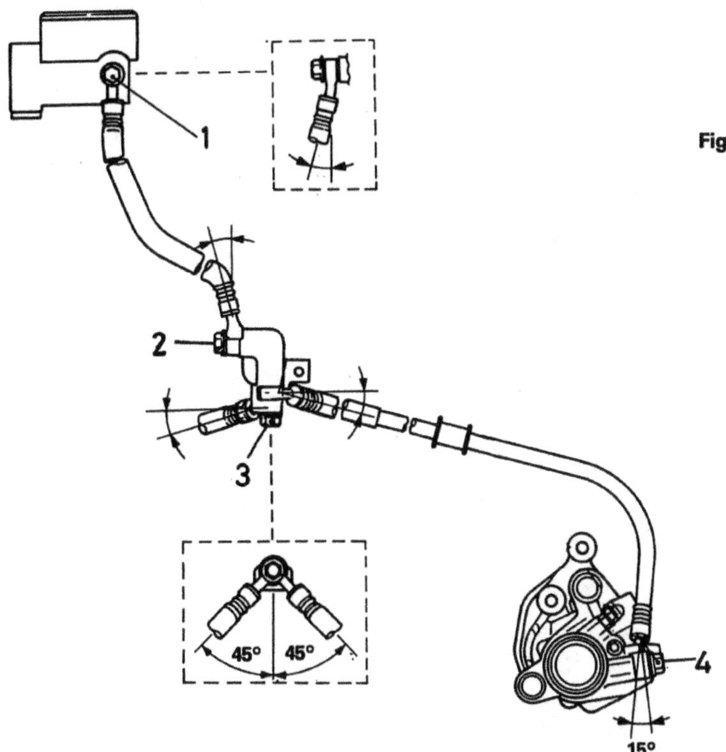

Fig. 6.10 Correct fitted position of front brake hydraulics

1 Master cylinder hose union bolt
2 Hose connection block top union bolt
3 Hose connection block lower union bolt
4 Caliper hose union bolt
5 Sealing washers

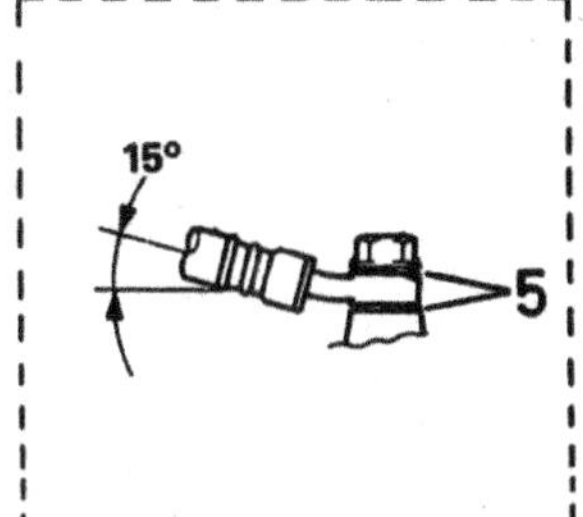

10 Rear brake master cylinder: removal, overhaul and refitting

1 The rear brake master cylinder can be dealt with in much the same way as has been described in Section 8, the main differences arising in the procedure for removing the assembly from the frame.
2 Remove the bolts securing the right-hand footrest plate to the frame and to the silencer, lifting it away to reveal the master cylinder. Connect a bleed tube to the caliper bleed nipple, open the nipple and pump the brake pedal to drain the hydraulic system. Once drained, slacken the hose clip which retains the hose from the reservoir to the master cylinder inlet and disconnect it.
3 Remove the circlip and clevis pin from the end of the master cylinder pushrod and free it from the brake pedal. Remove the two mounting bolts and lift the master cylinder away. Clean the cylinder carefully before dismantling commences. The internal arrangement of the cylinder differs only in detail from that of the front unit, and can be dealt with in a similar manner.

11 Brake discs: examination and removal

1 The brake discs are retained to the hub by six Allen-headed bolts. To remove the disc, first remove the relevant wheel as described later in this Chapter, then unscrew the Allen bolts evenly and in a diagonal sequence. When refitting the disc(s) check that the mounting surfaces are clean and apply Loctite to the Allen bolts. Tighten them evenly and progressively to 2.0 kgf m (14 lbf ft).
2 Examination of the disc can be carried out with the wheel installed. Look for signs of excessive scoring. Some degree of scoring is inevitable, but in severe cases renewal of the disc may prove necessary to restore full braking effect. Check the disc for warpage, which can often result from overheating or impact damage and may cause brake judder. This is best checked using a dial gauge mounted on the fork leg and should not exceed 0.15 mm (0.006 in).
3 The disc thickness should be measured using a vernier caliper or micrometer in several places around the disc surface. The nominal thickness is 5.0 mm (0.19 in) and the disc is in need of renewal if it is worn down to 4.5 mm (0.18 in) or less.

10.3 Rear master cylinder mounting (shown with wheel removed)

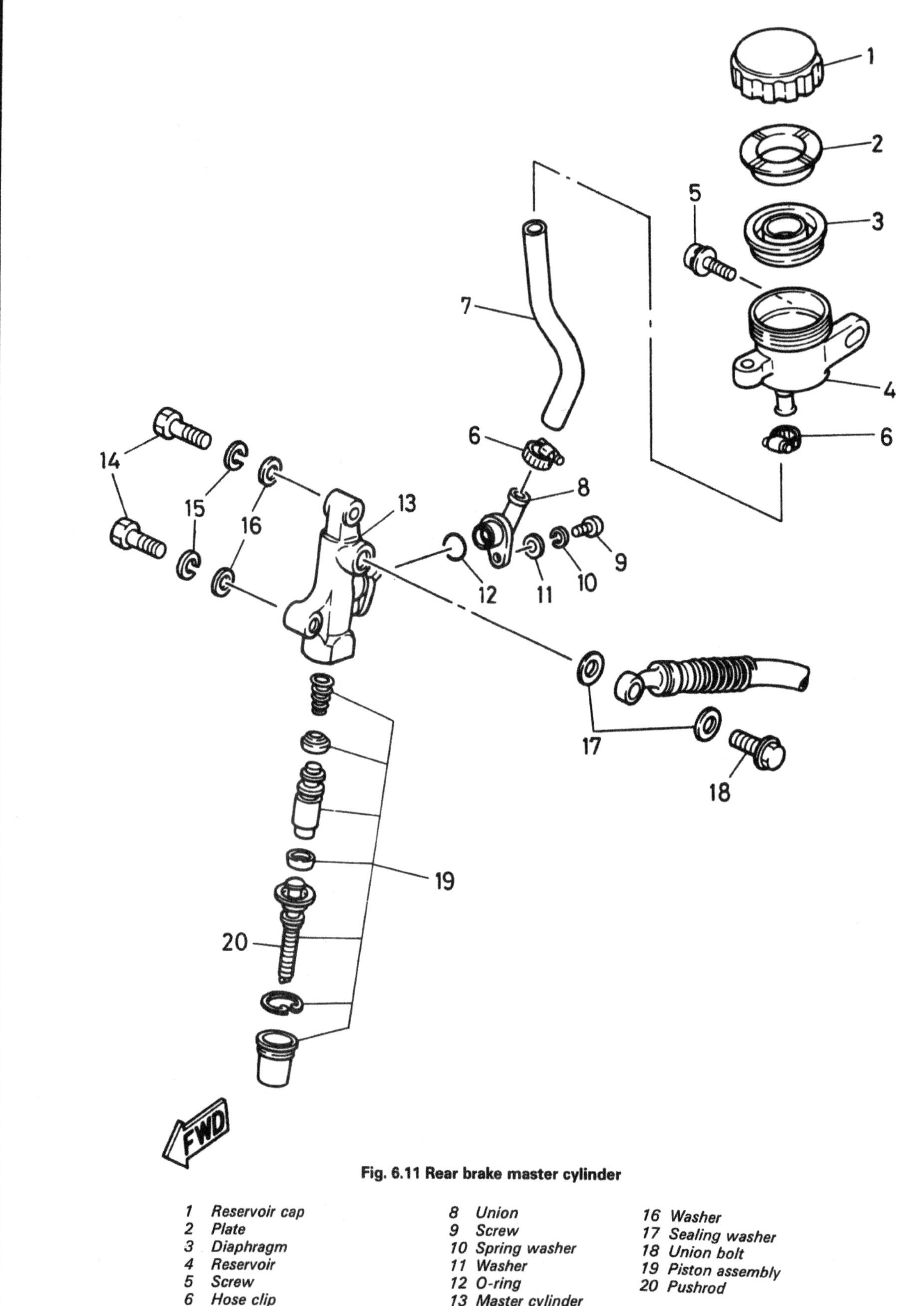

Fig. 6.11 Rear brake master cylinder

1 Reservoir cap
2 Plate
3 Diaphragm
4 Reservoir
5 Screw
6 Hose clip
7 Reservoir to master cylinder hose
8 Union
9 Screw
10 Spring washer
11 Washer
12 O-ring
13 Master cylinder
14 Bolt
15 Spring washer
16 Washer
17 Sealing washer
18 Union bolt
19 Piston assembly
20 Pushrod

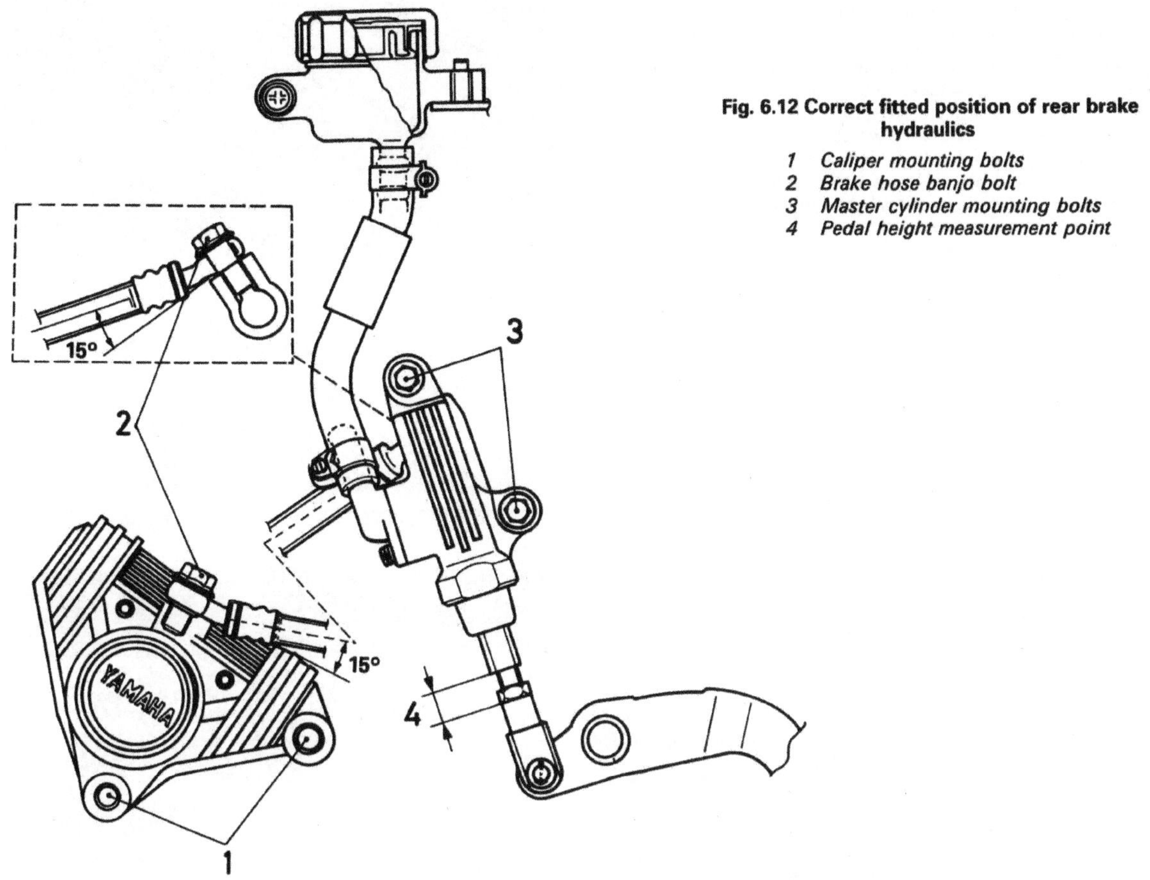

Fig. 6.12 Correct fitted position of rear brake hydraulics

1 *Caliper mounting bolts*
2 *Brake hose banjo bolt*
3 *Master cylinder mounting bolts*
4 *Pedal height measurement point*

12 Bleeding the hydraulic system

1 The method of bleeding a brake system of air and the procedure described below apply equally to either a front brake or rear brake of the hydraulically actuated type.
2 If the brake action becomes spongy, or if any part of the hydraulic system is dismantled (such as when a hose is replaced) it is necessary to bleed the system in order to remove all traces of air. The procedure for bleeding the hydraulic system is best carried out by two people.
3 Check the fluid level in the reservoir and top up with new fluid of the specified type if required. Keep the reservoir at least half full during the bleeding procedure; if the level is allowed to fall too far air will enter the system requiring that the procedure be started again from scratch. Screw the cap onto the reservoir to prevent the ingress of dust or the ejection of a spout of fluid.
4 Remove the dust cap from the caliper bleed nipple and clean the area with a rag. Place a clean glass jar below the caliper and connect a pipe from the bleed nipple to the jar. A clear plastic tube should be used so that air bubbles can be more easily seen. Place some clean hydraulic fluid in the glass jar so that the pipe is immersed below the fluid surface throughout the operation.
5 If parts of the system have been renewed, and thus the system must be filled, open the bleed nipple about one turn and pump the brake lever until fluid starts to issue from the clear tube. Tighten the bleed nipple and then continue the normal bleeding operation as described in the following paragraphs. Keep a close check on the reservoir level whilst the system is being filled.
6 Operate the brake lever as far as it will go and hold it in this position against the fluid pressure. If spongy brake operation has occurred it may be necessary to pump rapidly the brake lever a number of times until pressure is achieved. With pressure applied, loosen the bleed nipple about half a turn. Tighten the nipple as soon as the lever has reached its full travel and then release the lever. Repeat this operation until no more air bubbles are expelled with the fluid into the glass jar. When this condition is reached the air bleeding operation should be complete, resulting in a firm feel to the brake operation. If sponginess is still evident continue the bleeding operation; it may be that an air bubble trapped at the top of the system has yet to work down through the caliper.
7 When all traces of air have been removed from the system, top up the reservoir and refit the diaphragm and cap or cover, as appropriate. Check the entire system for leaks, and check also that the brake system in general is functioning efficiently before using the machine on the road.
8 Brake fluid drained from the system will almost certainly be contaminated, either by foreign matter or more commonly by the absorption of water from the air. All hydraulic fluids are to some degree hygroscopic, that is, they are capable of drawing water from the atmosphere, and thereby degrading their specifications. In view of this, and the relative cheapness of the fluid, old fluid should always be discarded.
9 Great care should be taken not to spill hydraulic fluid on any painted cycle parts; it is a very effective paint stripper. Also, the plastic glasses in the instrument heads, and most other plastic parts, will be damaged by contact with this fluid.
10 It should be noted that there have been some instances where a small air pocket has remained in the rear caliper after normal bleeding, causing spongy operation of the rear brake. If this problem occurs, remove the rear mounting bolt and slacken slightly the front mounting bolt. Tip the caliper upwards, leaving about 30% of the pad material in contact with the disc. Repeat the bleeding operation, which should succeed in removing the residual air. Refit the caliper and tighten the mounting bolts to the specified torque setting.

13 Front wheel: removal and refitting

1 Place the machine on its centre stand and raise the front wheel clear of the ground using wooden blocks or similar. Disconnect the speedometer drive cable from the gearbox by releasing the knurled retaining ring. Remove the two bolts which retain one of the two brake calipers to its fork leg. Lift the caliper clear of the wheel and forks, tying it to the frame to avoid placing undue strain on the hydraulic hose. Removal of one of the calipers allows the tyre sufficient clearance to permit removal; it is not necessary to remove both.

2 Straighten and remove the split pin which retains the wheel spindle nut and remove it. The wheel spindle can now be pulled out to free the wheel, which can be lowered to the ground and lifted clear of the forks. Before proceeding further, insert a wooden wedge between the brake pads in both calipers. This will prevent the pistons from being expelled if the brake lever is operated accidentally.

3 The wheel is refitted by reversing the above sequence. Before fitting the wheel, grease the oil seal lips and the speedometer drive gearbox. Make sure that the latter locates correctly over the lugs on the fork lower leg. Tighten the wheel spindle nut to 7.5 kgf m (54 lbf ft) and fit a new split pin to secure it. Refit the brake caliper and tighten the mounting bolts to 3.5 kgf m (25 lbf ft). Reconnect the speedometer drive cable.

13.1 Unscrew knurled ring to free the speedometer drive cable

13.3a Fit speedometer drive gearbox, ensuring that drive tangs engage in their slots

13.3b Slide wheel spindle into place and fit the plain washer ...

13.3c ... and tighten nut to the correct torque setting

13.3d The nut should be secured with a new split pin

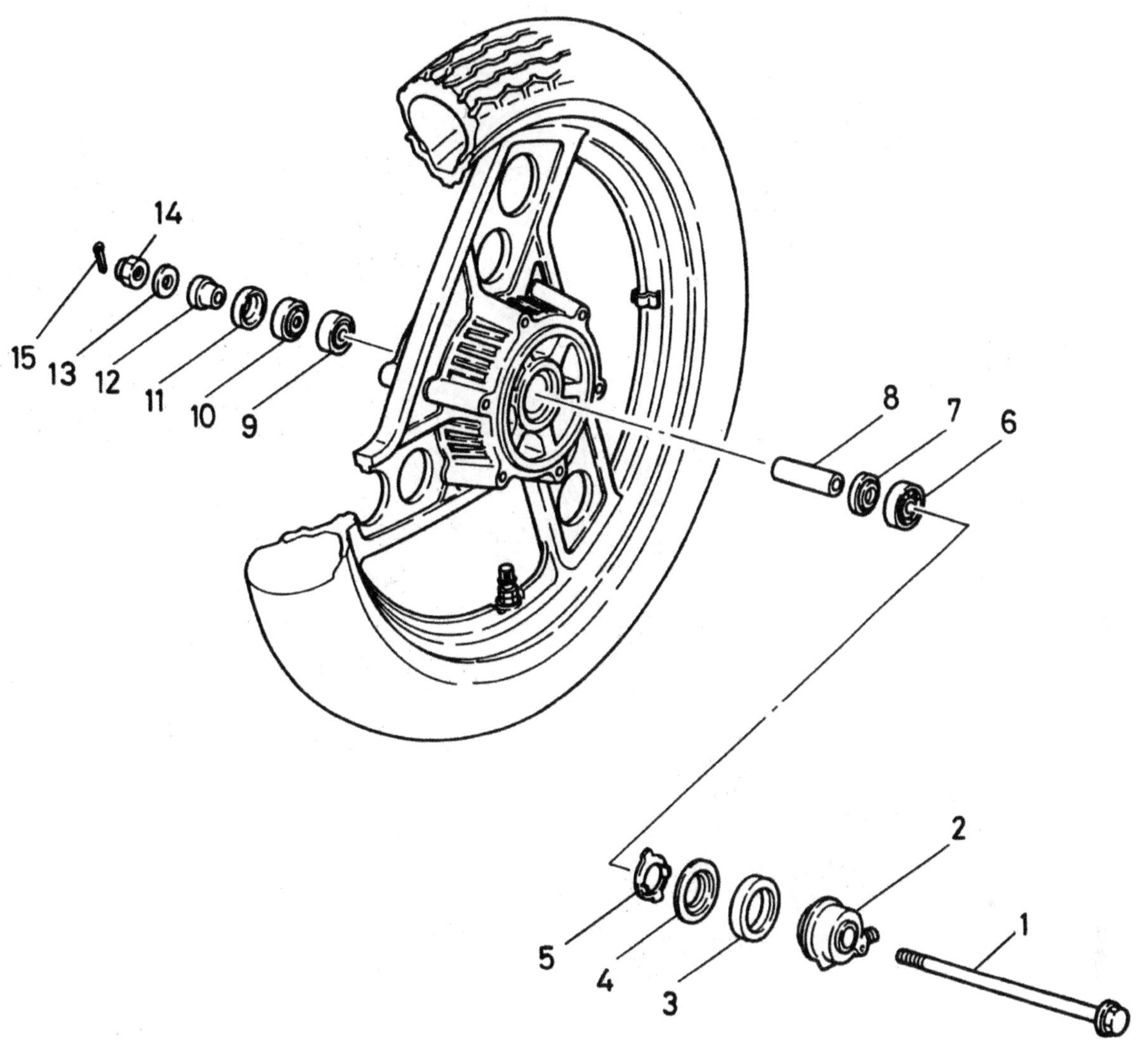

Fig. 6.13 Front wheel

1 *Spindle*
2 *Speedometer gearbox*
3 *Oil seal*
4 *Retainer*
5 *Speedometer drive plate*
6 *Left-hand wheel bearing*
7 *Spacer flange*
8 *Spacer*
9 *Right-hand wheel bearing*
10 *Oil seal*
11 *Collar – RD350 LC II only*
12 *Spacer*
13 *Washer*
14 *Nut*
15 *Split pin*

14 Rear wheel: removal and refitting

1 Place the machine on its centre stand so that the rear wheel is raised clear of the ground. On RD350 LC II models, slacken the chain adjuster locknuts and slacken the adjuster bolts. Push the adjusters clear of the swinging arm ends, then slide the wheel forward so that the chain is slack. In the case of the RD350 F and RD350 N a different design of adjuster is used. These should be slackened off and the wheel pushed fully forward. On all models, lift the chain away from the sprocket and allow it to hang around the wheel spindle.

2 Straighten and remove the split pin which retains the rear wheel spindle nut. Remove the nut, then tap the spindle through to free the wheel, the spacer on the sprocket side and the chain adjusters. Once the wheel has been pulled clear, place a wooden wedge between the brake pads to prevent the accidental expulsion of the caliper pistons if the pedal is depressed.

3 When installing the wheel, grease the oil seal lips. Do not omit the spacer on the left-hand side of the wheel. Check that the adjusters are fitted with the alignment marks outwards. Before tightening the wheel spindle nut, check the chain free play as described in Section 16. Tighten the wheel spindle nut to the figure specified in the Specifications.

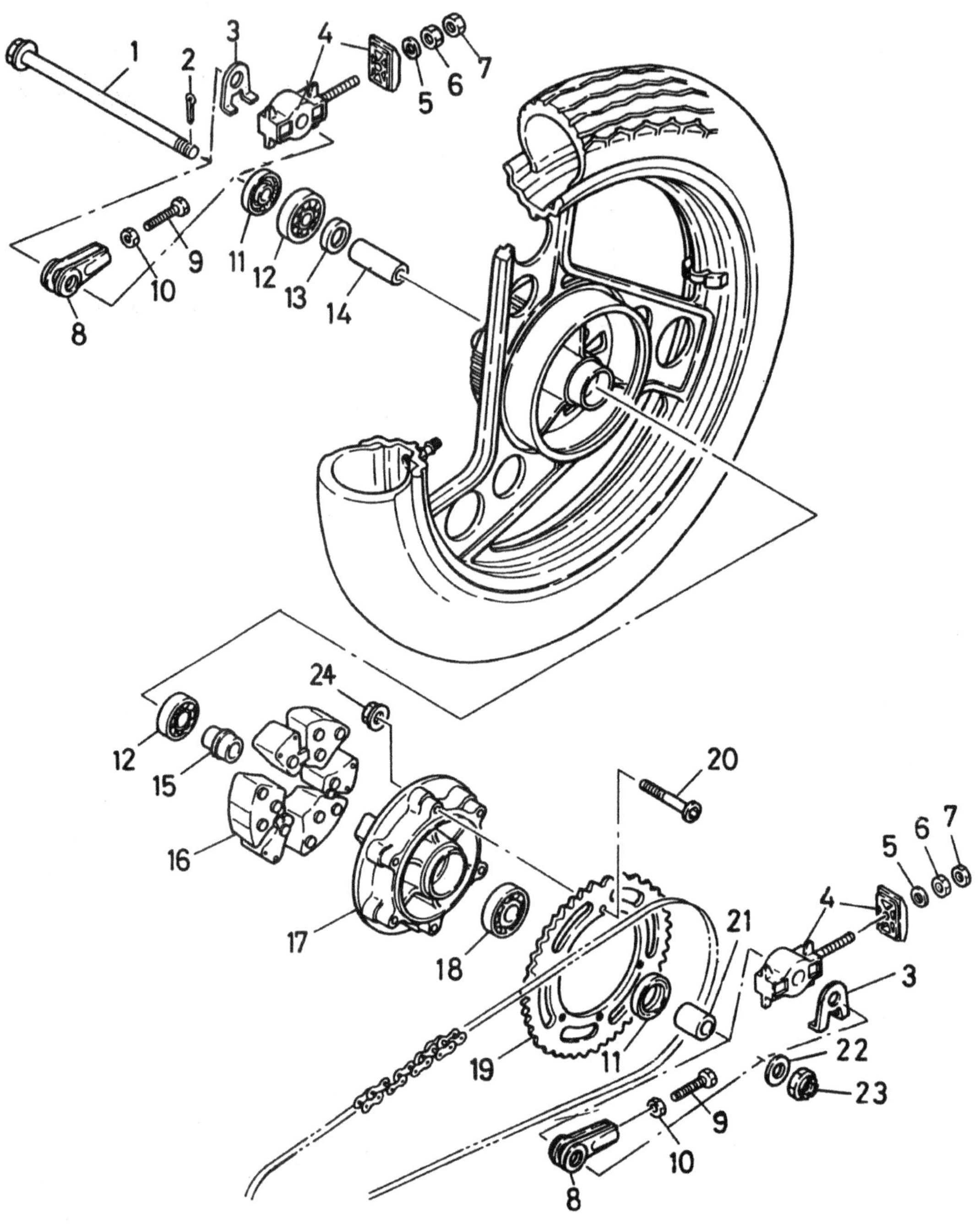

Fig. 6.14 Rear wheel

1 Spindle
2 Split pin
3 Wear indicator plate
4 Chain adjuster
5 Washer
6 Adjusting nut
7 Locknut
8 Chain adjuster
9 Adjusting bolt
10 Locknut
11 Oil seal
12 Wheel bearing
13 Spacer flange
14 Spacer
15 Collar
16 Cush drive rubbers
17 Cush drive flange
18 Bearing
19 Sprocket
20 Bolt – 6 off
21 Spacer
22 Washer
23 Nut
24 Nut – 6 off

Note: Items 3 – 7 fitted to RD350 F and N
Items 8 – 10 fitted to RD350 LC II

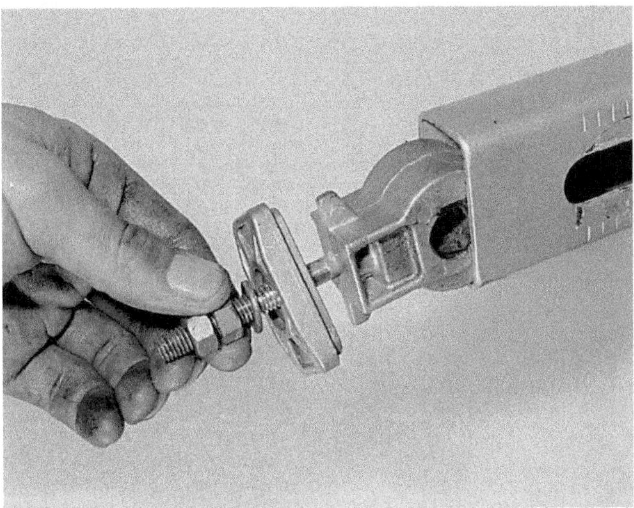
14.3a Later models use this type of chain tensioner

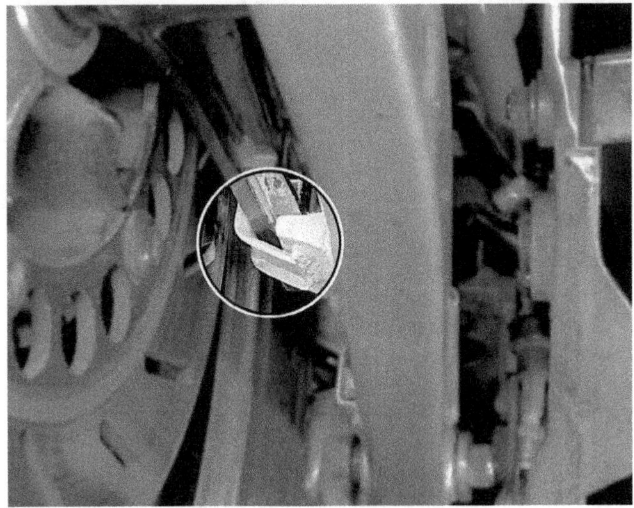
14.3b Note locating slot for rear caliper torque arm

15 Wheel bearings: examination and renewal

1 The front wheel bearings are an interference fit in the wheel hub, and can be removed by passing a long drift through the centre of one bearing and driving the remaining bearing out from the opposite side. It is advisable to support the wheel on wooden blocks to avoid damage to the disc, or to remove the disc from the wheel.

2 With the wheel suitably supported, pass the drift into position, displacing the spacer between the bearings so that the drift can bear on the inner race of the right-hand bearing. Drive the bearing out of the hub, and remove the spacer.

3 Invert the wheel and drive out the left-hand bearing by inserting a drift of the appropriate size, through the hub. During the removal of either bearing it may be necessary to support the wheel across an open-ended box so that there is sufficient clearance for the bearing to be displaced completely from the hub.

4 Remove all the old grease from the hub and bearings, giving the latter a final wash in petrol. Check the bearings for signs of play or roughness when they are turned. If there is any doubt about the condition of a bearing, it should be renewed.

5 Before replacing the bearings, first pack the hub with new grease. Then drive the bearings back into position, not forgetting the distance piece that separates them. Take great care to ensure that the bearings enter the housings perfectly squarely otherwise the housing surface may be broached.

6 The rear wheel bearing arrangement is much the same as that of the front wheel, and the same approach can be adopted if renewal is required. There is a further bearing housed in the cush drive hub, and this too can be removed using a drift and the new bearing tapped home using a large socket. For further information, see Section 16. Do not forget to renew the cush drive hub seal if this is worn or damaged.

15.5a The front wheel bearing arrangement

15.5b Place the headed spacer against the left-hand bearing ...

15.5c ... then install the greased right-hand bearing ...

15.5d ... tapping it home with a large socket

15.5e Fit the grease seal into the hub bore ...

15.5f ... and fit the headed spacer as shown

15.5g Place the speedometer drive dog in position ...

15.5h ... followed by its retainer ...

15.5i .. and the grease seal

15.6 Rear cush drive hub houses extra bearing and a spacer

16 Cush drive assembly: examination and renovation

1 The cush drive assembly is contained within the left-hand side of the rear wheel hub. It comprises a set of synthetic rubber buffers, housed within a series of vanes cast in the hub shell. A plate attached to the centre of the rear wheel sprocket has four cast-in dogs which engage with slots in these rubbers, when the wheel is replaced in the frame. The drive to the rear wheel is transmitted via these rubbers, which cushion any surges of roughness in the drive which would otherwise convey the impression of harshness.
2 Examine the rubbers periodically for signs of damage or general deterioration. Renew and fit the rubbers as a set if there is any doubt about their condition; there is no difficulty in removing or replacing them as they are not under compression when the drive plate is attached.

17 Rear wheel sprocket: examination and renovation

1 The rear wheel sprocket assembly can be removed as a separate unit after the rear wheel has been detached from the frame as described in Section 14 of this Chapter.
2 Check the condition of the sprocket teeth. If they are hooked, chipped or badly worn, the sprocket must be renewed. It is secured to the cush drive plate by six bolts.
3 It is considered bad practice to renew one sprocket on its own. The final drive sprockets should always be renewed as a pair and a new chain fitted, otherwise rapid wear will necessitate even earlier renewal on the next occasion.
4 An additional bearing is located within the cush drive plate, which supports the collar through which the rear wheel spindle fits. In common with the wheel bearings, this bearing is a journal ball and when wear occurs, the sprocket will give the appearance of being loose on its mounting bolts. The bearing is a push fit in the cush drive hub and is secured on the inside by a circlip.
5 Remove the circlip and bearing and wash out the latter to remove all traces of the old grease. If the bearing has any play or runs roughly, it must be renewed.
6 If the bearing has not been renewed it should be repacked with grease and refitted in its housing, followed by the circlip. Replace the rear wheel assembly by reversing whichever method was adopted for its removal.

18 Final drive chain: examination and maintenance

1 The final drive chain is fully exposed, with only a light chainguard over the top run. Periodically the tension will need to be adjusted, to compensate for wear. This is accomplished by placing the machine on the centre stand and slackening the wheel nut on the left-hand side of the rear wheel so that the wheel can be drawn backward by means of the drawbolt adjusters in the fork ends.
2 The chain is in correct tension if there is approximately 30 – 40 mm (1.2 – 1.6 in) slack at a point about 4 in forward of the rear wheel sprocket, on the lower run. Always check when the chain is at its tightest point as a chain rarely wears evenly during service. Note that the tension should be checked with the machine resting on its wheels, though it will be necessary to place it back on the centre stand to carry out adjustment.
3 Always adjust the drawbolts an equal amount in order to preserve wheel alignment. The fork ends are clearly marked with a series of horizontal lines above the adjusters, to provide a simple, visual check. If desired, wheel alignment can be checked by running a plank of wood parallel to the machine, so that it touches the side of the rear tyre. If wheel alignment is correct, the plank will be equidistant from each side of the front wheel tyre, when tested on both sides of the rear wheel. It will not touch the front wheel tyre because this tyre is of smaller cross section. See accompanying diagram.
4 Do not run the chain overtight to compensate for uneven wear. A tight chain will place undue stress on the gearbox and rear wheel bearings, leading to their early failure. It will also absorb a surprising amount of power.
5 After a period of running, the chain will required lubrication. Lack of oil will greatly accelerate the rate of wear of both the chain and the sprockets and will lead to harsh transmission. The application of engine oil will act as a temporary expedient, but it is preferable to remove the chain and clean it in a paraffin bath before it is immersed in a molten lubricant such as 'Linklife' or 'Chainguard'. These lubricants achieve better penetration of the chain links and rollers and are less likely to be thrown off when the chain is in motion.
6 To check whether the chain is due for replacement, lay it lengthwise in a straight line and compress it endwise so that all the play is taken up. Anchor one end and measure the length. Now pull the chain with one end anchored firmly, so that the chain is fully extended by the amount of play in the opposite direction. If there is a difference of more than 1/4 inch per foot in the two measurements, the chain should be renewed in conjunction with the sprockets. Note that this check should be made after the chain has been washed out, but before any lubricant is applied, otherwise the lubricant may take up some of the play.
7 When replacing the chain, make sure that the spring link is seated correctly, with the closed end facing the direction of travel.
8 Replacement chains are now available in standard metric sizes from Renold Limited, the British chain manufacturer. When ordering a new chain, always quote the size, the number of chain links and the type of machine to which the chain is to be fitted.

19 Tyres: removal and refitting

1 It is strongly recommended that should a repair to a tubeless tyre be necessary, the wheel is removed from the machine and taken to a tyre fitting specialist who is willing to do the job or taken to an official dealer. This is because the force required to break the seal between the wheel rim and tyre bead is considerable and considered to be beyond the capabilities of an individual working with normal tyre removing tools. Any abortive attempt to break the rim to bead seal may also cause damage to the wheel rim, resulting in an expensive wheel replacement. If, however, a suitable bead releasing tool is available, and experience has already been gained in its use, tyre removal and refitting can be accomplished as follows.

2 Remove the wheel from the machine by following the instructions for wheel removal as described in the relevant Section of this Chapter. Deflate the tyre by removing the valve insert and when it is fully deflated, push the bead of the tyre away from the wheel rim on both sides so that the bead enters the centre well of the rim. As noted, this operation will almost certainly require the use of a bead releasing tool.

3 Insert a tyre lever close to the valve and lever the edge of the tyre over the outside of the wheel rim. Very little force should be necessary; if resistance is encountered it is probably due to the fact that the tyre beads have not entered the well of the wheel rim all the way round the tyre. Should the initial problem persist, lubrication of the tyre bead and the inside edge and lip of the rim will facilitate removal. Use a recommended lubricant, a diluted solution of washing-up liquid or french chalk. Lubrication is usually recommended as an aid to tyre fitting but its use is equally desirable during removal. The risk of lever damage to wheel rims can be minimised by the use of proprietary plastic rim protectors placed over the rim flange at the point where the tyre levers are inserted. Suitable rim protectors may be fabricated very easily from short lengths (4 – 6 inches) of thick-walled nylon petrol pipe which have been split down one side using a sharp knife. The use of rim protectors should be adopted whenever levers are used and, therefore, when the risk of damage is likely.

4 Once the tyre has been edged over the wheel rim, it is easy to work around the wheel rim so that the tyre is completely free on one side.

5 Working from the other side of the wheel, ease the other edge of the tyre over the outside of the wheel rim, which is furthest away. Continue to work around the rim until the tyre is freed completely from the rim.

6 Refer to the following Section for details relating to puncture repair and the renewal of tyres. See also the remarks relating to the tyre valves in Section 21.

7 Refitting of the tyre is virtually a reversal of removal procedure. If the tyre has a balance mark (usually a spot of coloured paint), as on the tyres fitted as original equipment, this must be positioned alongside the valve. Similarly, any arrow indicating direction of rotation must face the right way.

8 Starting at the point furthest from the valve, push the tyre bead over the edge of the wheel rim until it is located in the central well. Continue to work around the tyre in this fashion until the whole of one side of the tyre is on the rim. It may be necessary to use a tyre lever during the final stages. Here again, the use of a lubricant will aid fitting. It is recommended strongly that when refitting the tyre only a recommended lubricant is used because such lubricants also have sealing properties. Do not be over generous in the application of lubricant or tyre creep may occur.

9 Fitting the upper bead is similar to fitting the lower bead. Start by pushing the bead over the rim and into the well at a point diametrically opposite the tyre valve. Continue working round the tyre, each side of the starting point, ensuring that the bead opposite the working area is always in the well. Apply lubricant as necessary. Avoid using tyre levers unless absolutely essential, to help reduce damage to the soft wheel rim. The use of the levers should be required only when the final portion of bead is to be pushed over the rim.

10 Lubricate the tyre beads again prior to inflating the tyre, and check that the wheel rim is evenly positioned in relation to the tyre beads. Inflation of the tyre may well prove impossible without the use of a high pressure air hose. The tyre will retain air completely only when the beads are firmly against the rim edges at all points and it may be found when using a foot pump that air escapes at the same rate as it is pumped in. This problem may also be encountered when using an air hose on new tyres which have been compressed in storage and by virtue of their profile hold the beads away from the rim edges. To overcome this difficulty, a tourniquet may be placed around the circumference of the tyre, over the central area of the tread. The compression of the tread in this area will cause the beads to be pushed outwards in the desired direction. The type of tourniquet most widely used consists of a length of hose closed at both ends with a suitable clamp fitted to enable both ends to be connected. An ordinary tyre valve is fitted at one end of the tube so that after the hose has been secured around the tyre it may be inflated, giving a constricting effect. Another possible method of seating beads to obtain initial inflation is to press the tyre into the angle between a wall and the floor. With the airline attached to the valve additional pressure is then applied to the tyre by the hand and shin, as shown in the accompanying illustration. The application of pressure at four points around the tyre's circumference whilst simultaneously applying the airhose will often effect an initial seal between the tyre beads and wheel rim, thus allowing inflation to occur.

11 Having successfully accomplished inflation, increase the pressure to 40 psi and check that the tyre is evenly disposed on the wheel rim. This may be judged by checking that the thin positioning line found on each tyre wall is equidistant from the rim around the total circumference of the tyre. If this is not the case, deflate the tyre, apply additional lubrication and reinflate. Minor adjustments to the tyre position may be made by bouncing the wheel on the ground.

12 Always run the tyre at the recommended pressures and never under or over-inflate. The correct pressures for various weights and configurations are given in the Specifications Section of this Chapter.

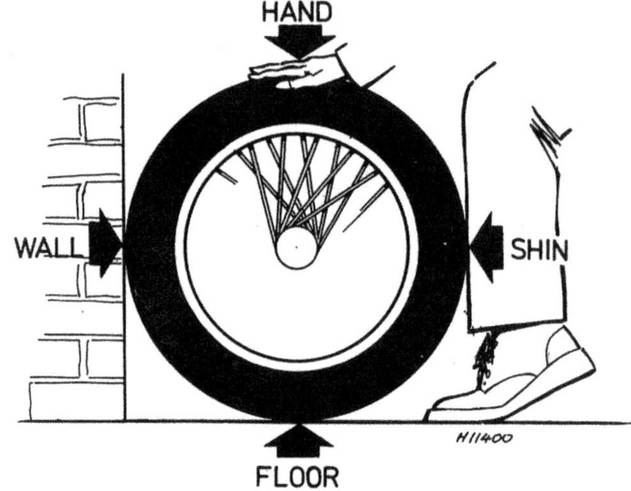

Fig. 6.15. Method of seating the beads on tubeless tyres

20 Puncture repair and tyre renewal

1 The primary advantage of the tubeless tyre is its ability to accept penetration by sharp objects such as nails etc without loss of air. Even if loss of air is experienced, because there is no inner tube to rupture, in normal conditions a sudden blow-out is avoided. If a puncture of the tyre occurs, the tyre should be removed for inspection for damage before any attempt is made at remedial action. The temporary repair of a punctured tyre by inserting a plug from the outside should not be attempted. Although this type of temporary repair is used widely on cars, the manufacturers strongly recommend that no such repair is carried out on a motorcycle tyre. Not only does the tyre have a thinner carcass, which does not give sufficient support to the plug, the consequences of a sudden deflation are often sufficiently serious that the risk of such an occurrence should be avoided at all costs.

2 The tyre should be inspected both inside and out for damage to the carcass. Unfortunately the inner lining of the tyre – which takes the place of the inner tube – may easily obscure any damage and some

TYRE CHANGING SEQUENCE - TUBELESS TYRES

Deflate tyre. After releasing beads, push tyre bead into well of rim at point opposite valve. Insert lever next to valve and work bead over edge of rim.

Use two levers to work bead over edge of rim. Note use of rim protectors.

When first bead is clear, remove tyre as shown.

Before installing, ensure that tyre is suitable for wheel. Take note of any sidewall markings such as direction of rotation arrows.

Work first bead over the rim flange.

Use a tyre lever to work the second bead over rim flange.

experience is required in making a correct assessment of the tyre condition.

3 There are two main types of tyre repair which are considered safe for adoption in repairing tubeless motorcycle tyres. The first type of repair consists of inserting a mushroom-headed plug into the hole from the inside of the tyre. The hole is prepared for insertion of the plug by reaming and the applications of an adhesive. The second repair is carried out by buffing the inner lining in the damaged area and applying a cold or vulcanised patch. Because both inspection and repair, if they are to be carried out safely, require experience in this type of work, it is recommended that the tyre be placed in the hands of a repairer with the necessary skills, rather than repaired in the home workshop.

4 In the event of an emergency, the only recommended 'get-you-home' repair is to fit a standard inner tube of the correct size. If this course of action is adopted, care should be taken to ensure that the cause of the puncture has been removed before the inner tube is fitted. It will be found that the valve hole in the rim is considerably larger than the diameter of the inner tube valve stem. To prevent the ingress of road dirt, and to help support the valve, a spacer should be fitted over the valve.

5 In the event of the unavailability of tubeless tyres, ordinary tubed tyres fitted with inner tubes of the correct size may be fitted. Refer to the manufacturer or a tyre fitting specialist to ensure that only a tyre and tube of equivalent type and suitability is fitted, and also to advise on the fitting of a valve nut to the rim hole.

21 Tyre valves: description and renewal

1 It will be appreciated from the preceding Sections that the adoption of tubeless tyres has made it necessary to modify the valve arrangement, as there is no longer an inner tube which can carry the valve core. The problem has been overcome by fitting a separate tyre valve which passes through a close-fitting hole in the rim, and which is secured by a nut and locknut. The valve is fitted from the rim well, and it follows that the valve can be removed and replaced only when the tyre has been removed from the rim. Leakage of air from around the valve body is likely to occur only if the sealing seat fails or if the nut and locknut become loose.

2 The valve core is of the same type as that used with tubed tyres, and screws into the valve body. The core can be removed with a small slotted tool which is normally incorporated in plunger type pressure gauges. Some valve dust caps incorporate a projection for removing valve cores. Although tubeless tyre valves seldom give trouble, it is possible for a leak to develop if a small particle of grit lodges on the sealing face. Occasionally, an elusive slow puncture can be traced to a leaking valve core, and this should be checked before a genuine puncture is suspected.

3 The valve dust caps are a significant part of the tyre valve assembly. Not only do they prevent the ingress of road dirt in the valve, but also act as a secondary seal which will reduce the risk of sudden deflation if a valve core should fail.

22 Wheel balancing

1 It is customary on all high performance machines to balance the wheels complete with the tyre. The out of balance forces which exist are eliminated and the handling of the machine is improved in consequence. A wheel which is badly out of balance produces through the steering a most unpleasant hammering effect at high speeds.

2 Some tyres have a balance mark on the sidewall, usually in the form of a coloured spot. This mark must be in line with the tyre valve, when the tyre is fitted. Even then the wheel may require the addition of balance weights, to offset the weight of the tyre valve itself.

3 If the wheel is raised clear of the ground and is spun, it will probably come to rest with the tyre valve or the heaviest part downward and will always come to rest in the same position. Balance weights must be added to a point diametrically opposite this heavy spot until the wheel will come to rest in ANY position after it is spun.

4 It should be noted that the front wheel must always be checked for balance with the brake disc in position, though it may prove necessary to remove the calipers to eliminate brake drag. The rear wheel can be balanced if required but this is unlikely to have much effect in practice. The balance weights used must be of the correct type for use with Yamaha rims, and to this end should be purchased from a Yamaha dealer. Balance weights are available in 10 gm, 20 gm and 30 gm sizes.

Chapter 7 Electrical system

For information relating to the RD350 F II, N II and R models, refer to Chapter 8

Contents

Specifications

Battery

Make	Nippon Denso
Type	Lead acid
Charge rate	0.55A for 10 hours
Specific gravity:	
RD350 LC II	1.260
Other models	1.280

Alternator

Make	Nippon Denso
Model:	
RD350 LC II	AVCC58
Other models	51L
Charging output	14V 14A @ 5000 rpm
Charging coil resistance:	
RD350 LC II	0.4 ohms ± 20% @ 20°C (68°F)
Other models	0.5 ohms ± 20% @ 20°C (68°F)

Voltage regulator

Type	Short circuit, combined with rectifier
Make	Shindengen Kougyou
Model	SH235-12C
Regulated voltage	14.5 ± 0.5V

Rectifier	
Type	3-phase, full wave, combined with regulator
Make	Shindengen Kougyou
Model	SH235-12C
Capacity	15A
Withstand voltage	200V
Horn	
Make	Nikko
Model	CF12
Max amperage	2.5A or less
Turn signal relay	
Make	Nippon Denso
Model	FJ245ED
Type	Transistor
Frequency	85 cpm
Capacity	12 volt 21W x2 plus 3.4W x1
Fuses	
Main	20A
Headlamp:	
RD350 LC II	10A
Other models	15A
Turn signal	15A
Ignition (RD350 LC II only)	5A
YPVS (not RD350 LC II)	5A
Spare:	
RD350 LC II	10A x2
Other models	20A x1, 15A x1, 5A x1
Temperature gauge sender	
Make	Nissei
Model	YA55901NO
Bulb wattages (all 12 volt)	
Headlamp	60/55W Quartz halogen
Tail/brake	5/21W (x2 except RD350 LC II model)
Turn signal	21W x4
Meter lamp	3.4W (x3, RD350 LC II – x5, other models)
Parking/city lamp	3.4W
Indicator lamps:	
Neutral	3.4W
High beam	3.4W
Oil warning	3.4W
Turn	3.4W x2

1 General description

The electrical system is powered by a crankshaft-mounted three-phase alternator located behind the left-hand engine casing. Output from the alternator is fed to a combined rectifier/regulator unit where it is converted from alternating current (ac) to direct current (dc) by the full-wave rectifier section, and the system voltage is regulated to 14.5 ± 0.5 volts by the electronic voltage regulator.

2 Testing the electrical system

1 Simple continuity checks, for instance when testing switch units, wiring and connections, can be carried out using a battery and bulb arrangement to provide a test circuit. For most tests described in this Chapter, however, a pocket multimeter should be considered essential. A basic multimeter capable of measuring volts and ohms can be bought for a very reasonable sum and will provide an invaluable tool. Note that separate volt and ohm meters may be used in place of the multimeter, provided those with the correct operating ranges are available.

2 Care must be taken when performing any electrical test, because some of the electrical components can be damaged if they are incorrectly connected or inadvertently shorted to earth. This is particularly so in the case of electronic components. Instructions

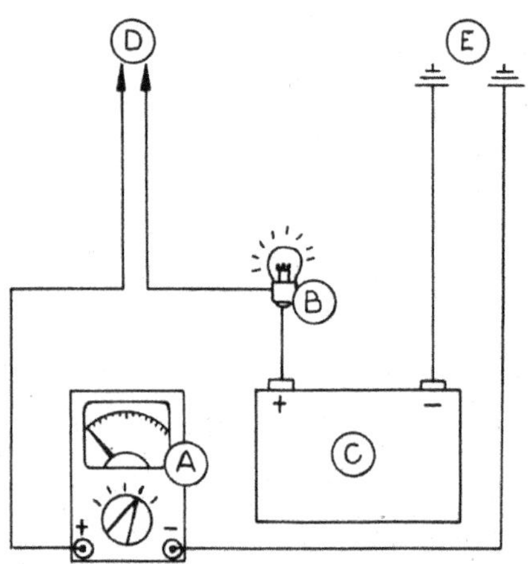

Fig. 7.1 Simple testing apparatus for electrical tests

A Multimeter
B Bulb
C Battery
D Positive probe
E Negative probe

regarding meter probe connections are given for each test, and these should be read carefully to preclude accidental damage occurring.
3 Where test equipment is not available, or the owner feels unsure of the procedure described, it is strongly recommended that professional assistance is sought. Errors made through carelessness or lack of experience can so easily lead to damage and the need for expensive replacement parts.
4 A certain amount of preliminary dismantling will be necessary to gain access to the components to be tested. Normally, removal of the seat and side panels will be required, with the possible addition of the fuel tank and headlamp unit to expose the remaining components.

3 Charging system: checking the output

1 In the event that the charging system fails or appears to be over- or under-charging the battery, the system voltage should be checked using a dc voltmeter or a multimeter set on the 0 – 20 volts dc scale. Remove the side panel to gain access to the battery terminals, noting that the battery leads must **not** be disconnected during the test. **Note:** If the machine is run with the battery disconnected the increased voltage across the alternator terminals will rise, causing damage to the regulator/rectifier unit or to the alternator windings.
2 Connect the positive (red) probe lead to the positive (+) battery terminal and the negative (black) probe lead to the negative (–) battery terminal. Start the engine and note the voltage reading at 2000 rpm. This should be 14.5 ± 0.5 volts if the system is operating correctly. If the voltage is outside this range it will be necessary to check the following, in the order shown below:

(a) Battery condition; see Sections 4 and 5
(b) Alternator windings; see Section 6
(c) Regulator/rectifier; see Section 7

4 Battery: examination and maintenance

1 The battery is housed in a tray on the right-hand side of the machine, behind the side panel. It is retained in the tray by a rubber strap. Note that the strap also holds a plastic shroud which is designed to cover the battery positive terminal. Care should be taken not to lose this; without it the battery may short out against the fuel tank.
2 The transparent plastic case of the battery permits the upper and lower levels of the electrolyte to be observed, without disturbing the battery, by removing the side cover. Maintenance is normally limited to keeping the electrolyte level between the prescribed upper and lower limits and making sure that the vent tube is not blocked.The lead plates and their separators are also visible through the transparent case, a further guide to the general condition of the battery. If electrolyte level drops rapidly, suspect over-charging and check the system.
3 Unless acid is spilt, as may occur if the machine falls over, the electrolyte should always be topped up with distilled water to restore the correct level. If acid is spilt onto any part of the machine, it should be neutralised with an alkali such as washing soda or baking powder and washed away with plenty of water, otherwise serious corrosion will occur. Top up with sulphuric acid of the correct specific gravity (1.260 to 1.280) only when spillage has occurred. Check that the vent pipe is well clear of the frame or any of the other cycle parts.
4 It is seldom practicable to repair a cracked battery case because the acid present in the joint will prevent the formation of an effective seal. It is always best to renew a cracked battery, especially in view of the corrosion which will be caused if the acid continues to leak.
5 If the machine is not used for a period of time, it is advisable to remove the battery and give it a 'refresher' charge every six weeks or so from a battery charger. The battery will require recharging when the specific gravity falls below 1.260 (at 29°C – 68°F). The hydrometer reading should be taken at the top of the meniscus with the hydrometer vertical. If the battery is left discharged for too long, the plates will sulphate. This is a grey deposit which will appear on the surface of the plates, and will inhibit recharging. If there is sediment on the bottom of the battery case, which touches the plates, the battery needs to be renewed. Prior to charging the battery refer to the following Section for correct charging rate and procedure. If charging from an external source with the battery on the machine, disconnect the leads, or the rectifier will be damaged.
6 Note that when moving or charging the battery, it is essential that the following basic safety precautions are taken:

(a) Before charging check that the battery vent is clear or, where no vent is fitted, remove the combined vent/filler caps. If this precaution is not taken the gas pressure generated during charging may be sufficient to burst the battery case, with disastrous consequences.
(b) Never expose a battery on charge to naked flames or sparks. The gas given off by the battery is highly explosive.
(c) If charging the battery in an enclosed area, ensure that the area is well ventilated.
(d) Always take great care to protect yourself against accidental spillage of the sulphuric acid contained within the battery. Eyeshields should be worn at all times. If the eyes become contaminated with acid they must be flushed with fresh water immediately and examined by a doctor at once. Similar attention should be given to a spillage of acid on the skin.

Note also that although, should an emergency arise, it is possible to charge the battery at a more rapid rate than that stated in the following Section, this will shorten the life of the battery and should therefore be avoided if at all possible.
7 Occasionally, check the condition of the battery terminals to ensure that corrosion is not taking place, and that the electrical connections are tight. If corrosion has occurred, it should be cleaned away by scraping with a knife and then using emery cloth to remove the final traces. Remake the electrical connections whilst the joint is still clean, then smear the assembly with petroleum jelly (NOT grease) to prevent recurrence of the corrosion. Badly corroded connections can have a high electrical resistance and may give the impression of complete battery failure.
8 It should be noted that it is almost impossible to test the electrical system with any degree of accuracy unless the battery is in sound condition and fully charged. Many apparent charging system faults can be attributed to an old and worn out battery which can no longer hold a charge. If the battery runs flat when the machine is left unused for a few days it is fairly safe to assume that its useful life has ended. To be certain, have the battery checked under load by an auto-electrical specialist, or check the electrical system using a battery known to be in good condition.

5 Battery: charging procedure

1 The normal charging rate for the 5.5 amp hour battery is 0.5 amps. A more rapid charge, not exceeding 1 amp can be given in an emergency. The higher charge rate should, if possible, be avoided since it will shorten the working life of the battery.
2 Make sure that the battery charger connections are correct, red to positive and black to negative. It is preferable to remove the battery from the machine whilst it is being charged and to remove the vent plug from each cell. When the battery is reconnected to the machine, the black lead must be connected to the negative terminal and the red lead to positive. This is most important, as the machine has a negative earth system. If the terminals are inadvertently reversed, the electrical system will be damaged permanently.

6 Alternator: winding resistance tests

1 The condition of the alternator windings can be checked by measuring their resistance. It should be noted that the resistance figure is very low and cannot be measured accurately using a very cheap multimeter with only a K ohms scale. In the absence of a good meter, take the machine to a Yamaha dealer to have the test performed, or check the fault by substituting a new stator assembly and noting the effect.
2 Trace the alternator wiring back to the four-pin connector and separate it. Measure the resistance between each pair of White leads; a total of three tests. A reading of 0.4 ohms ± 20% at 20°C (68°F) for the RD350 LC II or 0.5 ohms ± 20% at 20°C (68°F) for the remaining models should be indicated. A break in one or more windings will show infinite resistance. A short-circuit is less easy to distinguish, given the very low standard resistance of the windings.

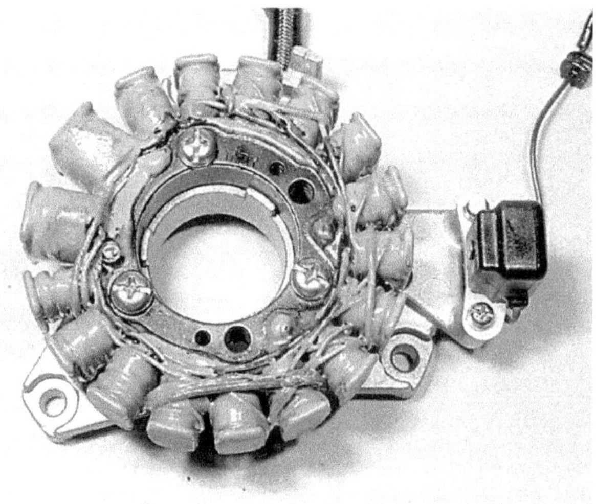

6.2 Alternator stator windings are encapsulated in resin – broken or shorted windings indicate renewal

7 Regulator/rectifier unit: testing

1 The regulator/rectifier unit is a sealed, finned alloy unit, mounted between the frame top tubes below the rear of the fuel tank. To remove the unit it is first necessary to release the electrical panel. The unit is attached to the underside of the panel. Note that it can be tested in position.
2 The regulator's operation is tested by performing the charging system output test, as described in Section 3. If the output voltage is incorrect, and the battery, alternator and rectifier are known to be operating correctly, it can be assumed that the unit is faulty and in need of renewal. It is worth having this checked by a Yamaha dealer though.
3 The rectifier is an arrangement of six diodes, connected in a bridge pattern to provide full-wave rectification. This means that the full output of the alternator is converted to dc, rather than half of it, as is the case with simple half-wave rectifiers as used on lightweight machines and mopeds.
4 The condition of the rectifier can be checked using a multimeter, set on its resistance scale, as a continuity tester. Each of the diodes acts as a one-way valve, allowing current to flow in one direction, but blocking it if the polarity is reversed. Perform the resistance check by following the table accompanying this Section. If any one test produces the wrong reading the rectifier will have to be renewed.

8 YPVS system: general description and fault diagnosis

1 The Yamaha Power Valve System (YPVS) comprises a mechanical valve and control cables, described in Chapter 1, and the associated electrical and electronic control equipment, these being covered in this Section.
2 The valve is opened and closed by the servomotor unit. This is mounted below the frame top tubes, near the steering head. Beneath the seat is the control unit which senses engine speed by picking up pulses from the CDI unit and uses this information to set the power valve to the correct position.
3 In the event of a suspected fault, always check first that the power valve cables are correctly adjusted as described in Section 35 of Chapter 1. Note also that it is essential that the recommended resistor-type spark plugs are used. If non-resistor plugs are fitted, spurious signals are fed to the control unit, causing the valve to flutter at some engine speeds. If the fault persists, proceed as described below.
4 Switch the ignition on and check that the valve opens and closes normally. The valve cycles automatically each time the ignition is switched on as a self-cleaning measure. If the valve fails to move, and assuming that it is not physically jammed, check that all wiring connectors between the servomotor, control unit and CDI unit are sound. If this fails to effect a cure, test the servomotor (see Section 9). If this fails to resolve the problem, check the control unit (see Section 10).
5 If the valve cycles normally, but the valve is not in its closed position when the engine is started, check the wiring connections between the CDI unit and the control unit. If this fails to find the problem, check the control unit (Section 10). If everything else seems to operate normally, the fault lies with the CDI unit which should be checked by substitution.

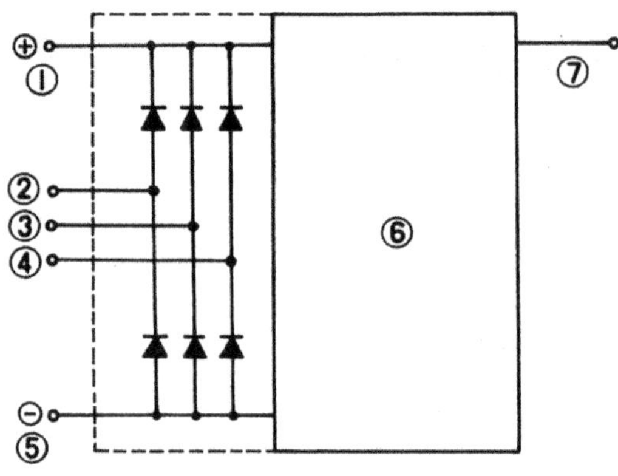

Checking element	Pocket test connecting point		Good	Replace (element shorted)	Replace (element opened)
	(+) (red)	(–) (black)			
D_1	B	U	O	O	×
	U	B	×	O	×
D_2	B	V	O	O	×
	V	B	×	O	×
D_3	B	W	O	O	×
	W	B	×	O	×
D_4	U	E	O	O	×
	E	U	×	O	×
D_5	V	E	O	O	×
	E	V	×	O	×
D_6	W	E	O	O	×
	E	W	×	O	×

O – Continuity
X – No continuity

Fig. 7.2 Rectifier test table

1 Red wire (B)
2 White wire (U)
3 White wire (V)
4 White wire (W)
5 Black wire (E)
6 Regulator unit
7 Brown wire (L)

8.2a The YPVS servomotor is located to the rear of the steering head

8.2b The YPVS control unit (arrowed) is mounted next to the fuse box

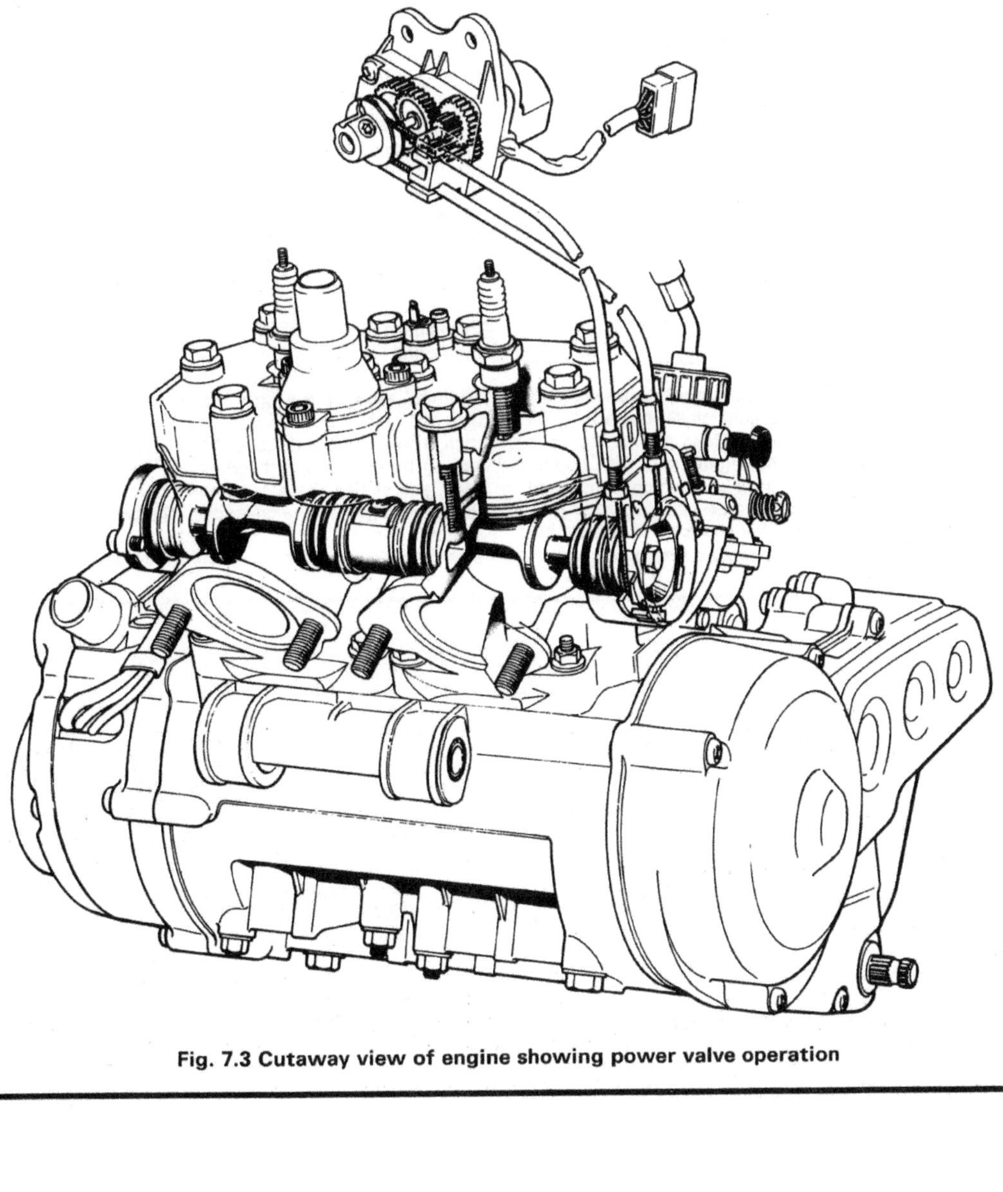

Fig. 7.3 Cutaway view of engine showing power valve operation

9 YPVS system: testing the servomotor

1 The tests described below require a good quality multimeter capable of reading on an ohms x 1 range, and also two insulated jumper leads with small crocodile clips on each end. Trace the servomotor wiring back to the 5-pin connector and separate it.

Motor operation

2 Connect a jumper lead between the negative battery terminal and the Black/Red pin in the connector block, and a second jumper lead between the positive battery terminal and the Black/Yellow pin in the connector. If the motor operates it can be considered serviceable. If not, renew the servomotor complete.

Potentiometer resistances

3 Using a multimeter, measure the potentiometer resistances as shown below. If the resistance of any one of the tests falls outside the specified figure of 7.5 K ohms ± 30% at 20°C (68°F), renew the servomotor assembly. The potentiometer test connections are as follows:

(a) Connect the Blue/Yellow lead to the White/Black lead
(b) Connect the Blue/Yellow lead to the White/Red lead
(c) Connect the White/Red lead to the White/Black lead

10 YPVS system: testing the control unit

1 Trace the wiring from the control unit and separate the connector. Using an electrical screwdriver, unclip the Black/white lead terminal and withdraw it from the connector, then join the two halves of the connector, leaving the Black/white lead free.

2 Using an insulated jumper lead, connect the Black/white lead from the control unit to the Black/white lead of the CDI unit. Start the engine, and open the throttle briefly so that the engine reaches just below 7000 rpm. If the valve fails to operate normally, the control unit can be considered faulty and should be renewed.

11 Fuses: location and renewal

1 The electrical system is protected by fuses, intentional weak links designed to break down before an electrical fault causes damage to the system or electrical components. The various fuse ratings for each model are shown in the Specifications at the beginning of the Chapter. Spare fuses are contained in the lid of the fuse box, which is mounted at the front of the rear mudguard, below the seat.

2 If a fuse blows, it should be replaced, after checking to ensure that no obvious short circuit has occurred. If the second fuse blows shortly afterwards, the electrical circuit must be checked thoroughly, to trace the fault.

3 When a fuse blows whilst the machine is running and no spare is available, a 'get you home' remedy is to remove the blown fuse and wrap it in metal foil before replacing it in the fuseholder. The foil will restore electrical continuity by bridging the broken fuse wire. This expedient should never be used if there is evidence of a short circuit or other major electrical fault, otherwise more serious damage will be caused. Replace the blown fuse at the earliest possible opportunity, to restore full circuit protection.

12 Headlamp: bulb renewal and beam alignment

1 On RD350 LC II models, remove the two bolts holding the fairing and tip it forward to gain access to the headlamp unit. On this and the RD350 N model, the headlamp unit should be detached from its shell by releasing the retaining screws and lifting the unit away. As it comes clear of the shell, disconnect the wiring from the headlamp and parking lamp. In the case of the RD350 F, the bulb and wiring can be reached from the back of the fairing, no preliminary dismantling being required.

2 Disconnect the wiring from the bulb terminals by pulling off the connector. The moulded plastic cover can now be peeled off the back of the unit to expose the bulb holder ring. Before the bulb is removed it should be noted that it is of the quartz halogen type. The bulb envelope will become permanently etched if touched, and so it is essential that it is handled by the metal part only. If the quartz envelope is touched accidentally it should be cleaned with methylated spirit and a soft cloth.

3 Remove the retaining ring by twisting it anticlockwise and lift the bulb away. Replacement is a straightforward reversal, the three locating tangs on the bulb flange ensure that it is correctly aligned.

4 The parking lamp bulb holder can be removed by twisting it anticlockwise. The bulb is a bayonet fitting type rated at 3.4 watts in the UK and 4.0 watts in other countries.

5 The headlamp can be adjusted for both horizontal and vertical alignment. Horizontal adjustment is made via the small screw which passes through the side of the headlamp rim. To set the vertical alignment locate the bolt which passes through the slotted bracket at the rear of the unit. This should be slackened and the headlamp moved to the required position. The bolt is then tightened to retain the setting. In the case of the RD350 F, headlamp adjustment is made from inside the fairing using the two control knobs provided for this purpose.

6 In the UK, regulations stipulate that the headlamp must be arranged so that the light will not dazzle a person standing at a distance greater than 25 feet from the lamp, whose eye level is not less than 3 feet 6 inches above that plane. It is easy to approximate this setting by placing the machine 25 feet away from a wall, on a level road, and setting the dipped beam height so that it is concentrated at the same height as the distance of the centre of the headlamp from the ground. The rider must be seated normally during this operation and also the pillion passenger, if one is carried regularly.

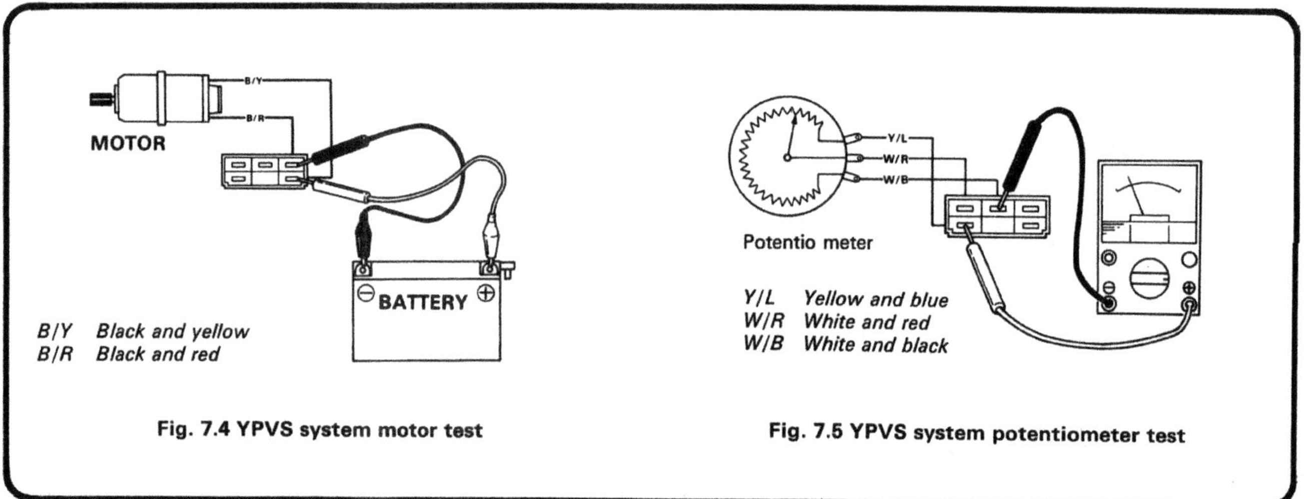

Fig. 7.4 YPVS system motor test

Fig. 7.5 YPVS system potentiometer test

11.1 Fuse box lid houses spare fuses – remember to replace them if used

12.1a On the 'F' model the headlamp bulb can be reached from inside the fairing. Pull off the wiring connector and rubber shroud ...

12.1b ... then twist the retaining ring and remove it ...

12.1c ... to free the bulb. Do not touch the envelope with fingers

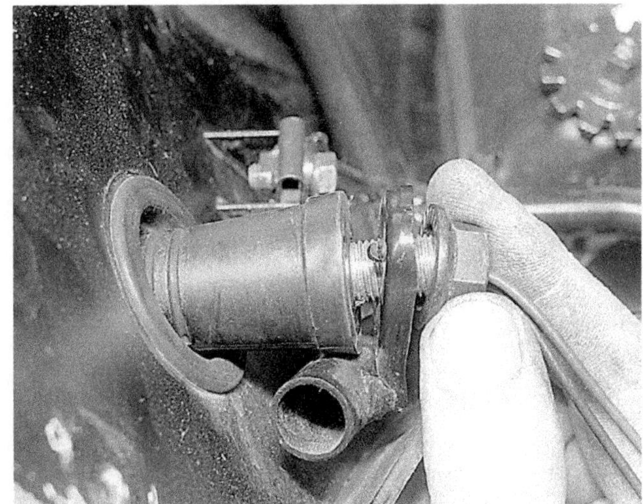

12.6a To remove the 'F' model headlamp unit, first remove the turn signal lamps ...

12.6b ... and the rear view mirrors

12.6c Slacken and remove the bolts holding the fairing to the subframe ...

12.6d ... taking care not to lose the rubber bushes and headed spacers

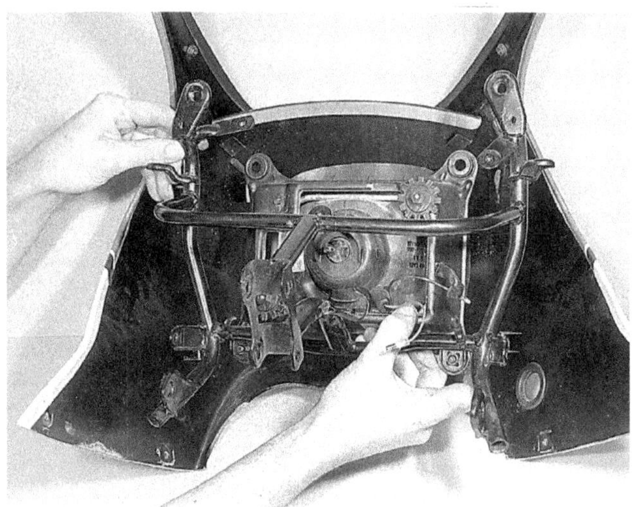

12.6e The subframe assembly can now be removed together with the headlamp

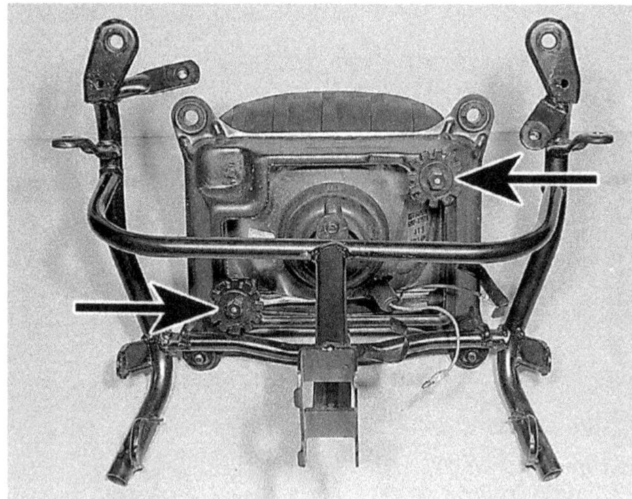

12.6f Headlamp support frame can be unbolted from fairing subframe if required. Note headlamp adjusters (arrowed)

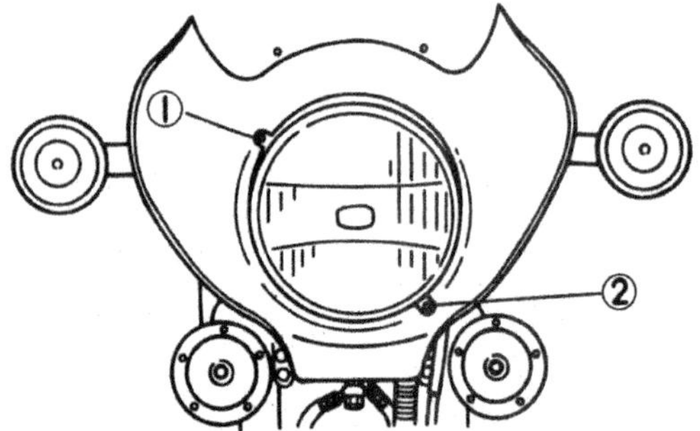

Fig. 7.6 Location of headlamp beam adjusting screws – RD350 LC II

1 *Horizontal adjustment screw*
2 *Vertical adjustment screw*

13 Stop/tail lamp: bulb renewal

1 The stop/tail lamp houses two twin filament bulbs. The bulb holders are a bayonet fit in the back of the lamp unit, and are accessible via the seat tail hump. The bulbs are a bayonet fit in the bulbholders and have offset pins to ensure that they are refitted correctly. Each bulb is rated at 12V 5/21W.

14 Turn signals: bulb renewal

1 The turn signal lamps are mounted on short stalks at the front and rear of the machine. To gain access to the bulbs, remove the plastic lens by removing the retaining screws or prising it away from the lamp body, depending on the model. Bulbs are a bayonet fit and are rated at 12V 21W.

13.1 Access to tail lamp bulbs is through the seat tail hump

14.1 Turn signal bulbs are a bayonet fitting

15 Turn signals: fault diagnosis and testing

1 If the indicators fail to operate, the nature of the fault gives a good indication of its cause. If the fault is restricted to one set of lamps only and yet the remaining set operate correctly the fault is almost certainly due to a blown bulb, or broken or shorted wiring on that side of the circuit. If the system fails totally, check that this is not caused by the self-cancelling system by unplugging the flasher cancelling unit (see Section 16). If it is found that the latter is at fault it can be left disconnected and the indicators used manually until a replacement unit is obtained.
2 If the fault cannot be attributed to any other cause it will be necessary to renew the flasher relay. It is located on the electrical panel below the fuel tank and is held in a rubber mounting. The relay is a sealed unit and cannot be repaired if it is faulty. Ensure that the new unit is of the same rating as the standard item. Any variation in its output will affect the flash rate.

3 In the event of a fault occurring in the indicator circuit, the following check list will prove helpful.
Indicators do not work

(a) Check bulb
(b) Right circuit:
 1 Check for 12V on dark Green wire to light.
 2 Check for ground on Black wire to light assembly.
(c) Left circuit:
 1 Check for 12V on dark Brown wire to light.
 2 Check for ground on Black wire to light assembly.
(d) Right and left circuits do not work:
 1 Check for 12V on Brown/white wire to flasher switch on left handlebar.
 2 Check for 12V on Brown wire to flasher relay.
 3 Replace flasher relay
 4 Replace flasher switch.
(e) Check flasher self cancelling system.
 (Refer to flasher self cancelling system).

15.1 Turn signal self-cancelling unit (arrowed) is mounted beneath the tank

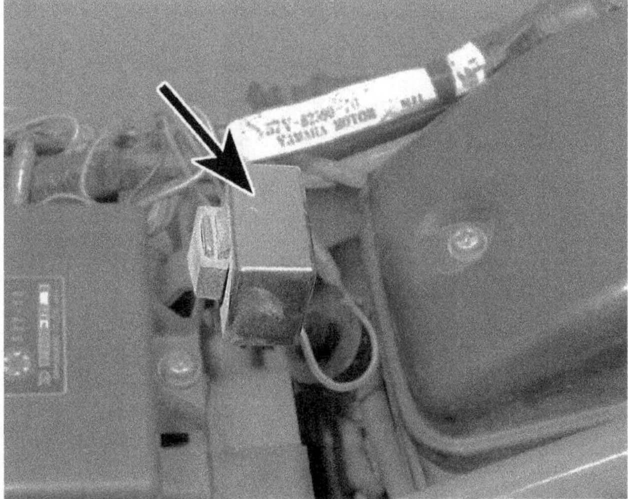
15.2 Rubber-mounted turn signal relay is fitted forward of the air filter casing

16 Turn signal self-cancelling circuit: fault diagnosis and testing

1 The RD350 YPVS models are equipped with self-cancelling indicators, this function being controlled by a timing circuit and a speedometer sensor which measures the distance travelled. In practice the indicators should switch off after 10 seconds or after 150 metres (164 yards) have been covered. Both systems must switch off before the indicators stop, thus at low speeds the system is controlled by distance, whilst at high speeds, elapsed time is the controlling factor.
2 A speedometer sensor measures the distance covered from the moment that the switch is operated. After the 150 metres have been covered, this part of the system will reset to off. The flasher cancelling unit starts a ten second countdown from the moment that the switch is operated. As soon as both sides of the system are at the off position, the flashers are cancelled. If required, the system may be overriden manually by depressing the switch inwards.
3 In the event of malfunction, refer to the accompanying figure which shows the circuit diagram for the self-cancelling system. The self-cancelling unit is located beneath the fuel tank, immediately to the rear of the steering head. Trace the output leads to the 6-pin connector and disconnect it. If the ignition switch is now turned on and the indicators will operate normally, albeit with manual cancelling, the flasher relay, bulbs, wiring and switch can be considered sound.
4 To check the speedometer sensor, connect a multimeter to the White/green and the black leads of the wiring harness at the 6-pin connector. Set the meter to the ohms x 100 scale. Release the speedometer cable at the wheel end and use the projecting cable end to turn the speedometer. If all is well, the needle will alternate between zero resistance and infinite resistance. If not, the sender or the wiring connections will be at fault.
5 Connect the meter probes between the Yellow/red lead and earth, again on the harness side of the 6-pin connector. Check the switch and associated wiring by turning the indicator switch on and off. In the off position, infinite resistance should be shown, with zero resistance in both on positions.

17 Instrument panel warning lamps: bulb renewal

1 The various warning and illumination bulbs are housed in rubber bulb holders which are a push fit into the underside of the instrument panel. The method of gaining access to the bulbs varies according to the model. In the case of the RD350 LC II, it is first necessary to remove the plastic cover from the underside of the panel, this being retained by four self-tapping screws.
2 The bulb holders can be pulled out of the panel to allow the bulbs to be checked or renewed. It is suggested that they are dealt with individually to avoid any risk of them becoming interchanged.

18 Coolant temperature gauge: description and testing

1 The coolant temperature is monitored by a gauge arrangement consisting of the meter unit mounted in the instrument panel, controlled by a sender unit which is screwed into the cylinder head water jacket. As the engine temperature rises, the resistance of the sender unit reduces, and this is used to control the position of the meter needle.
2 In the event of a fault, check the wiring connections between the meter and sender unit. Note that if the sender lead is broken or disconnected, the gauge will read cold all the time, whilst if it becomes shorted, the gauge will read hot all the time.
3 If the fault persists, check the sender resistances as described in Section 19. If this proves that the sender unit is working normally, the meter must be renewed.

16.2 This shows the reed switch fitted to the speedometer to sense distance travelled. It controls the self-cancelling system

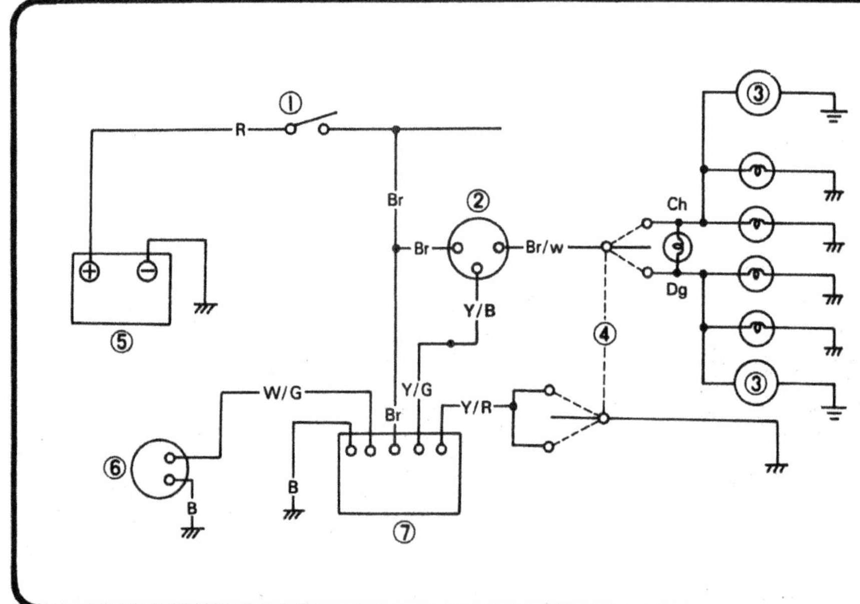

Fig. 7.7 Turn signal circuit diagram

1 *Ignition switch*
2 *Flasher unit*
3 *Turn signal lamps*
4 *Handlebar switch*
5 *Battery*
6 *Speedometer sensor*
7 *Cancelling unit*

B *Black*
Br *Brown*
Ch *Dark brown*
Dg *Dark green*
G *Green*
R *Red*
W *White*
Y *Yellow*

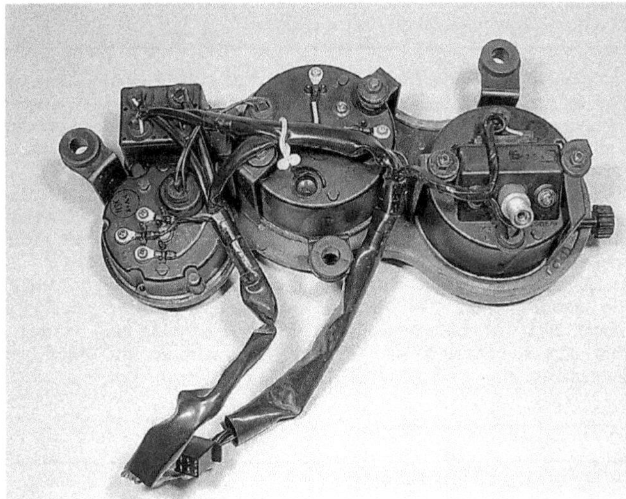

17.1a Instrument panel assembly removed to show wiring and bulb holders (F model shown)

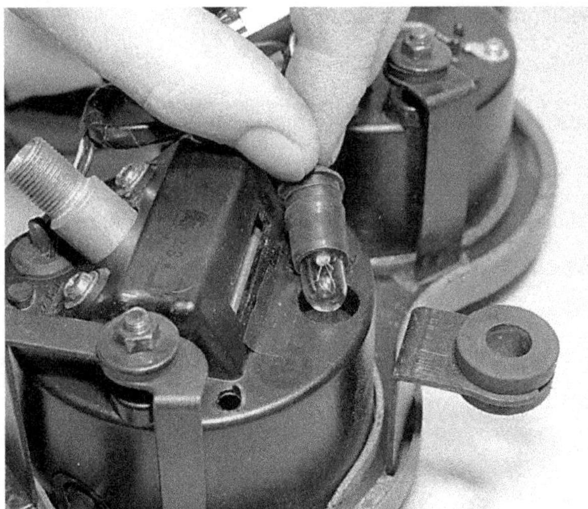

17.1b Bulb holders are a push-fit in the back of the instrument heads

17.1c Instrument heads are held in place by rubber-bushed studs and nuts

19 Coolant temperature gauge sender: testing

1 If the water temperature sender appears to be faulty it can be tested by measuring its resistance at various temperatures. To accomplish this it will be necessary to gather together a heatproof container into which the sender can be placed, a burner of some description (a small gas-powered camping burner would be ideal), a thermometer capable of measuring between 40°C and 120°C (122°F – 248°F) and an ohmmeter or multimeter capable of measuring 0 – 250 ohms with a reasonable degree of accuracy.
2 Fill the container with cold water and arrange the sender unit on some wire so that the probe end is immersed in it. Connect one of the meter leads to the sender body and the other to the terminal. Suspend the thermometer so that the bulb is close to the sender probe.
3 Start to heat the water, and make a note of the resistance reading at the temperature shown in the table below. If the unit does not give readings which approximate quite closely to those shown it must be renewed.

Water temperature	40°C 104°F	60°C 140°F	80°C 176°F	100°C 212°F
Sender resistances (ohms)	240	104±4	52.1±2	27.4

20 Horn: location and testing

Either a single or twin horns are fitted, depending on the model concerned. The horn or horns are bolted to a flexible mounting strip below the headlamp. In the event of the horn failing to operate it is usually necessary to renew it. It is of sealed construction and thus cannot be dismantled for repair. Before blaming the horn, check for battery voltage on the Brown lead when the horn button is pressed (ignition on). Check also that the Pink earth lead is securely connected.

21 Brake light circuit: testing

1 In the event of a fault in the brake light circuit follow the test sequence shown below. The switches must be renewed if they are faulty, but occasionally they can be persuaded to work after a good soaking in WD40 or similar.

(a) Check bulb and connections.
(b) Check for 12 volts on Yellow lead to brake lamp.
(c) Check for 12 volts on Brown lead to front and rear brake switches.
(d) Check Black earth lead from lamp unit to frame (continuity test)

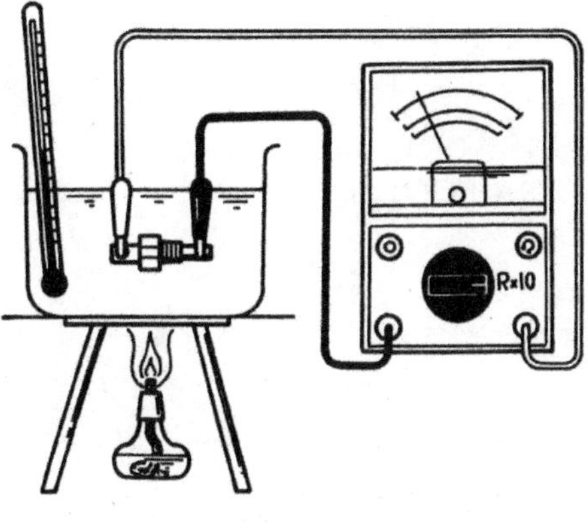

Fig. 7.8 Coolant temperature sender test

20.1 Typical horn mounting arrangement (F model shown)

22 Brake light switches: location and adjustment

1 All models have a stop lamp switch fitted to operate in conjunction with the rear brake pedal. The switch is located immediately to the rear of the crankcase, on the right-hand side of the machine. It has a threaded body giving a range of adjustment.
2 If the stop lamp is late in operating, slacken the locknuts and turn the body of the lamp in an anticlockwise direction so that the switch rises from the bracket to which it is attached. When the adjustment seems near correct, tighten the locknuts and test.
3 If the lamp operates too early, the locknuts should be slackened and the switch body turned clockwise so that it is lowered in relation to the mounting bracket.
4 As a guide, the light should operate after the brake pedal has been depressed by about 2 cm (¾ inch).
5 The front brake lever is also fitted with a switch. This is a non-adjustable unit and is of sealed construction. In the event of a fault, renew the switch.

22.1 Rear brake lamp switch is mounted inboard of the frame

23 Handlebar switches: maintenance

1 Generally speaking, the switches give little trouble, but if necessary they can be dismantled by separating the halves which form a split clamp around the handlebars. Note that the machine cannot be started until the ignition cut-out on the right-hand end of the handlebars is turned to the central 'Run' position.
2 Always disconnect the battery before removing any of the switches, to prevent the possibility of a short circuit. Most troubles are caused by dirty contacts, but in the event of the breakage of some internal part, it will be necessary to renew the complete switch.
3 Because the internal components of each switch are very small, and therefore difficult to dismantle and reassemble, it is suggested a special electrical contact cleaner be used to clean corroded contacts. This can be sprayed into each switch, without the need for dismantling.

24 Ignition switch: general

1 The combined ignition switch and steering lock is mounted on the underside of the top yoke. In the event of a fault, try soaking the switch in WD40 or similar. If this fails to get it working, it will have to be renewed. It is retained by two bolts.

25 Neutral switch: location and testing

1 The neutral indicator lamp is operated by a switch arrangement on the left-hand end of the gear selector drum. Access to the switch is straightforward once the left-hand engine casing has been removed. The switch consists of a triangular plastic cover which is retained by three screws. The inner face of the cover incorporates a circular track into which the fixed contact is set flush. A spring-loaded contact is fitted into the end of the selector drum.
2 If a fault is experienced it is not very likely that the switch will be the cause of it. The following check sequence should be followed:

(a) Check bulb and connections.
(b) Check for 12 volts on Sky blue lead at switch terminal.
(c) Check switch continuity. If faulty, clean or renew damaged parts.

22.5 Front brake switch (arrowed) is held by screw to brake lever assembly

23.3 Handlebar switches can be separated for examination and maintenance

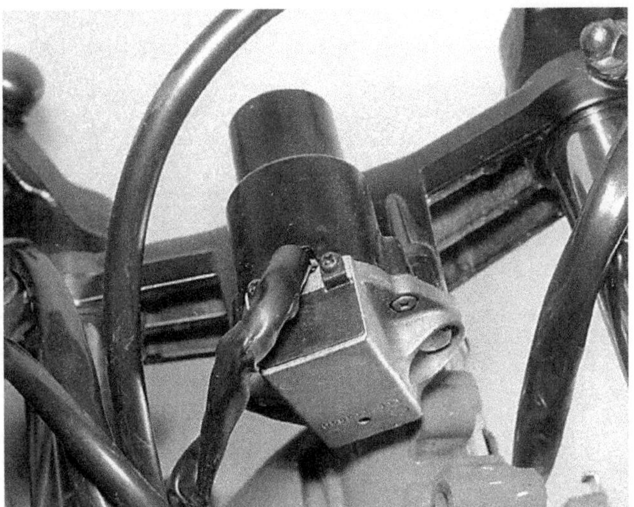

24.1 Ignition switch bolts to the underside of top yoke

26 Oil level warning lamp circuit: testing

1 The oil level warning lamp is operated by a float-type switch mounted in the oil tank. The circuit is wired through the neutral switch so that when the ignition is switched on and the machine is in neutral, the lamp comes on as a means of checking its operation. As soon as a gear is selected the lamp should go out unless the oil level is low.
2 In the event of a fault the bulb can be checked by switching the ignition on and selecting neutral. If this proves sound, check for 12 volts on the Black/Red lead to the switch. If the switch proves to be defective it can be unclipped from the tank and withdrawn.

27 Wiring: layout and examination

1 The wiring harness is colour-coded and will correspond with the accompanying wiring diagram. When socket connections are used, they are designed so that reconnection can be made in the correct position only.
2 Visual inspection will usually show whether there are any breaks or frayed outer covering which will give rise to short circuits. Occasionally a wire may become trapped between two components, breaking the inner core but leaving the more resilient outer cover intact. This can give rise to mysterious intermittent or total circuit failure. Another source of trouble may be the snap connectors and sockets, where the connector has not been pushed fully home in the outer housing, or where corrosion has occurred.
3 Intermittent short circuits can often be traced to a chafed wire that passes through or is close to a metal component such as a frame member. Avoid tight bends in the lead or situations where a lead can become trapped between casings.

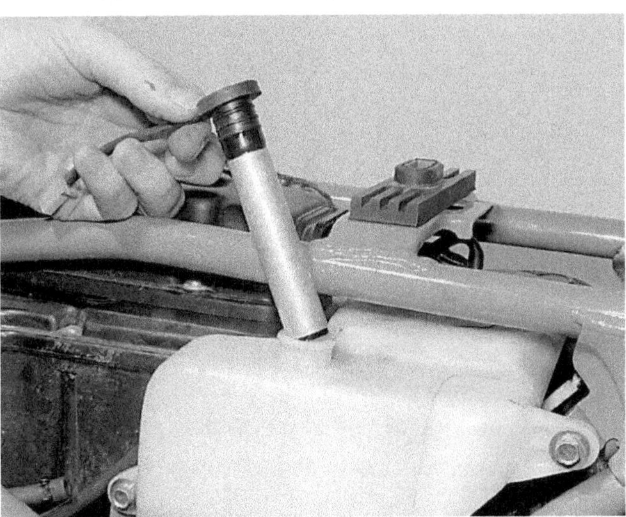

26.2 Oil level switch can be pulled out of tank for renewal

The RD350 F II model

Chapter 8
The RD350FII, NII and R models

Contents

Specifications

Note: *The additional specifications shown below relate to the RD350FII, RD350NII and RD350R models. Where no specification is given, refer to the specifications for the RD350F and N models in Chapters 1 to 7. Unless stated, the following information applies to all models.*

Model dimensions and weight

Overall length:	
RD350FII	2126 mm (83.7 in)
RD350NII	2120 mm (83.4 in)
RD350R	2095 mm (82.5 in)
Overall width:	
RD350FII and NII	700 mm (27.5 in)
RD350R	695 mm (27.3 in)
Overall height:	
RD350FII and R	1190 mm (46.8 in)
RD350NII	1070 mm (42.1 in)
Seat height:	
RD350FII and NII	790 mm (31.1 in)
RD350R	800 mm (31.5 in)
Wheelbase	1385 mm (54.5 in)
Ground clearance:	
RD350FII and NII	165 mm (6.5 in)
RD350R	175 mm (6.9 in)
Weight (with oil and full fuel tank):	
RD350FII	159 kg (350 lb)
RD350NII	155 kg (342 lb)
RD350R	166 kg (366 lb)

Specifications relating to Chapter 1

Torque wrench settings	kgf m	lbf ft
Cylinder barrel	2.8	20.2
Gearchange pedal	1.0	7.2
Thermosensor (temperature gauge sender unit)	1.5	10.8
Gear selector cam stopper plate	0.8	5.7
Stopper lever	1.0	7.2
Crankcase bolts:		
Lower	0.8	5.7
Upper	1.0	7.2

Specifications relating to Chapter 3

Fuel tank capacity

Overall	17 litres (3.7 Imp gal)
Reserve:	
RD350FII and NII	4 litres (0.8 Imp gal)
RD350R	3.5 litres (0.77 Imp gal)

Carburettors

ID mark	1UA 00
Main jet	185 (RD350R – 180*)
Air jet	0.8
Jet needle	5L20
Jet needle clip position (grooves from top)	2
Needle jet	N-8 (544)
Throttle valve cutaway	2.0
Pilot jet	27.5 (RD350R – 25)
Air screw setting (turns out)	1 ½
Fuel valve seat size	2.8
Starter jet	80
Power jet:	
Right-hand carburettor	60 (RD350R – 25*)
Left-hand carburettor	65 (RD350R – 25*)
Idle speed	1150 – 1250 rpm

**On RD350R models, a change of main jet size to 185 and power jet size to 55 can be made in conjunction with the exhaust system modification described in Section 5.*

Specifications relating to Chapter 4

Pulser (pickup) coil resistance

Pulser (pickup) coil resistance	94 – 140 ohms @ 20°C (68°F), white/red to white/green leads

Source coil resistances

Source coil 1	3.6 – 4.5 ohms @ 20°C (68°F), brown to red leads
Source coil 2	129 – 193 ohms @ 20°C (68°F), brown to green leads

Ignition coil

Make/type	Nippon Denso J0137
Primary winding resistance	0.28 – 0.38 ohm @ 20°C (68°F)
Secondary winding resistance	4.72 – 7.08 ohms @ 20°C (68°F)

CDI unit

Make/type	Nippon Denso QAB49

Spark plugs

Make/type	NGK BR9ES
Electrode gap	0.7 – 0.8 mm (0.027 – 0.031 in)

Specifications relating to Chapter 5

Front forks

Spring free length	416.6 mm (16.4 in)
Service limit	411.6 mm (16.2 in)
Oil capacity (per leg)	282 cc (9.9 Imp fl oz)
Oil level	128.7 mm (5.06 in)
Oil grade	SAE 10W fork oil
Air pressure:	
Standard	0.4 kg/cm² (5.7 psi)
Minimum	Atmospheric
Maximum	1.2 kg/cm² (17.0 psi)

Torque wrench settings

Swinging arm pivot	7.0 kgf m (50 lbf ft)

Specifications relating to Chapter 6

Brakes

Disk thickness	4.5 mm (0.18 in)
Service limit	4.0 mm (0.16 in)
Pad thickness	5.5 mm (0.22 in)
Service limit	0.5 mm (0.02 in)

Specifications relating to Chapter 7

Alternator

Make/model	Nippon Denso VCD88
Charging output	14 volt, 13A @ 5000 rpm
Charging coil resistance	0.44 – 0.66 ohm @ 20°C (68°F), white to white leads

Voltage regulator

Make/type/model	Shindengen, short circuit, combined with rectifier, SH569
No-load voltage	14.3 – 15.3 volts

Rectifier

Make/type/model	Shindengen, 3-phase full wave, combined with voltage regulator, SH569
Capacity	25 amps
Withstand voltage	200 volts

Horn

Make/model	Nikko YF3-12
Maximum amperage	2.5 amps

Turn signal relay

Make/type/model	Nippon Denso, condenser with self-cancelling device, FU249CD
Frequency	75 – 95 cpm
Wattage	21W x 2, plus 3.4W x 1

Thermosensor (temperature gauge sender)

Make/model	Nihon Seiki, 11H

1 Introduction

This Chapter covers the fully-faired RD350FII and RD350R, and the unfaired RD350NII.

The RD350FII and NII superseded the FI and NI models covered in the main text of this manual. The most noticeable changes were to the exhausts, which changed to the alloy end can type, the fuel tank design and general styling. The FII continued in production until December 1991, having received slight modification in 1988. The NII was discontinued in July 1987.

The RD350R is manufactured in Brazil. Apart from a new fairing, twin headlamps, remote-operated choke and tapered-roller steering head bearings, it is otherwise the same as the RD350FII.

Where procedures or specifications for these later models differ from the main text, they are given in this supplementary chapter. To aid model identification, the models covered in this chapter are listed below, together with their model code number and initial engine/frame number.

Model	Code No.	Engine/frame No.
RD350NII	1UA	1WT-005101
RD350FII (early)	1WT	1WT-000101
RD350FII (late)	3DH1	1WT-015101
RD350R	4CE6	4CE-0000101 (engine) 9C64CE00*N0000101 (frame)

2 Engine/gearbox unit — modifications

1 The engine/gearbox remained largely unchanged for the models covered in this Chapter, changes being confined to detail improvements to a well-proven design. Internally, there were alterations to the port timing and to the combustion chamber shape. These, together with the new carburettors and exhaust system, being responsible for the improvements in torque and power output.

2 There are also a number of minor detail changes to the engine and transmission components and assemblies. These have little effect on working procedure, but underline the need to quote in full the engine and frame numbers when ordering replacement parts.

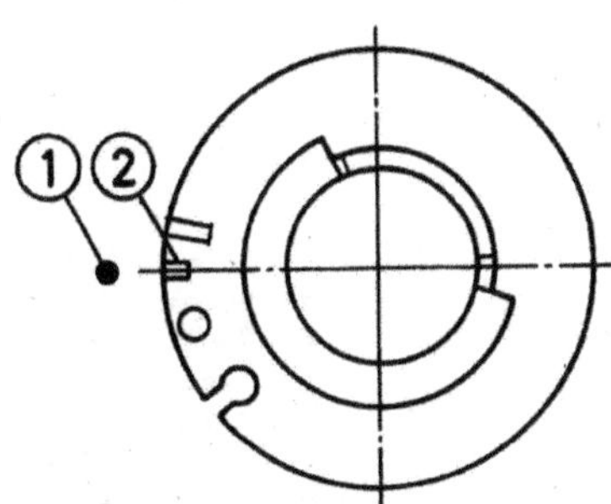

Fig 8.1 Oil pump pulley alignment mark (Sec 3)
1 Plunger pin 2 Pulley alignment mark

3 Oil pump: modification and cable adjustment

1 The oil pump remains similar to that used on the previous models, with the exception of a small lock washer added at the end of the pump drive shaft and secured by the drive pinion circlip.

2 When checking the pump cable adjustment proceed as shown in Routine Maintenance, noting that in the case of models covered in this chapter, the check is made with the throttle fully open. Check that the alignment mark shown in the accompanying illustration coincides with the pin, and if necessary reset the cable adjuster at the top of the casing to achieve this. Open and close the throttle twistgrip a few times, then re-check the setting.

4 Fuel tank and tap — modifications

Fuel tank — removal and refitting

1 Remove the fairing lower panels (see Section 10). Turn the fuel tap to ON (RD350NII and early RD350FII (1WT)) or to OFF (later RD350FII (3DH1) and all RD350R models). Remove the side panels (see Section 9) and disconnect the hose(s) from the fuel tap. Remove the seat.

2 Remove the single bolt at the rear of the tank and the screw on each side of the tank, then lift it off the frame.

3 When refitting the fuel tank, make sure that the breather pipe is not trapped at any point, and that it is properly routed.

Fuel tap — modifications

4 On the RD350NII and early RD350FII (1WT) models, there are detail changes to the fuel tap lever assembly; these are shown in the accompanying illustration.

5 The fuel tap fitted to later RD350FII (3DH1) and all RD350R models does not have a diaphragm unit, ie fuel is gravity-fed through the tap. The sediment bowl at the base of the tap incorporates a gauze filter which must be cleaned periodically. It is advised that filter cleaning is carried out under the 'Yearly, or every 8000 mile (12 000 km)' interval — *see Routine Maintenance*.

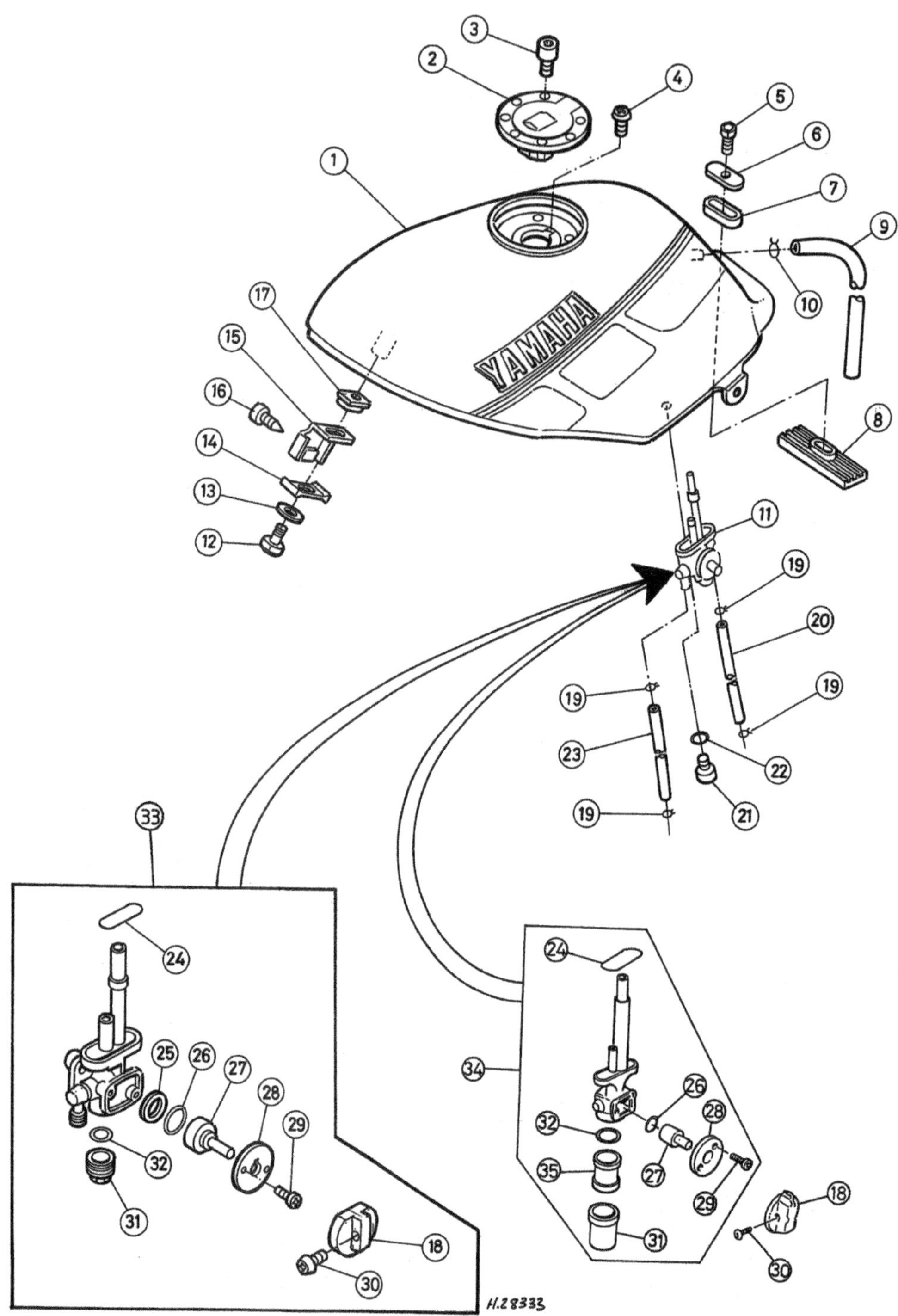

Fig. 8.2 Fuel tank and tap (Sec 4)

1 Fuel tank
2 Filler cap
3 Allen screw – 3 off
4 Screw
5 Bolt
6 Washer
7 Damping rubber
8 Damping rubber
9 Breather pipe
10 Clip
11 Fuel tap
12 Bolt – 2 off
13 Washer – 2 off
14 Plate – 2 off
15 Mounting bracket – 2 off
16 Screw – 2 off
17 Damping rubber – 2 off
18 Tap control knob
19 Clip – 4 off
20 Fuel pipe
21 Screw – 2 off
22 Washer – 2 off
23 Vacuum pipe – RD350NII and early RD350FII (1WT)
24 O-ring
25 Seal
26 O-ring
27 Rotor
28 Tap positioning disc
29 Screw – 2 off
30 Screw
31 Sediment bowl
32 O-ring
33 Fuel tap – RD350NII and early RD350FII (1WT)
34 Fuel tap – later RD350FII (3DH1) and RD350R
35 Filter

6.1 Fully unscrew retaining nut and withdraw choke plunger from left-hand carburettor

6.2 Choke cable can be withdrawn once retaining nut has been unscrewed

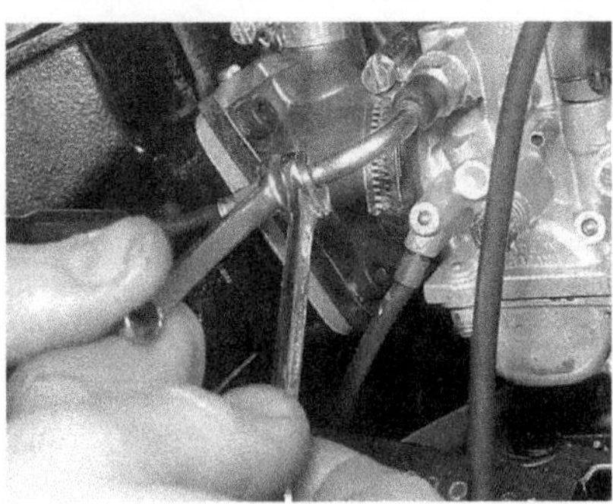
6.3 Make choke cable adjustment at the in-line adjuster

5 Carburettors: modifications

Power jets

1 In line with modified cylinder porting and the revised exhaust system, all later models are fitted with Mikuni power jet carburettors.
2 The power jet is a press fit (RD350FII and NII) or a screw fit (RD350R) in the float chamber, and is linked externally by a short length of pipe to the carburettor body.
3 Note that a modification is available for the RD350R which involves the removal of a restrictive device in the exhaust downpipes, coupled with a change of jetting for the main and power jets (see Carburettor Specifications). **Caution:** *if this modification is made, it is important that the modification is carried out in total, and not part. Owners are recommended to refer to a Yamaha dealer.*

Carburettor removal and refitting

4 Removal and refitting of the carburettors is unchanged from the procedure given in Chapter 3, Section 5 with the exception of the following.
5 On later RD350FII (3DH1) and all RD350R models, turn the fuel tap to the OFF position before disconnecting the fuel pipe from the carburettors. Note that there is no vacuum pipe to disconnect.
6 On RD350R models, unscrew the choke cable nut from the left-hand carburettor and slide the cable and choke plunger out.

6 Choke cable: removal and refitting – RD350R

1 The choke cable terminates in the left-hand carburettor. Remove the fairing left-hand lower panel for access (see Section 10). Unscrew the base nut and slide the choke plunger out of the carburettor body.
2 Remove the main fairing inner left-hand panel; it is retained by five screws. Fully unscrew the nut at the base of the mounting bracket and withdraw the choke knob and cable through the bracket aperture. Note the routing of the cable so that the new cable can be installed in the same path.
3 Reverse the removal procedure to fit the new cable. There should be a very small amount of freeplay at the choke knob before

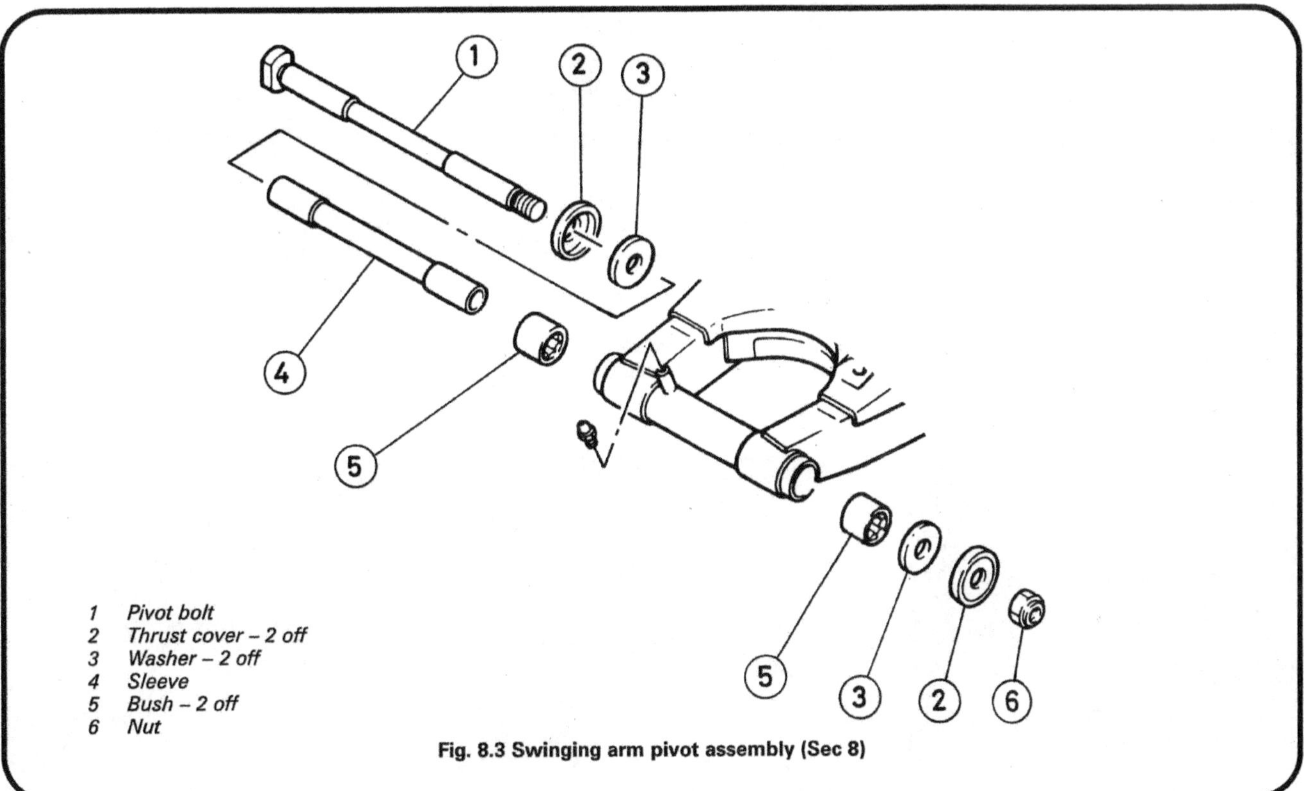

Fig. 8.3 Swinging arm pivot assembly (Sec 8)

Loading	Fork air pressure	Rear spring preload
Rider only	0.4 kg/cm² (5.7 psi)	2 - 4
Rider and luggage	0.6 kg/cm² (8.5 psi)	3 - 5
Rider and passenger	0.8 kg/cm² (11.0 psi)	4 - 6
Rider, passenger and luggage	0.8 kg/cm² (11.0 psi)	5 - 7
Standard rear suspension preload	4	
Standard fork air pressure	0.4 kg/cm² (5.7 psi)	
Maximum fork air pressure	0.8 kg/cm² (11.0 psi)	

Fig. 8.4 Suspension unit setting table (Sec 8)

the cable comes into operation. If necessary, make adjustment at the in-line adjuster at the carburettor end of the cable. Tighten the locknut on completion and slide the protective sleeve back into place.

7 Ignition system: modifications

1 Although remaining essentially similar in operation, many of the ignition system components have been revised. A new CDI unit provides a revised ignition advance curve, although this is of academic interest only and is not adjustable in any way.
2 The model of ignition coil has also been changed, and this will give different readings than those specified in Chapter 4. When testing the coil, refer to the specifications given at the beginning of this Chapter, using them in conjunction with Section 7 of Chapter 4.
3 Revised pulser (pickup) and source coil resistances are specified for these models, although the test procedures remain unaltered. For details, use the specifications which accompany this Chapter in conjunction with Chapter 4, Sections 5 and 6.

8 Cycle parts: modifications

1 The frame remained largely unchanged from that used on the previous models, with the exception of modified location points and brackets for the various ancillary and bodywork parts.
2 The rear suspension arrangement is unchanged, apart from the thrust cover and washer arrangement at each end of the swinging arm pivot shaft (see Fig. 8.3).
3 The rear suspension unit is similar to that used on the earlier models, but the remote adjustment system using a toothed belt and pulley was abandoned. In the case of the later models, adjustment is carried out conventionally, using a C-spanner on the adjustment collar at the top of the unit. Adjust the unit from the right-hand side of the machine, using the C-spanner supplied in the toolkit. Turn the adjuster clockwise to increase spring preload and anticlockwise to decrease it. Note the revised suspension unit settings given in Fig. 8.4.
4 Other changes include modifications to the front forks, steering head bearings and bodywork; these are discussed in the following Sections.

9 Side panels and tail fairing: removal and refitting

1 The styling changes made include a revised side panel/tail fairing arrangement. To gain access to components mounted below the side panels, the following procedure should be adopted.
2 Start by removing the seat to gain access to the storage area below the tail fairing. Remove the two screws which retain the centre section of the tail fairing, and slide it forward until it can be lifted away. Remove any tools or other items stored inside the compartment.
3 Release the four bolts which secure the grab rail to the frame tubes and lift it away. If removing the right-hand side panel, note that the screw at its front edge must be removed first. The side panels

10.2 Top rear nut of access panel has security wire attached

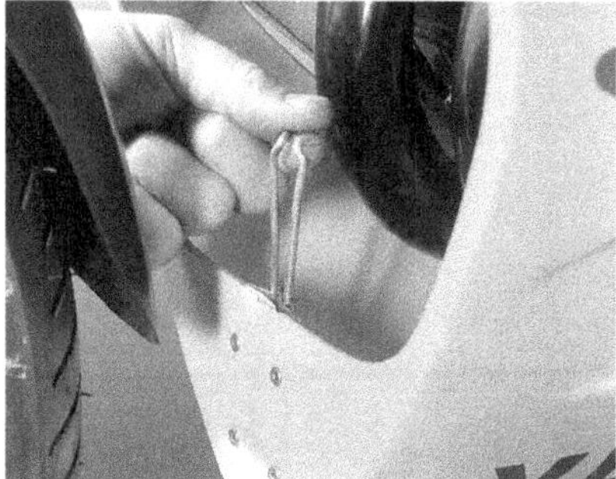

10.3 Fairing lower panels are held together by a spring clip front and rear

10.4a Each lower panel is retained by four screws (arrows) ...

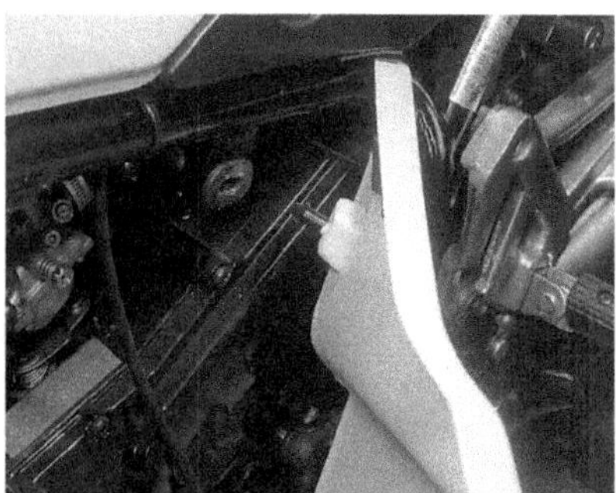

10.4b ... and by a peg and grommet

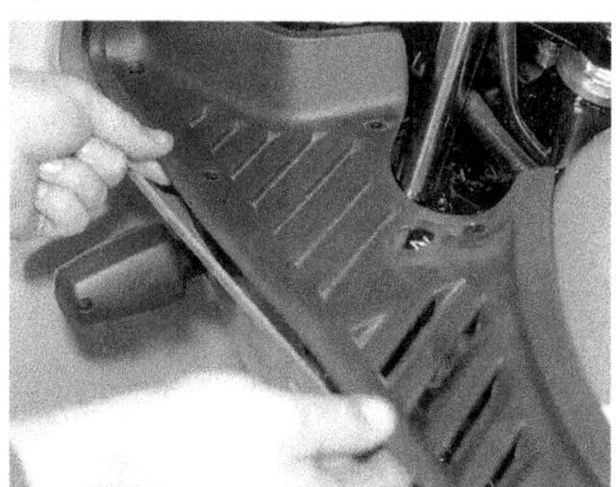

10.6 Remove the five screws to free fairing inner panels

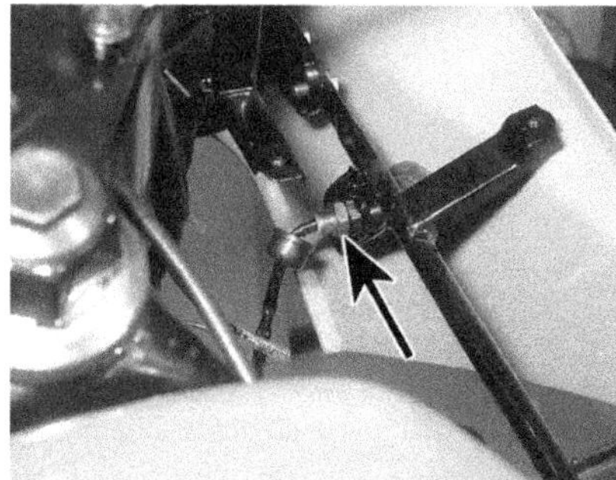

10.8 Unscrew the nut (arrow) to free turn signal from fairing bracket

10.10 Loop security wire over fairing mounting bracket stay

10.14 Fairing inner cover is retained by three screws

can be detached by pulling their lugs out of the frame grommets and hooks.
4 The side panels and centre section are refitted by reversing the removal sequence. Check that the grab rail bolts are tightened to 1.5 kgf m (11.0 lbf ft). When refitting the centre section of the tail fairing, make sure that the locating pegs in the centre section locate correctly in the holes in each side panel.

10 Fairing: removal and refitting – RD350R

1 Set the bike on its centrestand.

Lower fairing panels

2 Remove the access cover from both the left and right sides of the fairing. Each access cover is retained by four screws which locate in captive nuts on the fairing lower panels. Note that the top rear nut has a security wire attached; slip this nut off the fairing.
3 Reach in from the front opening of the fairing and pull out the two spring clips which hold the lower panels together.
4 Remove the four screws which retain the lower panel and work the peg out of its grommet in the frame. Take note of the washer locations to ensure correct reassembly.
5 Refitting is the reverse of removal ensuring the lower fairing panels are correctly hooked together at the bottom edge and secured by the spring clips. Tighten all screws securely, but do not overtighten otherwise the fairing material may crack.

Main fairing

6 Firstly remove the inner panels from each side. Each inner panel is retained by five screws which thread directly into the fairing mounting bracket.
7 Remove the two screws from the top edge of the fairing lower panels on each side.
8 From inside the fairing, trace and disconnect the wiring to the front turn signals. Peel back its rubber dust cover, and remove the nut which retains the turn signal to the fairing mounting bracket on each side. Pull the turn signals and wiring out of the fairing.
9 The fairing is retained to its mounting bracket at four points; at the top by the mirror mountings and at the bottom by a nut on each side (just below the headlamps). Have an assistant support the main fairing as the mountings are removed and take note of the washer and mounting damper locations.
10 Refitting is the reverse of removal. Ensure that the foam-rubber protection around the headlamp assembly seats correctly against the fairing aperture. Before refitting the inner panels, ensure that the security wire is looped over the mounting bracket stay.

Main fairing mounting bracket

11 Remove the lower fairing panels and main fairing as described above.
12 Detach the choke cable mounting from the left-hand carburettor (see Section 6).
13 Disconnect the wire connectors from the back of both headlamps and disconnect the wires to the parking lamp bulbholders at the base of each headlamp. Unscrew the four screws and separate the headlamp unit from the fairing bracket.
14 Remove the main fairing inner cover from the front of the instruments; it is retained by three screws.
15 Trace and disconnect the two block connectors for the instrument lighting, tachometer feed, warning lights and temperature gauge. Unscrew the speedometer cable knurled ring from the base of the meter, and free the cable. Remove the three retaining screws

10.16a Mounting bracket-to-steering head bolts (arrows) ...

10.16b ... and mounting bracket-to-frame bolt

10.16c Mounting bracket side sections can be detached by removing the two mounting bolts

and lift the instruments off the motorcycle; keep them upright to prevent damage to the meters.
16 Remove the mounting bracket-to-steering head bolts, and the bracket-to-frame top tube bolts and manoeuvre if off the motorcycle. Note that the side sections of the bracket can be detached by removing the two bolts on each side.
17 Refit the mounting bracket in a reverse of the removal sequence, tightening its bolts securely.

11 Front forks: modifications

1 As will be noted from the accompanying line drawing, the later type forks have a modified top plug arrangement. The general procedure for removing and overhauling the forks however, remains as described in Chapter 5, Sections 3, 5 and 6, noting the following points.
2 The design of the fork top plug is changed from a threaded type to a sealed, plain plug retained by a wire circlip. To remove the plug, unscrew and remove the dust cap, then depress the valve core to release air pressure.
3 Using a socket or piece of tubing, press down on the plug so that it is pushed into the top of the stanchion. This will reveal the retaining clip, which can then be displaced and removed using a small screwdriver. Once the clip has been freed, gradually release pressure on the plug, allowing the pressure from the fork spring to push it out of the stanchion. This operation is easier if two people are involved, but can be carried out unaided.
4 Before refitting the plug, which is carried out by reversing the above sequence, check the condition of the O-ring seal. If this is damaged or broken, renew it. Check that the wire circlip is located correctly before allowing the plug to rest against it. Check that the fork air pressures are set correctly, and in particular, that the pressure is equal in each leg, before refitting the dust cap.

12 Steering head: dismantling and reassembly – RD350R

Dismantling
1 Remove the main fairing (see Section 10 of this Chapter).
2 Remove the front forks (see Chapter 5).
3 Unbolt the horn and brake hose union from the lower yoke;

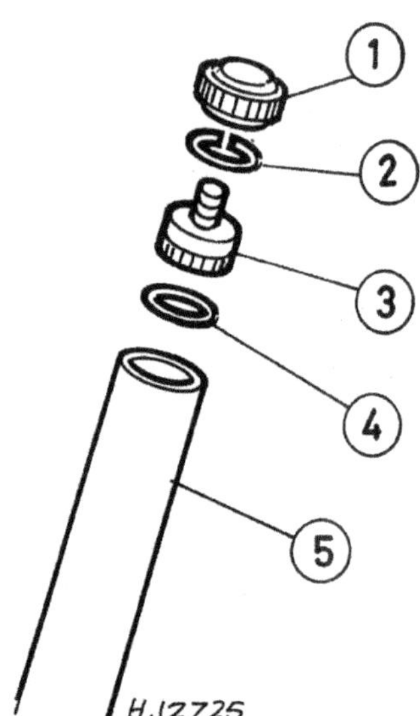

Fig. 8.5 Front fork top plug assembly (Sec 11)

1 Dust cap
2 Circlip
3 Top plug (incorporating air valve)
4 O-ring
5 Stanchion

there is no need to disconnect the hydraulic hoses, but ensure that no strain is placed on them.
4 Remove the fuel tank for access to the wiring block connectors (see Section 4), then trace and disconnect the wiring to the ignition switch.
5 Remove the steering stem top bolt, position the handlebars to one side, and lift off the upper yoke complete with ignition switch.
6 Support the base of the steering stem/lower yoke to prevent it falling as the steering stem nut is removed. Remove the nut and bearing dust cover, then lower the steering stem/lower yoke out of the steering head. Lift out the upper bearing from the top of the steering head.
7 Refer to the following section for details of bearing examination and renewal.

Reassembly
8 Apply fresh gease to the bearing races and rollers.
9 Insert the steering stem/lower yoke into the steering head and hold it in position whilst the upper bearing, dust cover and steering stem nut are installed. Tighten the nut enough to remove all freeplay from the bearings whilst still allowing free movement of the steering – do not overtighten. Make final adjustment once the steering head has been fully reassembled.
10 Install all other components in a reverse order of the tightening sequence, applying the torque settings where given.
11 Before tightening the steering stem top bolt, check the bearing freeplay. Grasp the lower fork legs and push and pull them to check for freeplay in the steering head bearings. If play exists, take this up by tightening the steering head nut with a C-spanner and recheck for freeplay. If correctly set, it should be possible to move the steering from lock-to-lock with the lightest pressure on the handlebar end. Tighten the steering stem top bolt to the specified torque setting when adjustment is complete.

13 Steering head bearings: examination and renewal – RD350R

1 Tapered roller steering head bearings are fitted instead of the ball bearing type used on earlier models.
2 Dismantle the steering head as described in the previous section.
3 Clean all parts in solvent and dry them thoroughly, using compressed air if available. If you do use compressed air, don't let the bearings spin as they are dried – it could ruin them. Wipe the old grease out of the frame steering head and bearing races.

Races
4 Examine the races in the steering head for cracks, dents and pits. If even the slightest amount of wear or damage is evident, the races must be renewed. The races and bearings are replaced as a set.
5 To remove the races, drive them out of the steering head by tapping around their inner edges with a drift passed through the steering head; tap evenly around the bearing edge to ensure that it leaves its housing squarely. If there is not enough access to the bearing edge for the drift, it will be necessary to use an internally-expanding bearing puller – a task best left to a Yamaha dealer.
6 Since the races are an interference fit in the steering head, fitting will be easier if the new races are left overnight in a refrigerator. This will cause them to contract and slip into place in the steering head with very little effort. When installing races, tap them gently into place with a hammer and large socket or tube which bears only on the outer edge of the race, and not on its working surface. Coat the races with grease.

Bearings
7 Examine the bearing rollers and cage for signs of damage. If renewal is required, the races must be renewed as well.
8 Whilst the upper bearing is easily removed during dismantling, the lower bearing should not be withdrawn from the steering stem unless renewal is required – it will most likely be damaged during removal.
9 Ideally, a bearing splitter and puller setup are needed for removal of the lower bearing, but if care is exercised, the bearing can be levered off with two flat-bladed screwdrivers.

16.10a RD350R headlamp beam adjusters are located at the base ...

16.10b ... and top of each headlamp unit

10 Tap the new lower bearing into position using a tubular drift over the steering stem which bears only on the inner edge of the bearing – it must not bear on the rollers or cage otherwise the bearing will be ruined.
11 Pack the bearings with grease.

14 Rear wheel: modification

1 All models feature restyled cast alloy wheels, these being somewhat lighter than the previous type. Mechanically, there is little difference between the two types, the only significant alteration being the inclusion of an O-ring on the left-hand side of the hub. This is located between item 12 and the wheel hub on Fig. 6.14, Chapter 6.

15 Alternator: modifications

1 On RD350NII and early RD350FII (1WT) models, the alternator is retained by a flanged nut. On the later RD350FII (3DH1) and all RD350R models it is retained by a flanged nut and plain washer.
2 Refer to the Specifications section of this Chapter for details of alternator test values.

16 Headlamp: bulb renewal and beam alignment

Bulb renewal – RD350FII

1 The headlamp bulb can be reached from inside the main fairing.
2 Remove the three instrument panel screws, and move the panel to one side to gain access to the headlamp bulbholder. Take care not to place any strain on the instrument panel wires or speedometer cable as this is done.
3 Renew the headlamp bulb as described in Chapter 7, Section 12.

Bulb renewal – RD350R

4 Each headlamp bulb can be reached from beneath the main fairing. Note that the main fairing inner panels can be removed (five screws) for improved access.
5 Pull off the wire connector from the back of the headlamp, and peel back the rubber dust cover.
6 The bulb is retained by a chrome retainer ring; twist the ring anti-clockwise to release the bulb. **Note:** *Handle the bulb by its terminals only, not by touching the glass – oil from your skin will cause the bulb to overheat and fail prematurely.*
7 Install the bulb and secure with the retainer ring, twisting it clockwise. Fit the dust cover, noting the TOP marking which must be positioned uppermost. Reconnect the wiring plug.

Beam alignment

8 On the RD350NII, beam alignment adjustment is made via the two long screws which pass through the headlamp rim. As viewed from the front of the motorcycle, the screw in the 10 o'clock position adjusts horizontal alignment, and the screw in the 5 o'clock position adjusts vertical alignment.
9 On the RD350FII, beam alignment adjustment is made by reaching inside the fairing to the wheel-type adjusters on the back of the headlamp. The upper right-hand adjuster controls horizontal alignment, and the lower left-hand adjuster controls vertical alignment.
10 On the RD350R, beam alignment adjustment is made via the modules at the base of each headlamp unit and the screw at the top of each unit. They can be accessed from beneath and inside the main fairing.

17 Side stand switch and control unit: function and testing

1 The purpose of the system is to prevent the machine from being started and ridden off while the side stand is down.
2 The switch, as might be expected, is mounted near to the side stand. The control unit will be found below the seat, just forward of the rear mudguard (early models) or under the fuel tank (later models). In the event of a suspected fault in the circuit, proceed as described below, referring to the wiring diagram at the end of this Chapter for details of the wiring colours and connections.
3 Disconnect the blue/yellow and black leads from the side stand switch. Using a multimeter set on the ohms x 1 scale, connect one probe to each of the switch terminals. With the side stand retracted, continuity (0 ohms) should be indicated by the meter. If a high resistance or an erratic reading is found, the switch is at fault. Try spraying WD40 or similar into the switch and operating it repeatedly. If this fails to resolve the problem, a new switch will be required.
4 If the switch worked normally in the above test, but the fault persists, check that the neutral switch is working normally. Separate the four-pin connector at the side stand control unit and identify the light blue neutral switch lead. Connect the multimeter positive (+) probe to the neutral switch lead and the negative (—) probe to a sound earth point on the frame. When neutral is selected, continuity (0 ohms) should be indicated by the meter. If a high resistance or an erratic reading is found, the switch is at fault.
5 To check the side stand control unit, disconnect the single brown lead and check whether this eliminates the ignition fault. If the system works normally with the lead disconnected, but reappears when it is connected, the control unit can be considered faulty and should be renewed. If the fault is evident with the brown lead disconnected, the fault must lie elsewhere, most probably with the CDI unit itself.

Component key – RD350 LC II model

1 Engine stop switch
2 Lighting switch
3 Front brake lamp switch
4 Temperature gauge sender unit
5 Frame earth point
6 Ignition coil
7 Servomotor
8 Turn signal relay
9 Rear brake lamp switch
10 Turn signal cancelling unit
11 Battery
12 Control unit
13 Oil level sender unit
14 Rear right-hand turn signal
15 Tail/stop lamp
16 Rear left-hand turn signal
17 Fuses
18 Neutral indicator switch
19 Alternator
20 CDI unit
21 Regulator/rectifier
22 Horn switch
23 Turn signal switch
24 Headlamp dip switch
25 Headlamp pass switch
26 Horn
27 Front left-hand turn signal
28 Ignition switch
29 Parking lamp
30 Headlamp
31 Instrument light
32 Instrument light
33 Instrument light
34 Temperature gauge
35 Reed switch
36 Neutral lamp
37 Oil level warning lamp
38 High beam indicator
39 Left-hand turn signal warning lamp
40 Right-hand turn signal warning lamp
41 Front right-hand turn signal
42 Horn

Component key – RD350 N and F models

1 Engine stop switch
2 Lighting switch
3 Front brake lamp switch
4 Temperature gauge sender unit
5 Ignition coil
6 Turn signal cancelling unit
7 Servomotor
8 Turn signal relay
9 Rear brake lamp switch
10 Battery
11 Control unit
12 Oil level sender unit
13 Rear right-hand turn signal
14 Tail/stop lamp
15 Rear left-hand turn signal
16 Fuses
17 Neutral indicator switch
18 Alternator
19 CDI unit
20 Regulator/rectifier unit
21 Horn switch
22 Turn signal switch
23 Headlamp dip switch
24 Headlamp pass switch
25 Front left-hand turn signal
26 Ignition switch
27 Parking lamp
28 Headlamp
29 Instrument light
30 Tachometer
31 Temperature gauge
32 Reed switch
33 Neutral indicator light
34 Oil level warning light
35 High beam indicator
36 Turn signal warning light
37 Front right-hand turn signal
38 Horn

Component key – RD350 N II and F II models

1 Lighting switch
2 Engine stop switch
3 Front brake lamp switch
4 Turn signal cancelling unit
5 Servomotor
6 CDI unit
7 Ignition coil
8 Spark plug
9 Rear brake lamp switch
10 Regulator/rectifier unit
11 Side stand control unit
12 Battery
13 Rear right-hand turn signal
14 Tail/stop lamp
15 Rear left-hand turn signal
16 Control unit
17 Sidestand switch
18 Oil level sender unit
19 Turn signal relay
20 Fuses
21 Neutral indicator switch
22 Alternator
23 Horn switch
24 Turn signal switch
25 Headlamp pass switch
26 Headlamp dip switch
27 Front left-hand turn signal
28 Ignition switch
29 Headlamp
30 Parking lamp
31 Turn signal warning light
32 High beam indicator
33 Oil level warning light
34 Neutral indicator light
35 Tachometer
36 Temperature gauge
37 Reed switch
38 Instrument lights
39 Front right-hand turn signal
40 Horn
41 Thermo switch

Wire colour key

B Black
Br Brown
Ch Dark brown
Dg Dark green
G Green
Gr Grey
L Blue
O Orange
P Pink
R Red
Sb Light blue
W White
Y Yellow

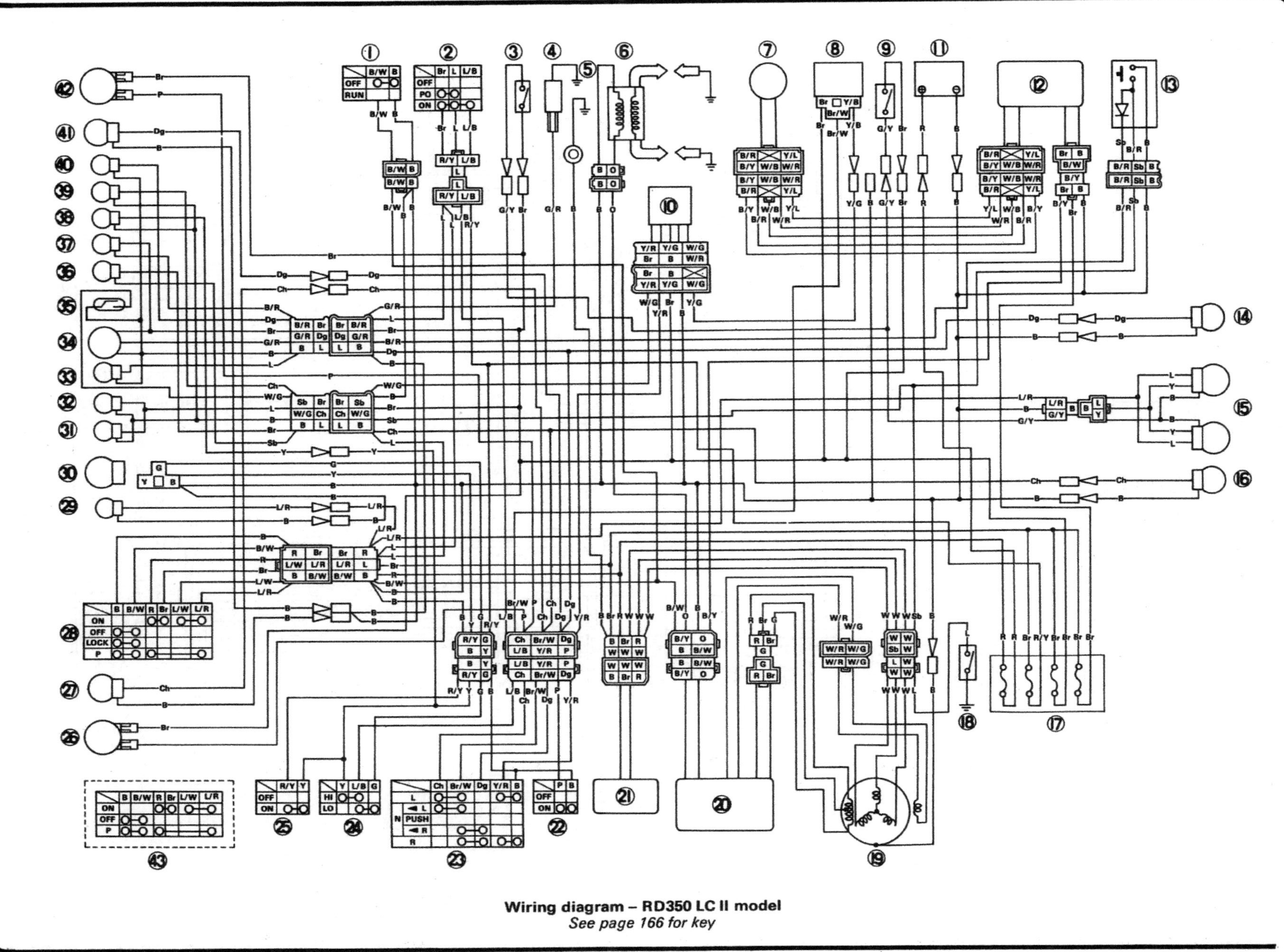

Wiring diagram – RD350 LC II model

See page 166 for key

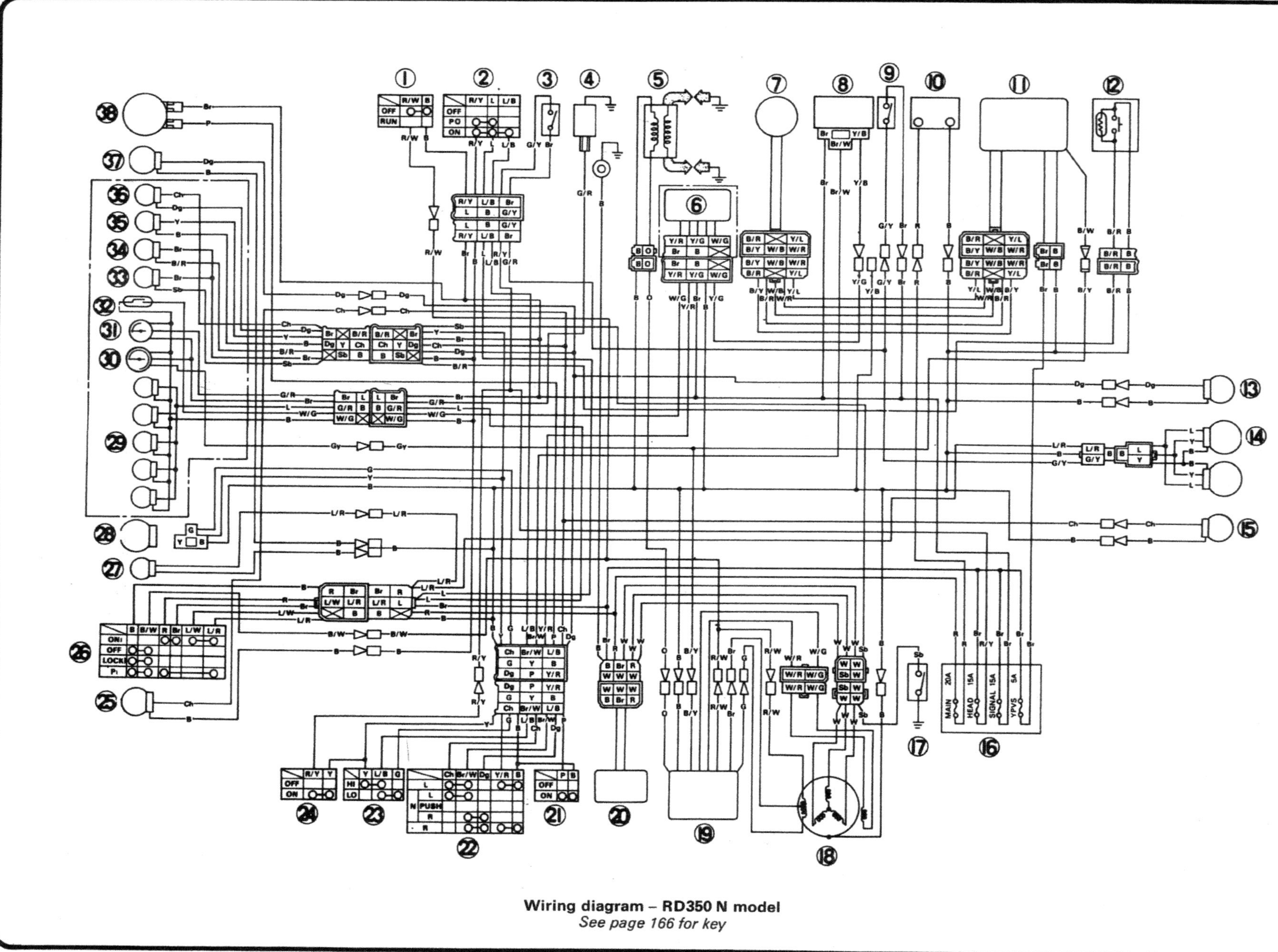

Wiring diagram – RD350 N model
See page 166 for key

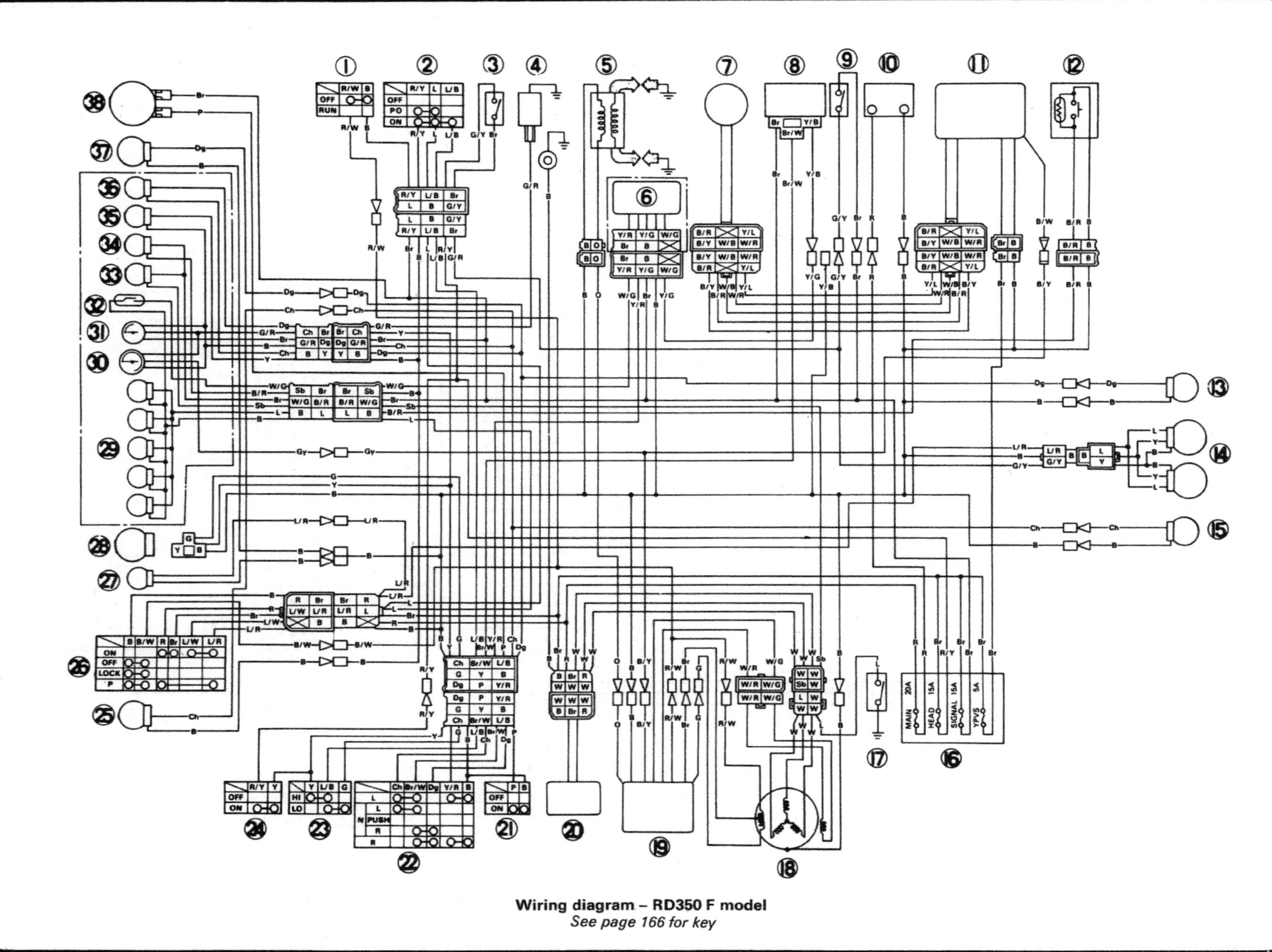

Wiring diagram – RD350 F model
See page 166 for key

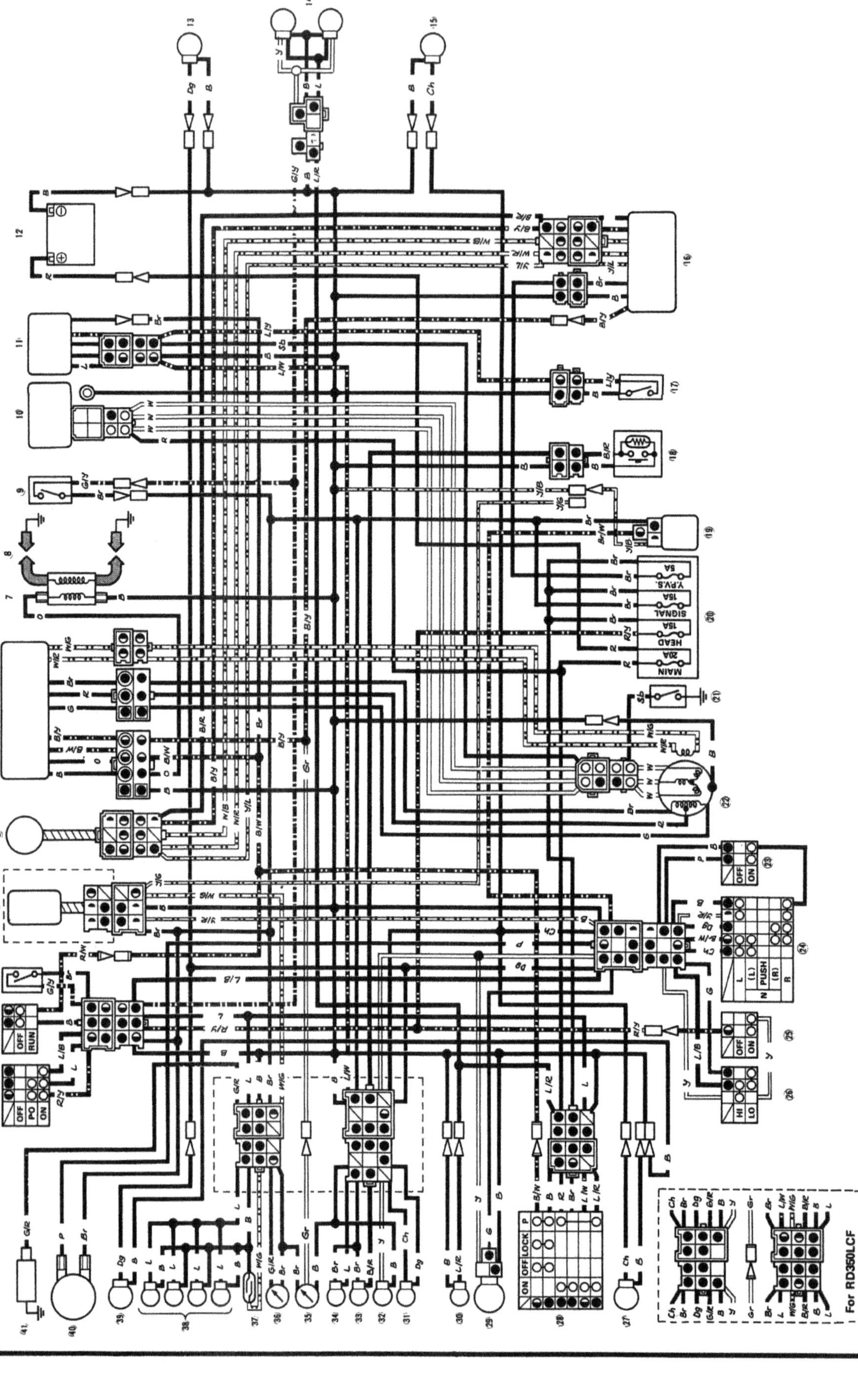

For RD350LCF

Wiring diagram – RD350 N II, F II and R models

See page 166 for key

Conversion factors

Length (distance)						
Inches (in)	X	25.4	= Millimetres (mm)	X	0.0394	= Inches (in)
Feet (ft)	X	0.305	= Metres (m)	X	3.281	= Feet (ft)
Miles	X	1.609	= Kilometres (km)	X	0.621	= Miles
Volume (capacity)						
Cubic inches (cu in; in^3)	X	16.387	= Cubic centimetres (cc; cm^3)	X	0.061	= Cubic inches (cu in; in^3)
Imperial pints (Imp pt)	X	0.568	= Litres (l)	X	1.76	= Imperial pints (Imp pt)
Imperial quarts (Imp qt)	X	1.137	= Litres (l)	X	0.88	= Imperial quarts (Imp qt)
Imperial quarts (Imp qt)	X	1.201	= US quarts (US qt)	X	0.833	= Imperial quarts (Imp qt)
US quarts (US qt)	X	0.946	= Litres (l)	X	1.057	= US quarts (US qt)
Imperial gallons (Imp gal)	X	4.546	= Litres (l)	X	0.22	= Imperial gallons (Imp gal)
Imperial gallons (Imp gal)	X	1.201	= US gallons (US gal)	X	0.833	= Imperial gallons (Imp gal)
US gallons (US gal)	X	3.785	= Litres (l)	X	0.264	= US gallons (US gal)
Mass (weight)						
Ounces (oz)	X	28.35	= Grams (g)	X	0.035	= Ounces (oz)
Pounds (lb)	X	0.454	= Kilograms (kg)	X	2.205	= Pounds (lb)
Force						
Ounces-force (ozf; oz)	X	0.278	= Newtons (N)	X	3.6	= Ounces-force (ozf; oz)
Pounds-force (lbf; lb)	X	4.448	= Newtons (N)	X	0.225	= Pounds-force (lbf; lb)
Newtons (N)	X	0.1	= Kilograms-force (kgf; kg)	X	9.81	= Newtons (N)
Pressure						
Pounds-force per square inch (psi; lbf/in²; lb/in²)	X	0.070	= Kilograms-force per square centimetre (kgf/cm²; kg/cm²)	X	14.223	= Pounds-force per square inch (psi; lbf/in²; lb/in²)
Pounds-force per square inch (psi; lbf/in²; lb/in²)	X	0.068	= Atmospheres (atm)	X	14.696	= Pounds-force per square inch (psi; lbf/in²; lb/in²)
Pounds-force per square inch (psi; lbf/in²; lb/in²)	X	0.069	= Bars	X	14.5	= Pounds-force per square inch (psi; lbf/in²; lb/in²)
Pounds-force per square inch (psi; lbf/in²; lb/in²)	X	6.895	= Kilopascals (kPa)	X	0.145	= Pounds-force per square inch (psi; lbf/in²; lb/in²)
Kilopascals (kPa)	X	0.01	= Kilograms-force per square centimetre (kgf/cm²; kg/cm²)	X	98.1	= Kilopascals (kPa)
Millibar (mbar)	X	100	= Pascals (Pa)	X	0.01	= Millibar (mbar)
Millibar (mbar)	X	0.0145	= Pounds-force per square inch (psi; lbf/in²; lb/in²)	X	68.947	= Millibar (mbar)
Millibar (mbar)	X	0.75	= Millimetres of mercury (mmHg)	X	1.333	= Millibar (mbar)
Millibar (mbar)	X	0.401	= Inches of water (inH_2O)	X	2.491	= Millibar (mbar)
Millimetres of mercury (mmHg)	X	0.535	= Inches of water (inH_2O)	X	1.868	= Millimetres of mercury (mmHg)
Inches of water (inH_2O)	X	0.036	= Pounds-force per square inch (psi; lbf/in²; lb/in²)	X	27.68	= Inches of water (inH_2O)
Torque (moment of force)						
Pounds-force inches (lbf in; lb in)	X	1.152	= Kilograms-force centimetre (kgf cm; kg cm)	X	0.868	= Pounds-force inches (lbf in; lb in)
Pounds-force inches (lbf in; lb in)	X	0.113	= Newton metres (Nm)	X	8.85	= Pounds-force inches (lbf in; lb in)
Pounds-force inches (lbf in; lb in)	X	0.083	= Pounds-force feet (lbf ft; lb ft)	X	12	= Pounds-force inches (lbf in; lb in)
Pounds-force feet (lbf ft; lb ft)	X	0.138	= Kilograms-force metres (kgf m; kg m)	X	7.233	= Pounds-force feet (lbf ft; lb ft)
Pounds-force feet (lbf ft; lb ft)	X	1.356	= Newton metres (Nm)	X	0.738	= Pounds-force feet (lbf ft; lb ft)
Newton metres (Nm)	X	0.102	= Kilograms-force metres (kgf m; kg m)	X	9.804	= Newton metres (Nm)
Power						
Horsepower (hp)	X	745.7	= Watts (W)	X	0.0013	= Horsepower (hp)
Velocity (speed)						
Miles per hour (miles/hr; mph)	X	1.609	= Kilometres per hour (km/hr; kph)	X	0.621	= Miles per hour (miles/hr; mph)
Fuel consumption*						
Miles per gallon, Imperial (mpg)	X	0.354	= Kilometres per litre (km/l)	X	2.825	= Miles per gallon, Imperial (mpg)
Miles per gallon, US (mpg)	X	0.425	= Kilometres per litre (km/l)	X	2.352	= Miles per gallon, US (mpg)

Temperature

Degrees Fahrenheit = (°C x 1.8) + 32

Degrees Celsius (Degrees Centigrade; °C) = (°F − 32) x 0.56

** It is common practice to convert from miles per gallon (mpg) to litres/100 kilometres (l/100km), where mpg (Imperial) x l/100 km = 282 and mpg (US) x l/100 km = 235*

Index

W

Y